Lecture Notes in Mathematics 2208

More information about this series at http://www.springer.com/series/304

Gunther Schmidt • Michael Winter

Relational Topology

Springer

Gunther Schmidt
Fakultät für Informatik
Universität der Bundeswehr München
Neubiberg, Germany

Michael Winter
Department of Computer Science
Brock University
St. Catharines, Ontario, Canada

ISSN 0075-8434 ISSN 1617-9692 (electronic)
Lecture Notes in Mathematics
ISBN 978-3-319-74450-6 ISBN 978-3-319-74451-3 (eBook)
https://doi.org/10.1007/978-3-319-74451-3

Library of Congress Control Number: 2018942705

Mathematics Subject Classification (2010): 54-XX, 03E20, 54E05, 54E17, 97E60

Printed on acid-free paper

This Springer imprint is published by the registered company Springer International Publishing AG part of Springer Nature.
The registered company address is: Gewerbestrasse 11, 6330 Cham, Switzerland

Preface

Over the years, the authors have encountered a multitude of topics that are ultimately related to general topology and the logics of spatial reasoning. On the other hand, they have long been working on and with relational methods in fields around computer science. Finally, programming was their daily lecturing task. They became increasingly unsatisfied with the many—but slightly diverging—approaches to the topics mentioned and decided to work on a unifying presentation.

Yet another stimulus was the idea to lift concepts to a relational level making them point-free as well as quantifier-free, thus liberating them from the style of first-order predicate logic and approaching the clarity of algebraic reasoning. For this, a calculus had already been invented, since the 1970s, introducing heterogeneous relations (i.e., relations between possibly different sets). Also the important domain construction steps of forming the direct product, direct sum, or direct power had in the meantime been given birth to, characterizing them uniquely *up to isomorphism*.

Treating a topic algebraically means to work with algebraic rules that are lastly based on axioms. As we know from Euclid's axioms of geometry, an axiomatic theory may admit not just one model. As early as in the 1980s, the problem of *sharp factorization* or *unsharpness* has been raised. One may best characterize it with the statement that the concept of predicate logic is insufficient in treating relations satisfactorily since it restricts us to just one model, namely the Boolean matrix model. There exist others that seem more appropriate when—more generally—considering processes.

In recent years, this relational approach has been extended introducing the constructs of a Kronecker operator, together with a strict fork and strict join operator. Axiomatic characterizations have been developed; the tool kit of rules and formulae is beginning to stabilize, and the effectivity of computing with them increases steadily.

Given this context, it was highly welcome that several concepts of topology, such as neighborhoods, transition to the open kernel, contact relations, proximity, etc., qualify for being typical application fields to be integrated under one common relational roof. All the transitions between such concepts may be formulated by concise relation-algebraic terms or rules. Any proof necessary lends itself to being executed algebraically, and in the near future possibly with machine assistance as earlier with RALF, if not via proof systems such as Isabelle/HOL and Coq.

First steps in this direction have already been made with the relational language TITUREL. When one is about to solve topological problems computationally, one often has to be able to convert the given topology to a suitable or favorable form which means to apply some step of transition that needs to be justified. Such justifications are here given for nearly all conceivable version switches.

Quotient topologies, product topologies, as well as relative topologies on a subset are handled in this way. It turned out that in all three cases, one comes close to the sharpness effect, which makes the intended point- and quantifier-free proofs unexpectedly complicated. Only when looking at these situations in full detail, one will recognize why. The typical situation is that an algebraic reasoning is allowed only via some additionally available relation seemingly peripheral to the problem proper and not even mentioned in its statement. We consider this as a deeper insight obtained during our work on the topic.

Furthermore, a study of several approaches to spatial reasoning on discreteness, proximity, nearness, apartness, betweenness, and Aumann contacts is presented, which are frequently performed by logicians. These concepts are heavily interrelated which we exhibit expressing one by means of the respective other concept. This would have hardly been possible when not with the relational shorthand expressions. We prove that these transitions are correct. In case of apartness, we had the opportunity to identify properties which to demand seems counterproductive.

Another point to be explained is that we do not make an overly detailed use of categories. Category theory has proved to be extremely versatile in studying concepts. Here, however, we also aim at computation and/or computational proofs. It is absolutely clear that in this context category theory is hardly used in its deeper sense. We go ahead and strip off overly detailed category theory, mentioning it just to the extent that typing is clarified.

Finally, some ideas about how to work relationally on simplicial complexes are demonstrated at least in examples. This differs from the approach taken for the algebraic transitions between related topics. Here, it seems possible to work practically using the computer. We could, of course, only give a very slight idea of how this might work.

This booklet rests on decades of work with colleagues and students, to whom we owe our sincere thanks. Without all the discussions, it would not have emerged. Special thanks are due to all those working and contributing to the by now well-established "intercontinental" RAMiCS conference series (Relational and Algebraic Methods in Computer Science). Also the European COST action 274 TARSKI (Theory and Applications of Relational Structures as Knowledge Instruments) from 2001 to 2005 with its meetings all over Europe and sometimes also in Canada gave many background ideas.

Direct input and repeated discussions and contributions have always been provided by Rudolf Berghammer and Wolfram Kahl—after earlier common work on such topics.

The authors are grateful to the publisher for having included this booklet in his program. They thank in particular for the agreeable cooperation with Ute McCrory.

Their deeply felt thanks go in particular to the anonymous reviewers. The sheer number of their suggestions made us feel that they were really interested in getting the authors to improve the text.

The second author gratefully acknowledges support from the Natural Sciences and Engineering Research Council of Canada.

Neubiberg, Germany
St. Catharines, ON, Canada
April 27, 2018

Gunther Schmidt
Michael Winter

Symbols

Sets

Union and intersection are denoted as $M \cup N$ and $M \cap N$—in the same way as later for relations. The complement is $\overline{M}$, provided the ground set is tacitly given. For a one-element set, we provide $\mathbb{1}$ as standard notation. The Cartesian product of sets is $M \times N$.

Logic

For metalanguage consequence, equivalence, and definition, "$\Longrightarrow$", "$\Longleftrightarrow$", and "$:\Longleftrightarrow$" are used. Definitional equality is denoted as "$:=$". The set of Boolean truth values is $\mathbb{B} = \{\, \mathbf{0}\,,\ \mathbf{1}\,\}$. In the context of propositional logic, "$\wedge$" , "$\vee$" are used for "and" and "or," together with "$\rightarrow$" for "if... then" and "$\leftrightarrow$" for "precisely when." In the context of predicate logic, "$\exists$" and "$\forall$" denote the existential quantifier and the universal quantifier.

Relations

$R : X \longrightarrow Y$	Relation with source and target	7
$\mathbb{1}$	One-element set	10
$\mathcal{P}(X)$	Powerset of X	14
$\mathbf{2}^X$	Powerset of X, variant	14
$R \cup S$	Union	7
$R \cap S$	Intersection	7

$\perp\!\!\!\perp$	Empty relation	provided the	7
$\top\!\!\!\top$	Universal relation	ground sets	7
$\mathbb{I}$	Identity	are tacitly given	7
R^T	Transposed relation, converse		7
$R ; S$	Product, composition		7
R^*	Reflexive-transitive closure of R		173
H_B	Hasse relation of ordering B		10
$S \backslash R$	Right residual		9
S/R,	Left residual		9
$\mathtt{syq}(R, S)$	Symmetric quotient		10
$(R \otimes S)$	Strict fork operator		26
$(R \otimes S)$	Strict join operator		26
$(R \otimes S)$	Kronecker product		26
$\mathtt{gre}_E(R)$	Greatest upper bounds functional		11
$\mathtt{greR}_E(R)$	Greatest upper bounds functional—row-wise		11
$\mathtt{ubd}_R(S)$	Upper bound cone functional		10
$\mathtt{lbd}_R(S)$	Lower bound cone functional		10
$\mathtt{lub}_E(t)$, $\mathtt{glb}_E(t)$	Least upper, greatest lower bounds		11
$\bigwedge_1$	Lower cone of an element	with regard to	45
$\bigwedge_2$	Lower cone of 2 elements	some tacitly given	45
$\bigwedge$	Lower cone of a set of elements	relation R	45
$\bigvee_1$	Upper cone of an element	with regard to	47
$\bigvee_2$	Upper cone of 2 elements	some tacitly given	47
$\bigvee$	Upper cone of a set of elements	relation R	47
$\mathfrak{J}$	Lifted join		48
$\mathfrak{J}_2$	Lifted binary join		48
$\mathfrak{M}$	Lifted meet		48
$\mathfrak{M}_2$	Lifted binary meet		48
ϑ_R	Existential image of relation R		17
ϑ_{R^T}	Inverse image of relation R		17
ε	Membership		14
σ	Singleton injection		15
E	Arbitrary ordering		10
Ω	Powerset ordering		15
$\mathcal{N}$	Powerset negation		15
$\mathcal{U}$	Neighborhood topology		72
$\mathcal{K}$	Open kernel-mapping topology		74
$\mathcal{O}_D$	Open diagonal topology		80
$\mathcal{O}_V$	Open set topology		79

Contents

Chapter 1
Introduction

There exist lots of concepts around topology: open sets, neighborhoods, transitions to their open kernels, proximity, nearness, betweenness, apartness, different concepts of contact and so on. Although, they are all heavily interrelated, this is often hard to recognize, because they are discussed in quite different settings resp. terminology. We are going to identify the core concepts of those ideas and to show how they may be mutually deduced from one another.

In contrast to what the title may insinuate, the authors don't claim to be topologists. So it need be explained why they felt entitled to write a text named *Relational Topology*. They have developed *relational methods* to quite some extent; but developing *methods* is triggered by work in application fields where these may be used. Among those was in particular topology where they may be used very effectively.

1.1 Lifting to Relational Style

People using relation algebra have early identified that *elements* of a set correspond to relations satisfying certain laws. These specific relations have for long been called *points* by the respective researchers. When using such points, it is a rather simple task to reformulate all the standard mathematics in terms of relations.

This, however, is not the really satisfactory approach to relational mathematics. When, e.g., formulating that a relation $M : X \longrightarrow Y$ is a partial function, one may say

$$\forall x \in X : \forall y_1, y_2 \in Y : (x, y_1) \in M \wedge (x, y_2) \in M \rightarrow y_1 = y_2,$$

which we sometimes refer to as being in *predicate-logic style*.

G. Schmidt, M. Winter, *Relational Topology*, Lecture Notes in Mathematics 2208, https://doi.org/10.1007/978-3-319-74451-3_1

There is, however, another form (anticipating that ^{T} is transposition, $;$ is composition, and $\mathbb{I}$ is the respective identity relation)

$$M^{\mathsf{T}};M \subseteq \mathbb{I},$$

which we refer to as having been expressed in *relational style*. It may be seen as a shorthand version saying indeed, that when going back from an image to one of its arguments and going forward again, one will always arrive at the same image.

We refer to the step going from the predicate-logic to the relational style as *lifting*. A very first transition in this direction might end with something like

$$\forall x \text{ element/point in } X: \quad \forall y_1, y_2 \text{ elements/points in } Y:$$
$$x;y_1^{\mathsf{T}} \subseteq M \quad \wedge \quad x;y_2^{\mathsf{T}} \subseteq M \quad \rightarrow \quad y_1;y_2^{\mathsf{T}} \subseteq \mathbb{I},$$

incorporating already earliest relational operations. But this is only half of the way intended, leaving us still with quantification. Working really in relational style, one will hardly ever quantify over elements/points of a set—look at the abovementioned end of the road $M^{\mathsf{T}};M \subseteq \mathbb{I}$. The notation for any quantification is hidden and stays deeply incorporated in the typing of the relational operations (here: transposition, composition) and their rules. Since elements correspond to the relational points mentioned earlier, people often speak of a *point-free* formulation when *quantifier-free* is the intended meaning. Being quantifier-free indicates algebraic reasoning.

To even further demonstrate the difference of predicate-logic style as opposed to relational style, we look ahead at two items handled later in this text. Firstly, a part of the definition[1] of a topology via a neighborhood system demands that

for every neighborhood $U \in \mathcal{U}(p)$ there exists a neighborhood $V \in \mathcal{U}(p)$

so that $U \in \mathcal{U}(y)$ for all $y \in V$.

This is a quantifier-prone verbose text which to process in a computer system seems hard. We give preference to the relation-algebraic condition

$$\mathcal{U} \subseteq \mathcal{U};\overline{\varepsilon^{\mathsf{T}};\overline{\mathcal{U}}}$$

with ε the membership relation. Secondly, in the condition[2] for $f : X \longrightarrow X'$ to be a continuous mapping:

For every point $p \in X$ and every neighborhood $V \in \mathcal{U}'(f(p))$,

there exists a neighborhood $U \in \mathcal{U}(p)$ such that $f(U) \subseteq V$

---

[1] See Definition 5.2.1.iv.

[2] See Definition 5.6.1.

as opposed to the relation-algebraic form with ϑ_{f^T} the inverse image mapping

$$f{:}\mathcal{U}' \subseteq \mathcal{U}{:}\vartheta_{f^\mathsf{T}}^\mathsf{T}.$$

Mathematicians have indeed been able to handle the complex textually quantifying form. Some would, however, vote for the simpler, i.e. relation-algebraic, conditions. These may also be handled efficiently by the state-of-the-art relational proof systems.

Some may blame the relational style to be too abstract, but we answer with a remark by Barthel Leendert van der Waerden[3]: *Das Ziel der Abstraktion in der Modernen Algebra ist nämlich nicht nur die größtmögliche Allgemeinheit. Sondern dadurch, dass man sich von allen Besonderheiten des gerade untersuchten Problems frei macht, trennt man das Wesentliche vom Unwesentlichen und macht die ganzen Zusammenhänge durchsichtig.*

1.2 Equational vs. Implicational Style

There is another aspect that needs to be explained when lifting from the quantifier-prone predicate-logic style to the relational level with its algebraic flavor:

Traditionally, algebraists work with universally quantified equational formulae over a given signature, such as in the simplest case $\forall a, b : a \wedge b = b \wedge a$, look for their free term algebra and divide out the congruence according to the universally quantified equation. Algebraists stay more or less completely in this area, should they confine their studies to algebraic extensions, quotients, etc.

A slight deviation took place when artificial intelligence emerged and also Horn formulae were made use of. Implicitly quantified propositional Horn formulae such as

$$\neg p \vee \neg q \vee \neg r \vee s, \quad \text{or else} \quad p \wedge q \wedge r \rightarrow s,$$

may be seen as implications (subjunctions) with several terms combined in conjunction on the left side, but just one on the right. This was fine for automated reasoning by first-order resolution, but made it more difficult to formally prove the existence of a model: The standard way of dividing out a congruence did no longer work in the same way.

In our relational work, we also use $A \subseteq B$ for relations to be seen as element-wise *implication* (subjunction). As for Horn formulae, it is important on which side

[3]Reported in [RK07, p. 153]: The goal of abstraction in Modern Algebra is *not* just to obtain the utmost generality. When one is unburdened from of all the peculiarities of the actual problem, one will more easily detach the essential from the unessential and make transparent how things are mutually connected.

a term happens to occur. Such conditions might—of course—also be written in an equational style, since

$$A \subseteq B \iff A \cap B = A,$$

but this is avoided since it does no longer open the clear view on the chance for chaining such implications. For the ordering of, e.g., numbers, we have a firm feeling that

$$x \le y \,\wedge\, y \le z \;\rightarrow\; x \le z,$$

which applies in an analogous way for relational containment. Although it is via the aforementioned transition still equational, one might better call this style *implicational*. Looking later at the Schröder equivalences, e.g., it is even more clear that both sides enforce completely different actions, which must be immediately visible when discussing such topics.

1.3 Chapter Organization

This monograph is organized as follows:

Chapter 2 It is a requirement for such research to be acquainted with the relation-algebraic methods. In Chap. 2, we collect what has to be mentioned from known relational methods to make this text sufficiently self-contained. It reveals the omnipresent membership $x \in U \subseteq X$ as a relation $\varepsilon : X \longrightarrow \mathbf{2}^X$ that holds for the pair (x, U). Starting therefrom, it recalls the concepts of an existential and an inverse image, frequently referred to in theoretical computer science, and develops the algebraic apparatus to efficiently work with them. Later, one will in retrospect see that this is a first example of *lifting* a concept to a relational level.

Chapter 3 The following Chap. 3 begins by recalling familiar constructions: the direct product with its projections, the—less commonly known—direct sum with its injections, as well as the direct quotient, dividing out an equivalence, with its natural projection.

Then we develop the calculus of the binary Kronecker $\otimes$, fork $\circledless$, and join $\circledgtr$ operators out of a rigorous relational axiomatization. It seems that this has so far never been systematically developed and, thus, is mainly novel material. It gives opportunity to discuss sharp factorization with its difficult model question to be presented here.

The chapter then ends by introducing also for binary mappings a fully lifted general form. Even such generally accepted concepts as commutativity, distributivity, and associativity assume new and very concise quantifier-free formulations.

Chapter 4 The aforementioned binary mappings are omnipresent in mathematics, not least when adding, multiplying, or forming joins and meets according to some ordering. Following our general idea, we lift in Chap. 4 also such binary mappings. It comes as quite a surprise that the algebraic formulae thus obtained turn out to be acceptably simple. This will then facilitate reasoning on this upper level—notwithstanding the fact that they are novel in style and require getting accustomed to them.

The lifted concepts include standard properties of (partial) functions, binary mappings such as cone mappings, join, meet, and other operations. All their manifold interdependencies will get an implicational flavor.

Chapter 5 Many known concepts of topology and continuity are recalled in Chap. 5. They are then lifted to a quantifier-free relational form, thus opening them to being handled relationally, using the existential image and the inverse image. Separability is then handled relationally. The chapter ends with a relational treatment of continuity. This in particular leads us to concise relation-algebraic formulations for each of in the diverse forms of topology.

Chapter 6 This chapter shows how to build new topologies from given ones. These techniques are known in principle: product topology, relative topology, and quotient topology. Nevertheless, each requires specific relational methods to be brought to the intended concise relational level.

Chapter 7 Chapter 7 mentions the less known Aumann contact relation—that in later publications by other authors resurrected as betweenness—and its connections with topology. One may go from an Aumann contact to a topology as well as back from a topology to a *different* Aumann contact. These transitions are, thus, not inverses of one another.

Chapter 8 Several concepts of the border zone between topological concepts and logical reasoning are recalled in Chap. 8, in the highly diverse forms in which they frequently appear. They are then brought to relational style and many of their interrelationships are exhibited and proved formally. This includes proximity, nearness, and apartness.

Chapter 9 Yet another aspect is covered by Chap. 9 when the reasoning about processes via just finite and partial observation is investigated as to its logical basis which consists of topological systems and frames. It elaborates on how one may obtain a topology from a frame and vice versa.

Chapter 10 To somehow complete our endeavor, we study concepts of homology such as orientation, boundary operators, etc., in Chap. 10. Again, these are brought to a relational form without quantifiers and then applied to simplicial complexes. Several examples illustrate how this might work.

1.4 Final Remarks

Quite often, one will detect similarities between approaches that come from absolutely different areas. Some researchers study all these minor differences in ever new papers. Our approach is definitely different: Can we—led by the ideas of these differing approaches—find some relational 'girder' carrying all the intertwined theories that shows us a basis with several sound anchoring supports that are relationally related in a simple way and that may provide a firm starting point for research? Such a girder should serve as a reference for further study; it should also be the measure against which any strengthening or weakening of the axioms should be discussed.

The present text is completely based on relational methods. A secondary objective while working on it was to further grind, sharpen, and edge our relational tools. We hope to have shown the effectiveness of these relational methods. Proofs have been given so detailed that one may see how it will be possible to execute them in a theorem proving environment that has a detailed type control—as it has already been done experimentally by the authors themselves.

All examples we provide are finite and discrete. While this might persuade a reader to believe that our approach is restricted to finiteness, it is not so: The relational formulae are just shorthand or abbreviated versions of the predicate logic formulae with which topology is traditionally defined. The many computer-produced examples of finite discrete topologies are generated using the language TITUREL[4] to interpret relational terms and formulae. The well-known RELVIEW[5] system would considerably scale up the size of problems that may be tackled.

A remark in [Die74] supports this discrete approach even in view of the general situation: There, Jean Dieudonné reconsiders what René Thom said concerning superiority of "continuous" considerations as compared with "discrete" ones: He rightly criticizes Kronecker for his one-sided view on mathematics as fully based on the concept of a number. ... But then Thom himself, says that the *continuum needs to be discretized*, and that since Poincaré the only way to understand topology somehow is the ever increasing application of algebra leading to topological invariants as objects of study.

While working on relational topology, we developed some hope that in addition the concept of *dimension* may be underpinned with sound relational arguments. This seems to have been successful, but had regrettably to be postponed to future work. The landmark books by Karl Menger [Men28], Witold Hurewicz and Henry Wallman [HW41], and Ryszard Engelking [Eng78]—extending over half a century—demonstrate convincingly how difficult these topics are. Also the concept of a matroid seems to lend itself to being treated relationally, which has not yet been achieved.

---

[4]http://www.titurel.org/TituRel/indexTituRel.html.

[5]http://www.informatik.uni-kiel.de/~progsys/relview/.

Chapter 2
Prerequisites

Relational methods are not yet broadly known and, thus, need a detailed introduction. We develop all the necessary methodology; it originates in particular from [SS89, SS93, Sch11, SW14]. There, full proofs may be found. In addition it is shown how everything is based on a concise axiomatic basis. However, some of the following results are new, and therefore given together with their proof.

2.1 Preliminaries

The basic prerequisites presented routinely for relational work are by now fairly well-known: Boolean operations and predicates $\cup, \cap, \overline{}, \subseteq$, together with the least $\mathbb{⊥}$ and the greatest elements $\mathbb{T}$; then the monoid operation of relational composition[1] $;$ together with the identities $\mathbb{I}$, and finally transposition or conversion (Fig. 2.1). Composition $;$ binds stronger than the Boolean operations. The most immediate interpretation—not the only one, however—is that of Boolean matrices, i.e., $\mathbf{0}$, $\mathbf{1}$-matrices; therefore we explain effects sometimes via rows, columns, and diagonals. Also, we sometimes refer to the entry of matrix R in row r and column c as R_{rc}.

It should be stressed that we treat heterogeneous relations, i.e. relations between possibly different sets. In theoretically oriented articles this is achieved by utilizing the typing found in categories—even when category theory is only used to an utterly moderate extent. When looking at Fig. 2.2, we better speak of typing when sets X, Y, Z are mentioned as source, resp. target, of a relation; they are otherwise considered being objects of a category. With $A : X \longrightarrow Y$, e.g., we denote a relation from X to Y.

[1] We have chosen this rather tiny symbol as a compromise with researchers who do not denote matrix multiplication at all and those using a rather dominant „ ;".

G. Schmidt, M. Winter, *Relational Topology*, Lecture Notes
in Mathematics 2208, https://doi.org/10.1007/978-3-319-74451-3_2

$$\begin{array}{c|cccc} & a & b & c & d \\ \hline a & 1 & 0 & 0 & 0 \\ b & 0 & 1 & 0 & 0 \\ c & 0 & 0 & 1 & 0 \\ d & 0 & 0 & 0 & 1 \end{array} \quad \begin{array}{c|ccccc} & 1 & 2 & 3 & 4 & 5 \\ \hline a & 1 & 0 & 0 & 1 & 0 \\ b & 0 & 0 & 0 & 0 & 0 \\ c & 1 & 0 & 0 & 0 & 1 \\ d & 0 & 1 & 0 & 0 & 0 \end{array} \quad \begin{array}{c|ccccc} & 1 & 2 & 3 & 4 & 5 \\ \hline 1 & 1 & 0 & 0 & 0 & 0 \\ 2 & 0 & 1 & 0 & 0 & 0 \\ 3 & 0 & 0 & 1 & 0 & 0 \\ 4 & 0 & 0 & 0 & 1 & 0 \\ 5 & 0 & 0 & 0 & 0 & 1 \end{array} \quad \begin{array}{c|cccc} & a & b & c & d \\ \hline 1 & 1 & 0 & 1 & 0 \\ 2 & 0 & 0 & 0 & 1 \\ 3 & 0 & 0 & 0 & 0 \\ 4 & 1 & 0 & 0 & 0 \\ 5 & 0 & 0 & 1 & 0 \end{array}$$

Fig. 2.1 A heterogeneous relation, with identities on either side, and its transpose

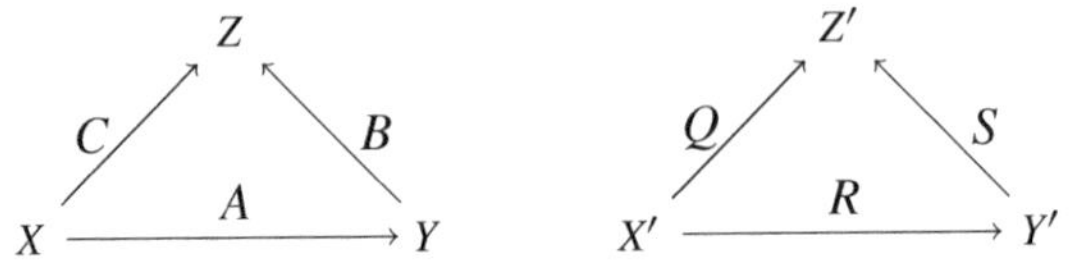

Fig. 2.2 Typing of Schröder equivalences and Dedekind rule

Still widely unknown are the Dedekind rule and the Schröder equivalences which composition and transposition in combination with the Boolean operations obey:

Dedekind rule:

$$R{:}S \cap Q \subseteq (R \cap Q{:}S^{\mathsf{T}}){:}(S \cap R^{\mathsf{T}}{:}Q)$$

Schröder equivalences:

$$A{:}B \subseteq C \quad\Longleftrightarrow\quad A^{\mathsf{T}}{:}\overline{C} \subseteq \overline{B} \quad\Longleftrightarrow\quad \overline{C}{:}B^{\mathsf{T}} \subseteq \overline{A}$$

The two rules are equivalent in the context mentioned. One will observe that the typing is valid also for the respective other side of these formulae.

Many other formulae are immediate consequences of this setting. We first mention those for the concepts of a function.

The most well-known properties of a relation Q are being *univalent*, i.e., a (possibly partial) *function*, ($Q^{\mathsf{T}}{:}\,Q \subseteq \mathbb{I}$), being *injective* (when Q^{T} is univalent), being *total* ($\mathbb{I} \subseteq Q{:}\,Q^{\mathsf{T}}$ or equivalently $Q{:}\,\mathbb{T} = \mathbb{T}$), being *surjective*, (when Q^{T} is total), and finally being a *mapping* (when univalent as well as total). We restrict ourselves in this text to use the latter word for a totally defined function. A mapping will always satisfy $f{:}\overline{A} = \overline{f{:}A}$, i.e., it may *slip below negation* from the left side.

There are three frequently applied rules that we recall here for convenience: When f is a mapping, always

$$A{:}f \subseteq B \quad\Longleftrightarrow\quad A \subseteq B{:}f^{\mathsf{T}},$$

a transition we refer to as *shunting*. When we call a transition *destroy and append*, we mean

$$(A{:}Q^{\mathsf{T}} \cap B){:}Q = A \cap B{:}Q$$

which holds for univalent Q. Of course, a univalent Q multiplied from the left side acts distributively $Q{:}(R \cap S) = Q{:}R \cap Q{:}S$.

Yet another rule is *masking* with a row-constant relation

$$(A \cap B{:}\mathbb{T}){:}C = A{:}C \cap B{:}\mathbb{T},$$

which says that one may extract rows according to $B{:}\mathbb{T}$ before or after composition of A with C. The two universal relations may here be differently typed concerning their codomain.

There remains an important point to mention concerning mappings: Whenever one has a mapping $f : X \longrightarrow Y$ in the presence of some "structure", conceived simply as a relation R_X on X, resp. R_Y on Y, it is interesting whether f somehow respects these given structures. When $R_X{:}f \subseteq f{:}R_Y$ we will speak of f being a *homomorphism*.[2] When f^{T} is also a homomorphism, i.e., f^{T} is a mapping with $R_Y{:}f^{\mathsf{T}} \subseteq f^{\mathsf{T}}{:}R_X$, we will speak of an *isomorphism*.

From time to time, we will speak of *rolling a homomorphism* as in Prop. 5.45 of [Sch11].[3] This shall express that via the homomorphism properties always

$$R_X{:}f \subseteq f{:}R_Y \iff R_X \subseteq f{:}R_Y{:}f^{\mathsf{T}} \iff f^{\mathsf{T}}{:}R_X \subseteq R_Y{:}f^{\mathsf{T}} \iff f^{\mathsf{T}}{:}R_X{:}f \subseteq R_Y.$$

Subsets are modelled as row-constant relations. When modelling subsets as row-constant relations, it is immaterial whether by the relation $v : X \longrightarrow Y$ or by the relation $v : X \longrightarrow Z$. In Fig. 2.3 it is indicated that matrices "of different breadths", related via composition with a suitable universal relation, may be used to characterize " the same" subset conceived as a vector.

Another way to represent what we conceive as a subset of a set is a subidentity, or partial diagonal $s \subseteq \mathbb{I}$. The vector $v = v{:}\mathbb{T}$ and the subidentity s may easily be converted into one another via $v = s{:}\mathbb{T}$ and $s = \mathbb{I} \cap v{:}\mathbb{T}$.

As we also have the concept of being univalent or surjective, we now introduce *points*, the algebraic counterpart of *elements* of a set as a relation p that is *row-constant* ($p = p{:}\mathbb{T}$), injective ($p{:}p^{\mathsf{T}} \subseteq \mathbb{I}$), and surjective ($\mathbb{T}{:}p = \mathbb{T}$).

When a non-commutative composition is available, one usually looks for the right and the left residual, defined via application of the Schröder rule as

$$A{:}B \subseteq C \iff B \subseteq \overline{A^{\mathsf{T}}{:}\overline{C}} =: A\backslash C \quad \text{and}$$

$$A{:}B \subseteq C \iff A \subseteq \overline{\overline{C}{:}B^{\mathsf{T}}} =: C/B.$$

[2] This concept has shown to be applicable to algebraic as well as to relational structures.

[3] At other occasions, the concept of a *cryptomorphism* will be used. Then it is assumed that we have two different relation algebraic concepts, each with an axiomatization, together with mappings forward and backwards between them, so that it is possible to prove all the axioms of concept 2 via the map_{12} and axioms of concept 1, and correspondingly vice versa. A most trivial example is the concept of a finite lattice in case 1 via its ordering and in case 2 using meet and join operations.

heart suit: $\begin{array}{c} \\ \spadesuit \\ \heartsuit \\ \diamondsuit \\ \clubsuit \end{array} \begin{array}{c} 1 \\ \begin{pmatrix} 0 \\ 1 \\ 0 \\ 0 \end{pmatrix} \end{array} \approx \begin{array}{c} \\ \spadesuit \\ \heartsuit \\ \diamondsuit \\ \clubsuit \end{array} \begin{array}{c} \mathtt{a}\ \mathtt{b}\ \mathtt{c} \\ \begin{pmatrix} 0 & 0 & 0 \\ 1 & 1 & 1 \\ 0 & 0 & 0 \\ 0 & 0 & 0 \end{pmatrix} \end{array}$ black suits: $\begin{array}{c} \\ \spadesuit \\ \heartsuit \\ \diamondsuit \\ \clubsuit \end{array} \begin{array}{c} 1\ \ 2 \\ \begin{pmatrix} 1 & 1 \\ 0 & 0 \\ 0 & 0 \\ 1 & 1 \end{pmatrix} \end{array} \approx \begin{array}{c} \\ \spadesuit \\ \heartsuit \\ \diamondsuit \\ \clubsuit \end{array} \begin{array}{c} \mathtt{a} \\ \begin{pmatrix} 1 \\ 0 \\ 0 \\ 1 \end{pmatrix} \end{array}$ $\begin{array}{c} \\ \spadesuit \\ \heartsuit \\ \diamondsuit \\ \clubsuit \end{array} \begin{array}{c} \spadesuit\ \heartsuit\ \diamondsuit\ \clubsuit \\ \begin{pmatrix} 1 & 0 & 0 & 0 \\ 1 & 1 & 0 & 0 \\ 1 & 1 & 1 & 0 \\ 1 & 1 & 1 & 1 \end{pmatrix} \end{array}$

Fig. 2.3 Point, subset, and the value ordering E on bridge card suits

The relation $A \backslash C$ describes which columns of A are contained in which columns of C

$$\left[A \backslash C\right]_{wz} = \forall v \in V : A_{vw} \rightarrow C_{vz}.$$

Among the standard tool kit for any mathematical work are orderings, denoted $a \leq b$ or $c < d$. Here however, we will work point-free without quantifiers and have to be very precise in notation distinguishing between an order E and the corresponding strictorder $C := E \cap \overline{\mathbb{I}}$. Then, of course, in the other direction $E = C \cup \mathbb{I}$. To be qualified as an order,[4] E must be *reflexive* ($\mathbb{I} \subseteq E$), *transitive* ($E ; E \subseteq E$), and *antisymmetric* ($E \cap E^{\mathsf{T}} \subseteq \mathbb{I}$).

Several order-theoretic operations will be used frequently. The first helps to obtain the set of immediate successors of a strictorder C, known as its *Hasse relation* $H := C \cap \overline{C ; C}$. The second gets the lower bounds of X with regard to some order E on U, i.e.

$$\mathtt{lbd}_E(X) := \overline{\overline{E} ; X} = E / X^{\mathsf{T}}$$

the *minorant set* of X wrt. the ordering E.

In predicate-logic style for a subset $X : U \longrightarrow \mathbb{1}$ (with $\mathbb{1}$ indicating a one-element set), this would read

$$\forall x : X_x \rightarrow E_{bx}$$

and could then be so interpreted that for any element b of the lower bound set all the x belonging to X have to satisfy $b \leq x$.

This works analogously for the other direction, providing the upper bound set, i.e.

$$\mathtt{ubd}_E(X) := \overline{\overline{E^{\mathsf{T}}} ; X} = E^{\mathsf{T}} / X^{\mathsf{T}}$$

the *majorant set* of X wrt. the ordering E.

[4]Sometimes also termed being a *partial order*,

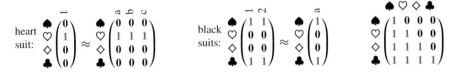

$$A = \begin{array}{c} \\ \spadesuit \\ \heartsuit \\ \diamondsuit \\ \clubsuit \end{array} \!\!\begin{array}{c} \begin{array}{cccccc} \text{Mon} & \text{Tue} & \text{Wed} & \text{Thu} & \text{Fri} & \text{Sat} \end{array} \\ \begin{pmatrix} 0 & 1 & 1 & 0 & 1 & 0 \\ 1 & 0 & 1 & 0 & 0 & 1 \\ 0 & 0 & 0 & 1 & 1 & 0 \\ 1 & 0 & 0 & 1 & 0 & 1 \end{pmatrix} \end{array} \quad B = \begin{array}{c} \\ \spadesuit \\ \heartsuit \\ \diamondsuit \\ \clubsuit \end{array} \!\!\begin{array}{c} \begin{array}{cccc} \text{red} & \text{green} & \text{blue} & \text{orange} \end{array} \\ \begin{pmatrix} 1 & 0 & 0 & 0 \\ 1 & 0 & 0 & 1 \\ 0 & 1 & 0 & 0 \\ 0 & 1 & 0 & 1 \end{pmatrix} \end{array} \quad \begin{array}{c} \\ \text{Mon} \\ \text{Tue} \\ \text{Wed} \\ \text{Thu} \\ \text{Fri} \\ \text{Sat} \end{array} \!\!\begin{array}{c} \begin{array}{cccc} \text{red} & \text{green} & \text{blue} & \text{orange} \end{array} \\ \begin{pmatrix} 0 & 0 & 0 & 1 \\ 0 & 0 & 0 & 0 \\ 1 & 0 & 0 & 0 \\ 0 & 1 & 0 & 0 \\ 0 & 0 & 0 & 0 \\ 0 & 0 & 0 & 1 \end{pmatrix} \end{array} = \mathtt{syq}(A,B)$$

Fig. 2.4 Column comparison via syq

Another functional delivers for a relation E and a subset X the—possibly empty—subset of its greatest elements $\mathtt{gre}_E(X) := X \cap \mathtt{ubd}_E(X)$ column-wise or row-wise as $\mathtt{greR}_E(X) := \left[\mathtt{gre}_E(X^\mathsf{T})\right]^\mathsf{T}$.

Having these operations available, one may proceed in the standard way; however, now in a point- as well as quantifier-free fashion, by introducing

$$\mathtt{lub}_E(X) := \mathtt{ubd}_E(X) \cap \mathtt{lbd}_E(\mathtt{ubd}_E(X)) \quad \textit{least upper bound} \text{ set of } X,$$

$$\mathtt{glb}_E(X) := \mathtt{lbd}_E(X) \cap \mathtt{ubd}_E(\mathtt{lbd}_E(X)) \quad \textit{greatest lower bound} \text{ set.}$$

The least upper bound—if it exists—is an upper bound, however, the uniquely determined least among all these. The relational construct will always exist, but may have an empty column, modelling non-existence.

By intersecting the residuals already introduced,

$$\mathtt{syq}(R,S) := \overline{R^\mathsf{T} ; \overline{S}} \cap \overline{\overline{R}^\mathsf{T} ; S} = R \backslash S \cap S^\mathsf{T} / R^\mathsf{T},$$

the *symmetric quotient* $\mathtt{syq}(R,S) : W \longrightarrow Z$ of two relations $R : V \longrightarrow W$ and $S : V \longrightarrow Z$ is defined. Symmetric quotients serve the purpose of *column comparison* when row types coincide (Fig. 2.4)

$$\left[\mathtt{syq}(R,S)\right]_{wz} = \forall v \in V : R_{vw} \leftrightarrow S_{vz}.$$

Two obvious rules for the symmetric quotient are

$$\left[\mathtt{syq}(A,B)\right]^\mathsf{T} = \mathtt{syq}(B,A) \quad \text{and} \quad \mathtt{syq}(\overline{A},\overline{B}) = \mathtt{syq}(A,B).$$

Symmetric quotients, cf. [Sch11, pages 46 and 174], are so named because they allow several cancelling rules which we here simply recall without proof. These rules are broadly unknown, but will easily be recognized as precise and powerful ones via their analogues.

The following has an analogy in the well-known $a \cdot \frac{b}{a} = b$:

Proposition 2.1.1 *Two relations A, B with common source satisfy*

$$A;\mathtt{syq}(A,B) = B \cap \mathbb{T};\mathtt{syq}(A,B), \quad \textit{therefore}$$
$$A;\mathtt{syq}(A,B) = B \quad \textit{when } \mathtt{syq}(A,B) \textit{ is surjective.} \qquad \square$$

The side-condition is to be expected: Also real-valued matrices can only be inverted when non-singular, e.g. Also the rule $\frac{b}{a} \cdot \frac{c}{b} = \frac{c}{a}$ has an analogous one for symmetric quotients:

Proposition 2.1.2 *Three relations A, B, C with common source satisfy*

$$\begin{aligned}
\mathtt{syq}(A,B);\mathtt{syq}(B,C) &= \mathtt{syq}(A,C) \cap \mathtt{syq}(A,B);\mathbb{T} \\
&= \mathtt{syq}(A,C) \cap \mathbb{T};\mathtt{syq}(B,C), \quad \textit{therefore} \\
\mathtt{syq}(A,B);\mathtt{syq}(B,C) &= \mathtt{syq}(A,C) \qquad \textit{if } \mathtt{syq}(A,B) \textit{ is total, } \textbf{\textit{or}} \\
&\qquad\qquad\qquad\qquad\quad \textit{if } \mathtt{syq}(B,C) \textit{ is surjective.}
\end{aligned}$$

The next proposition reminds us of $\frac{z}{x} : \frac{y}{x} = \frac{z}{y}$.

Proposition 2.1.3 *Relations X, Y, Z with common source always satisfy*

i) $\mathtt{syq}(X,Y) \backslash \mathtt{syq}(X,Z) \supseteq \mathtt{syq}(Y,Z)$
ii) $\mathtt{syq}(\mathtt{syq}(X,Y), \mathtt{syq}(X,Z)) \supseteq \mathtt{syq}(Y,Z)$
iii) $\mathtt{syq}(\mathtt{syq}(X,Y), \mathtt{syq}(X,Z)) = \mathtt{syq}(Y,Z)$
when both, $\mathtt{syq}(X,Y)$ ***and*** $\mathtt{syq}(X,Z)$, *are surjective.*

We present a novel and useful rule for composition of a univalent relation with a symmetric quotient. For total Q, i.e. when one has $Q;\mathbb{T} = \mathbb{T}$, the first reduces to Prop. 8.16.ii of [Sch11]:

Proposition 2.1.4 *Let f be a mapping. Then*

i) $f;\mathtt{syq}(X,Y) = \mathtt{syq}(X;f^\mathsf{T}, Y)$, *generalized to*
$Q;\mathtt{syq}(X,Y) = \mathtt{syq}(X;Q^\mathsf{T}, Y) \cap Q;\mathbb{T}$ *when Q is just univalent*
ii) $f^\mathsf{T};\mathtt{syq}(V,W) \subseteq \mathtt{syq}(V;f, W)$ *when $V = V;f;f^\mathsf{T}$*
iii) $f^\mathsf{T};\mathtt{syq}(V,W) = \mathtt{syq}(V;f, W)$ *when $V = V;f;f^\mathsf{T}$ and f is surjective*
iv) $\mathtt{syq}(A,B) \subseteq \mathtt{syq}(C;A, C;B)$ *for every relation C*

Proof

i) We show this for univalent Q, using that in such case always $Q;\overline{X} = Q;\mathbb{T} \cap \overline{Q;X}$. The mapping case f is a trivial consequence.

$$\begin{aligned}
Q;\mathtt{syq}(A,B) &= Q;\left[\overline{\overline{A}^\mathsf{T};B} \cap \overline{A^\mathsf{T};\overline{B}}\right] && \text{by definition} \\
&= Q;\overline{\overline{A}^\mathsf{T};B} \cap Q;\overline{A^\mathsf{T};\overline{B}} && \text{since } Q \text{ is univalent} \\
&= \left[Q;\mathbb{T} \cap \overline{Q;\overline{A}^\mathsf{T};B}\right] \cap \left[Q;\mathbb{T} \cap \overline{Q;A^\mathsf{T};\overline{B}}\right]
\end{aligned}$$

$$
\begin{aligned}
&= Q;\mathbb{T} \cap \overline{Q;\overline{A^{\mathsf{T}}};B} \cap \overline{Q;A^{\mathsf{T}};\overline{B}} \\
&= Q;\mathbb{T} \cap \overline{[Q;\mathbb{T} \cap \overline{Q;A^{\mathsf{T}}}];B} \cap \overline{Q;A^{\mathsf{T}};\overline{B}} \\
&= Q;\mathbb{T} \cap \overline{Q;\mathbb{T} \cap \overline{Q;A^{\mathsf{T}}};B} \cap \overline{Q;A^{\mathsf{T}};\overline{B}} \quad \text{masking} \\
&= Q;\mathbb{T} \cap \left[\overline{Q;\mathbb{T}} \cup \overline{\overline{Q;A^{\mathsf{T}}};B}\right] \cap \overline{Q;A^{\mathsf{T}};\overline{B}} \\
&= \left[Q;\mathbb{T} \cap \overline{Q;\mathbb{T}} \cap \overline{Q;A^{\mathsf{T}};\overline{B}}\right] \cup \left[Q;\mathbb{T} \cap \overline{\overline{Q;A^{\mathsf{T}}};B} \cap \overline{Q;A^{\mathsf{T}};\overline{B}}\right] \\
&= \perp\!\!\!\perp \cup \left[Q;\mathbb{T} \cap \mathtt{syq}(A;Q^{\mathsf{T}}, B)\right]
\end{aligned}
$$

ii) $f^{\mathsf{T}};\mathtt{syq}(V, W) = f^{\mathsf{T}};\mathtt{syq}(V;f;f^{\mathsf{T}}, W)$ by assumption
$= f^{\mathsf{T}};f;\mathtt{syq}(V;f, W)$ following (i)
$\subseteq \mathtt{syq}(V;f, W)$ since f is univalent

iii) follows because in the last step $f^{\mathsf{T}};f = \mathbb{I}$ when f is surjective.

iv) We show, e.g.

$$
C;B \subseteq C;B \iff C^{\mathsf{T}};\overline{C;B} \subseteq \overline{B} \iff B \subseteq \overline{C^{\mathsf{T}};\overline{C;B}}. \qquad \square
$$

Case (iii) is particularly relevant in a special setting to be encountered later: Assume an arbitrary equivalence $\Xi : X \longrightarrow X$ on a set X and consider the mapping f onto the set of classes X_Ξ modulo this equivalence, usually called the *natural projection*. Such a natural projection may be characterized by $\Xi = f;f^{\mathsf{T}}$ and $f^{\mathsf{T}};f = \mathbb{I}$. For a given equivalence Ξ, there may exist several mappings η satisfying these requirements. They are, however, all "the same" meaning that $f^{\mathsf{T}};\eta$ will always be an isomorphism.

Here are other basic rules together with their proof:

Proposition 2.1.5 *Let f be a mapping. Then*

i) f surjective	$\Longrightarrow$	$\mathtt{syq}(X, f;Y) \subseteq \mathtt{syq}(f^{\mathsf{T}};X, Y)$
ii) f surjective	$\Longrightarrow$	$\mathtt{syq}(X, Y) = \mathtt{syq}(f;X, f;Y)$
iii) f injective	$\Longrightarrow$	$\mathtt{syq}(X, f;Y) \supseteq \mathtt{syq}(f^{\mathsf{T}};X, Y)$
iv) f injective	$\Longrightarrow$	$\mathtt{syq}(X, Y) = \mathtt{syq}(f^{\mathsf{T}};X, f^{\mathsf{T}};Y)$

Proof Expanding the left and right term common for (i) and (iii) gives

$$
\overline{\overline{X^{\mathsf{T}}};f;Y} \cap \overline{X^{\mathsf{T}};\overline{f;Y}} \qquad \text{resp.} \qquad \overline{\overline{X^{\mathsf{T}};f};Y} \cap \overline{X^{\mathsf{T}};f;\overline{Y}}.
$$

The respective second parts are equal since f is a map, so that $\overline{f;Y} = f;\overline{Y}$.

i) Containment of the first ones follows from surjectivity:

$$
\mathbb{T} = \mathbb{T};f = X^{\mathsf{T}};f \cup \overline{X^{\mathsf{T}}};f \quad \Longrightarrow \quad \overline{X^{\mathsf{T}};f} \subseteq \overline{X^{\mathsf{T}}};f
$$

iii) We use the Schröder rule and injectivity in

$$X^{\mathsf{T}} ; f ; f^{\mathsf{T}} \subseteq X^{\mathsf{T}} \quad \Longleftrightarrow \quad \overline{X^{\mathsf{T}}} ; f \subseteq \overline{X^{\mathsf{T}} ; f}$$

ii) From $f^{\mathsf{T}} ; f = \mathbb{I}$ follows

$$\overline{\overline{X^{\mathsf{T}} ; f^{\mathsf{T}}} ; f ; Y} \cap \overline{X^{\mathsf{T}} ; f^{\mathsf{T}} ; \overline{f ; Y}} = \overline{\overline{X^{\mathsf{T}}} ; f^{\mathsf{T}} ; f ; Y} \cap \overline{X^{\mathsf{T}} ; f^{\mathsf{T}} ; f ; \overline{Y}} = \overline{\overline{X^{\mathsf{T}}} ; Y} \cap \overline{X^{\mathsf{T}} ; \overline{Y}}.$$

iv) This proof starts with Proposition 2.1.4.iv and then uses $f ; f^{\mathsf{T}} = \mathbb{I}$. □

Exercises

Exercise 2.1 Prove that Dedekind rule and Schöder equivalences are equivalent.

Exercise 2.2 Prove that $\mathtt{lbd}_E(\mathtt{ubd}_E(\mathtt{lbd}_E(X))) = \mathtt{lbd}_E(X)$.

Exercise 2.3 Prove that, given any equivalence Ξ, all the possible surjective mappings f, g satisfying $\Xi = f ; f^{\mathsf{T}} = g ; g^{\mathsf{T}}$ will be isomorphic, i.e. will satisfy $f ; \varphi \subseteq \varphi ; g$ and vice versa—for some adequately defined φ.

2.2 Power Operations

An important construction is the direct power accompanied by a membership relation. The symmetric quotient allows to characterize this *membership relation* $\varepsilon : A \longrightarrow \mathbf{2}^A$ between a set A and its powerset $\mathbf{2}^A$ or $\mathcal{P}(A)$ (Fig. 2.5).

The process is fairly intuitive and easy to understand from Fig. 2.6. The basic purpose is to make set arguments work together with more advanced algebraic mechanisms.

Definition 2.2.1 Given any set A, we define another set $\mathbf{2}^A$, its **direct power**, related to it by the

i) **membership** relation $\varepsilon : A \longrightarrow \mathbf{2}^A$ characterized by
$\mathtt{syq}(\varepsilon, \varepsilon) \subseteq \mathbb{I}$ and surjectivity of $\mathtt{syq}(\varepsilon, X)$ for all relations X.

$$\varepsilon = \begin{array}{c} \\ \mathrm{a} \\ \mathrm{b} \\ \mathrm{c} \end{array} \begin{array}{cccccccc} \{\} & \{a\} & \{b\} & \{a,b\} & \{c\} & \{a,c\} & \{b,c\} & \{a,b,c\} \\ \end{array}$$

$$\varepsilon = \begin{pmatrix} 0 & 1 & 0 & 1 & 0 & 1 & 0 & 1 \\ 0 & 0 & 1 & 1 & 0 & 0 & 1 & 1 \\ 0 & 0 & 0 & 0 & 1 & 1 & 1 & 1 \end{pmatrix}$$

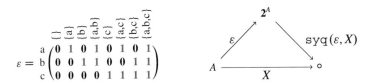

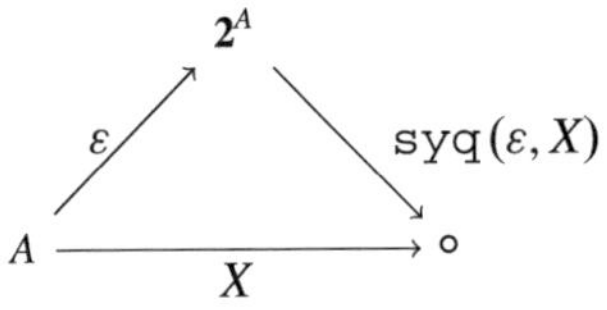

Fig. 2.5 An example of a membership and its typing

$$\varepsilon = \begin{array}{c} \\ a \\ b \\ c \\ d \end{array}\!\!\begin{array}{cccccccccccccccc} \{\} & \{a\} & \{b\} & \{a,b\} & \{c\} & \{a,c\} & \{b,c\} & abc & \{d\} & \{a,d\} & \{b,d\} & abd & \{c,d\} & \{a,c,d\} & \{b,c,d\} & \{a,b,c,d\} \\ 0&1&0&1&0&1&0&1&0&1&0&1&0&1&0&1 \\ 0&0&1&1&0&0&1&1&0&0&1&1&0&0&1&1 \\ 0&0&0&0&1&1&1&1&0&0&0&0&1&1&1&1 \\ 0&0&0&0&0&0&0&0&1&1&1&1&1&1&1&1 \end{array}$$

$$\sigma = \begin{array}{c} \\ a \\ b \\ c \\ d \end{array}\!\!\begin{array}{cccccccccccccccc} \{\} & \{a\} & \{b\} & \{a,b\} & \{c\} & \{a,c\} & \{b,c\} & abc & \{d\} & \{a,d\} & \{b,d\} & abd & \{c,d\} & \{a,c,d\} & \{b,c,d\} & \{a,b,c,d\} \\ 0&1&0&0&0&0&0&0&0&0&0&0&0&0&0&0 \\ 0&0&1&0&0&0&0&0&0&0&0&0&0&0&0&0 \\ 0&0&0&0&1&0&0&0&0&0&0&0&0&0&0&0 \\ 0&0&0&0&0&0&0&0&1&0&0&0&0&0&0&0 \end{array}$$

$$\mathbb{T}{:}\sigma^{\mathsf{T}}{:}\sigma = (0\ 1\ 1\ 0\ 1\ 0\ 0\ 0\ 1\ 0\ 0\ 0\ 0\ 0\ 0\ 0)$$

$$\begin{array}{c|cccccccccccccccc} & \{\} & \{a\} & \{b\} & \{a,b\} & \{c\} & \{a,c\} & \{b,c\} & \{a,b,c\} & \{d\} & \{a,d\} & \{b,d\} & abd & \{c,d\} & \{a,c,d\} & \{b,c,d\} & \{a,b,c,d\} \\ \hline \{\} & 0&0&0&0&0&0&0&0&0&0&0&0&0&0&0&0 \\ \{a\} & 0&1&0&0&0&0&0&0&0&0&0&0&0&0&0&0 \\ \{b\} & 0&0&1&0&0&0&0&0&0&0&0&0&0&0&0&0 \\ \{a,b\} & 0&0&0&0&0&0&0&0&0&0&0&0&0&0&0&0 \\ \{c\} & 0&0&0&0&1&0&0&0&0&0&0&0&0&0&0&0 \\ \{a,c\} & 0&0&0&0&0&0&0&0&0&0&0&0&0&0&0&0 \\ \{b,c\} & 0&0&0&0&0&0&0&0&0&0&0&0&0&0&0&0 \\ \{a,b,c\} & 0&0&0&0&0&0&0&0&0&0&0&0&0&0&0&0 \\ \{d\} & 0&0&0&0&0&0&0&0&1&0&0&0&0&0&0&0 \\ \{a,d\} & 0&0&0&0&0&0&0&0&0&0&0&0&0&0&0&0 \\ \{b,d\} & 0&0&0&0&0&0&0&0&0&0&0&0&0&0&0&0 \\ abd & 0&0&0&0&0&0&0&0&0&0&0&0&0&0&0&0 \\ \{c,d\} & 0&0&0&0&0&0&0&0&0&0&0&0&0&0&0&0 \\ \{a,c,d\} & 0&0&0&0&0&0&0&0&0&0&0&0&0&0&0&0 \\ \{b,c,d\} & 0&0&0&0&0&0&0&0&0&0&0&0&0&0&0&0 \\ \{a,b,c,d\} & 0&0&0&0&0&0&0&0&0&0&0&0&0&0&0&0 \end{array}$$

Fig. 2.6 Membership ε, singleton injection σ, and atoms, shown as vector $\sigma^{\mathsf{T}}{:}\sigma{:}\mathbb{T}$ and also as diagonal $\sigma^{\mathsf{T}}{:}\sigma$

Then one has also the derived constructs

ii) **singleton injection** $\sigma := \mathtt{syq}(\mathbb{I}, \varepsilon)$,

iii) **powerset ordering** $\Omega := \overline{\varepsilon^{\mathsf{T}}{:}\overline{\varepsilon}} = \varepsilon\backslash\varepsilon$,

iv) **powerset negation** $\mathcal{N} := \mathtt{syq}(\varepsilon, \overline{\varepsilon})$. □

Membership, thus algebraically characterized, is known to be determined uniquely up to isomorphism; cf. [Sch11, p. 141]. The equivalent version $\Omega = \varepsilon\backslash\varepsilon$ of the powerset order makes indeed clear that columns of ε—each of them representing a subset—are investigated as to whether they are contained in columns of ε. Any expression $\mathtt{syq}(X, \varepsilon)$ will turn out to be a mapping. For any relation $R : B \longrightarrow A$, we will get a mapping

$$\mathtt{syq}(R^{\mathsf{T}}, \varepsilon) : B \longrightarrow \mathbf{2}^A.$$

Then trivially $\varepsilon{:}\mathcal{N} = \overline{\varepsilon}$. We may also introduce from singleton injection σ the *atoms* $a := \sigma^{\mathsf{T}}{:}\sigma$. From Fig. 2.6, one may derive intuition as to the recursive formation of ε, σ, and a.

Some useful technicalities around $\varepsilon, \sigma, \Omega$ follow.

Lemma 2.2.2

i) $\varepsilon{:}\Omega = \varepsilon \qquad \varepsilon{:}\sigma^{\mathsf{T}} = \mathbb{I} \qquad \sigma^{\mathsf{T}}{:}\varepsilon \subseteq \Omega$

ii) $\sigma{:}\Omega^{\mathsf{T}} = \overline{\overline{\mathbb{I}}{:}\varepsilon} = \sigma \cup \overline{\mathbb{T}{:}\varepsilon}$

iii) $\sigma{:}\Omega = \varepsilon$

iv) $\varepsilon = \sigma \cup (\varepsilon \cap \overline{\mathbb{T}{:}\sigma})$

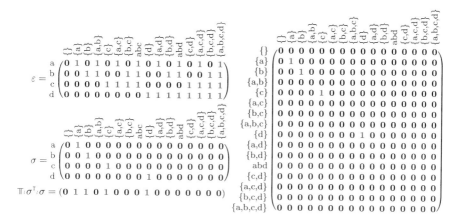

Proof

i) The following last step is equivalent with $\varepsilon^{\mathsf{T}};\overline{\varepsilon} \subseteq \overline{\Omega}$, which is true:

$$\varepsilon \subseteq \varepsilon;\Omega \subseteq \varepsilon$$

$$\varepsilon;\sigma^{\mathsf{T}} = \varepsilon;\mathtt{syq}(\varepsilon, \mathbb{I}) = \mathbb{I} \quad \text{cancellation Proposition 2.1.1}$$

$$\sigma;\overline{\Omega} = \sigma;\varepsilon^{\mathsf{T}};\overline{\varepsilon} = \overline{\varepsilon} \quad \Longrightarrow \quad \sigma^{\mathsf{T}};\varepsilon \subseteq \Omega \quad \text{with second of (i) and Schröder rule}$$

ii)

$$\sigma \cup \overline{\mathbb{T};\varepsilon} = \left[\overline{\overline{\mathbb{I}};\varepsilon} \cap \varepsilon\right] \cup \overline{(\overline{\mathbb{I}} \cup \mathbb{I});\varepsilon} = \overline{\overline{\mathbb{I}};\varepsilon} \cap (\varepsilon \cup \overline{\varepsilon}) = \overline{\overline{\mathbb{I}};\varepsilon}$$

$$\sigma;\Omega^{\mathsf{T}} = \sigma;\overline{\overline{\varepsilon^{\mathsf{T}}};\varepsilon} = \overline{\overline{\sigma;\varepsilon^{\mathsf{T}}};\varepsilon} = \overline{\overline{\mathbb{I}};\varepsilon}, \quad \text{using (i)}$$

iii)

$$\sigma;\Omega = \sigma;\overline{\varepsilon^{\mathsf{T}};\overline{\varepsilon}} = \overline{\sigma;\varepsilon^{\mathsf{T}};\overline{\varepsilon}} = \overline{\mathbb{I};\overline{\varepsilon}} = \varepsilon, \quad \text{using (i)}$$

iv) "$\supseteq$" is obvious. For "$\subseteq$", it suffices to prove

$$\mathbb{T};\sigma \cap \varepsilon \subseteq (\mathbb{T} \cap \varepsilon;\sigma^{\mathsf{T}});(\sigma \cap \mathbb{T};\varepsilon) = \mathbb{I};(\sigma \cap \mathbb{T};\varepsilon) \subseteq \sigma \text{ using (i).} \qquad \square$$

Note in particular the intuition captured by (ii): Going with σ to a singleton set and then to sets contained therein, one will arrive at singleton sets and the empty set; these don't contain other elements, via $\overline{\mathbb{I}}$, than the given one.

The following result often allows to reduce the size of a complicated formula.

Proposition 2.2.3 *Membership ε obeys these two simplifying rules, to be referred later as* membership deletions*:*

i) $\varepsilon;\overline{\varepsilon^{\mathsf{T}};Z} = \overline{Z}$ *and* $\overline{\varepsilon};\overline{\overline{\varepsilon}^{\mathsf{T}};Z} = \overline{Z}$ *for arbitrary* Z
ii) $\mathtt{syq}(\varepsilon^{\mathsf{T}};\overline{\varepsilon};X, \varepsilon^{\mathsf{T}};\overline{\varepsilon};Y) = \mathtt{syq}(\overline{\varepsilon};X, \overline{\varepsilon};Y)$

Proof

i) We deduce equality from the following chain of containments

$$Z = \varepsilon;\mathtt{syq}(\varepsilon, Z) \quad \text{cancellation Proposition 2.1.1 and definition of membership}$$

$$= \varepsilon;(\overline{\overline{\varepsilon}^{\mathsf{T}};Z} \cap \overline{\varepsilon^{\mathsf{T}};\overline{Z}}) \quad \text{expanding the symmetric quotient}$$

$$\subseteq \varepsilon;\overline{\varepsilon^{\mathsf{T}};\overline{Z}}$$

$$\subseteq Z \quad \text{Schröder rule}$$

For the second result, $\mathcal{N}$ can be used together with the former:

$$\overline{\varepsilon};\overline{\overline{\varepsilon}^{\mathsf{T}};Z} = \varepsilon;\mathcal{N};\overline{\mathcal{N};\varepsilon^{\mathsf{T}};Z} = \varepsilon;\overline{\varepsilon^{\mathsf{T}};Z} = Z$$

ii) The result (i) is applied twice in the expansion

$$\mathtt{syq}(\varepsilon^{\mathsf{T}};\overline{\varepsilon};X, \varepsilon^{\mathsf{T}};\overline{\varepsilon};Y) = \overline{\overline{X^{\mathsf{T}};\overline{\varepsilon}^{\mathsf{T}}};\varepsilon;\varepsilon^{\mathsf{T}};\overline{\varepsilon};Y} \cap \overline{X^{\mathsf{T}};\overline{\varepsilon}^{\mathsf{T}};\varepsilon;\overline{\varepsilon^{\mathsf{T}};\overline{\varepsilon};Y}}. \qquad \square$$

Following such ideas, we develop an interesting interrelationship between relations and their counterparts that hold between the corresponding powersets. It offers the possibility to work algebraically in situations in which this has so far not been the classical approach.

Definition 2.2.4 Let any relation $R : X \longrightarrow Y$ be given together with membership relations $\varepsilon : X \longrightarrow \mathbf{2}^X, \varepsilon' : Y \longrightarrow \mathbf{2}^Y$. Then the **existential image mapping** for R is defined as

$$\vartheta_R := \mathtt{syq}(R^{\mathsf{T}};\varepsilon, \varepsilon').$$

One may correspondingly study the existential image mapping for R^{T}, also called the **inverse image mapping** for R, defined as

$$\vartheta_{R^{\mathsf{T}}} = \mathtt{syq}(R;\varepsilon', \varepsilon). \qquad \square$$

Both constructs, see e.g. Fig. 2.8, are necessarily mappings. Forming the existential image is known to be a multiplicative operation, $\vartheta_R ; \vartheta_S = \vartheta_{R;S}$. The existential image satisfies in addition $\vartheta_{\mathbb{I}_X} = \mathbb{I}_{\mathbf{2}^X}$.[5]

We further recall the interesting facts concerning the existential and the inverse image; see [Sch11]. One may state them roughly by saying that the diagram Fig. 2.7 is commutative.

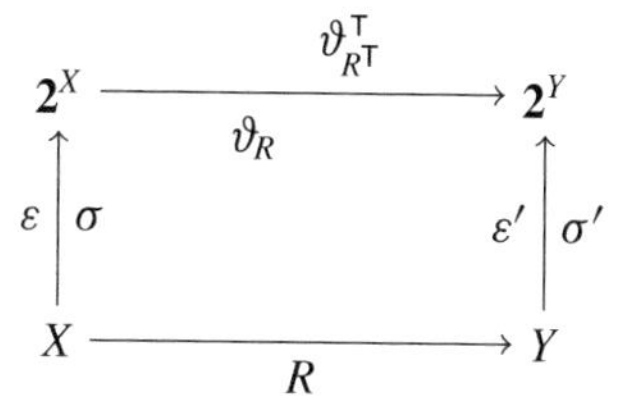

Fig. 2.7 Typing in case of the existential and the inverse image

[5] Therefore, the existential image is a monoid morphism as well as a functor of the category of relations into the subcategory of mappings.

$R =$

	a	b	c	d
1	0	1	0	1
2	1	0	0	0
3	0	0	1	0
4	0	0	0	1
5	1	0	1	0

$\vartheta_R =$

	{}	{a}	{b}	{a,b}	{c}	{a,c}	{b,c}	{a,b,c}	{d}	{a,d}	{b,d}	{a,b,d}	{c,d}	{a,c,d}	{b,c,d}	{a,b,c,d}
{}	1	0	0	0	0	0	0	0	0	0	0	0	0	0	0	0
{1}	0	0	0	0	0	0	0	0	0	0	1	0	0	0	0	0
{2}	0	1	0	0	0	0	0	0	0	0	0	0	0	0	0	0
{1,2}	0	0	0	0	0	0	0	0	0	0	0	1	0	0	0	0
{3}	0	0	0	0	1	0	0	0	0	0	0	0	0	0	0	0
{1,3}	0	0	0	0	0	0	0	0	0	0	0	0	0	0	1	0
{2,3}	0	0	0	0	0	1	0	0	0	0	0	0	0	0	0	0
{1,2,3}	0	0	0	0	0	0	0	0	0	0	0	0	0	0	0	1
{4}	0	0	0	0	0	0	0	0	1	0	0	0	0	0	0	0
{1,4}	0	0	0	0	0	0	0	0	0	0	1	0	0	0	0	0
{2,4}	0	0	0	0	0	0	0	0	0	1	0	0	0	0	0	0
{1,2,4}	0	0	0	0	0	0	0	0	0	0	0	1	0	0	0	0
{3,4}	0	0	0	0	0	0	0	0	0	0	0	0	1	0	0	0
{1,3,4}	0	0	0	0	0	0	0	0	0	0	0	0	0	0	1	0
{2,3,4}	0	0	0	0	0	0	0	0	0	0	0	0	0	1	0	0
{1,2,3,4}	0	0	0	0	0	0	0	0	0	0	0	0	0	0	0	1
{5}	0	0	0	0	0	1	0	0	0	0	0	0	0	0	0	0
{1,5}	0	0	0	0	0	0	0	0	0	0	0	0	0	0	0	1
{2,5}	0	0	0	0	0	1	0	0	0	0	0	0	0	0	0	0
{1,2,5}	0	0	0	0	0	0	0	0	0	0	0	0	0	0	0	1
{3,5}	0	0	0	0	0	1	0	0	0	0	0	0	0	0	0	0
{1,3,5}	0	0	0	0	0	0	0	0	0	0	0	0	0	0	0	1
{2,3,5}	0	0	0	0	0	1	0	0	0	0	0	0	0	0	0	0
{1,2,3,5}	0	0	0	0	0	0	0	0	0	0	0	0	0	0	0	1
{4,5}	0	0	0	0	0	0	0	0	0	0	0	0	0	1	0	0
{1,4,5}	0	0	0	0	0	0	0	0	0	0	0	0	0	0	0	1
{2,4,5}	0	0	0	0	0	0	0	0	0	0	0	0	0	1	0	0
{1,2,4,5}	0	0	0	0	0	0	0	0	0	0	0	0	0	0	0	1
{3,4,5}	0	0	0	0	0	0	0	0	0	0	0	0	0	1	0	0
{1,3,4,5}	0	0	0	0	0	0	0	0	0	0	0	0	0	0	0	1
{2,3,4,5}	0	0	0	0	0	0	0	0	0	0	0	0	0	1	0	0
{1,2,3,4,5}	0	0	0	0	0	0	0	0	0	0	0	0	0	0	0	1

$\vartheta_{R^\mathsf{T}} =$

	{}	{1}	{2}	{1,2}	{3}	{1,3}	{2,3}	{1,2,3}	{4}	{1,4}	{2,4}	{1,2,4}	{3,4}	{1,3,4}	{2,3,4}	{1,2,3,4}	{5}	{1,5}	{2,5}	{1,2,5}	{3,5}	{1,3,5}	{2,3,5}	{1,2,3,5}	{4,5}	{1,4,5}	{2,4,5}	{1,2,4,5}	{3,4,5}	{1,3,4,5}	{2,3,4,5}	{1,2,3,4,5}
{}	1	0	0	0	0	0	0	0	0	0	0	0	0	0	0	0	0	0	0	0	0	0	0	0	0	0	0	0	0	0	0	0
{a}	0	0	0	0	0	0	0	0	0	0	0	0	0	0	0	0	0	0	1	0	0	0	0	0	0	0	0	0	0	0	0	0
{b}	0	1	0	0	0	0	0	0	0	0	0	0	0	0	0	0	0	0	0	0	0	0	0	0	0	0	0	0	0	0	0	0
{a,b}	0	0	0	0	0	0	0	0	0	0	0	0	0	0	0	0	0	0	0	1	0	0	0	0	0	0	0	0	0	0	0	0
{c}	0	0	0	0	0	0	0	0	0	0	0	0	0	0	0	0	0	0	0	0	1	0	0	0	0	0	0	0	0	0	0	0
{a,c}	0	0	0	0	0	0	0	0	0	0	0	0	0	0	0	0	0	0	0	0	0	0	1	0	0	0	0	0	0	0	0	0
{b,c}	0	0	0	0	0	0	0	0	0	0	0	0	0	0	0	0	0	0	0	0	0	1	0	0	0	0	0	0	0	0	0	0
{a,b,c}	0	0	0	0	0	0	0	0	0	0	0	0	0	0	0	0	0	0	0	0	0	0	0	1	0	0	0	0	0	0	0	0
{d}	0	0	0	0	0	0	0	0	0	1	0	0	0	0	0	0	0	0	0	0	0	0	0	0	0	0	0	0	0	0	0	0
{a,d}	0	0	0	0	0	0	0	0	0	0	0	0	0	0	0	0	0	0	0	0	0	0	0	0	0	0	0	1	0	0	0	0
{b,d}	0	0	0	0	0	0	0	0	0	1	0	0	0	0	0	0	0	0	0	0	0	0	0	0	0	0	0	0	0	0	0	0
{a,b,d}	0	0	0	0	0	0	0	0	0	0	0	0	0	0	0	0	0	0	0	0	0	0	0	0	0	0	0	1	0	0	0	0
{c,d}	0	0	0	0	0	0	0	0	0	0	0	0	0	0	0	0	0	0	0	0	0	0	0	0	0	0	0	0	0	1	0	0
{a,c,d}	0	0	0	0	0	0	0	0	0	0	0	0	0	0	0	0	0	0	0	0	0	0	0	0	0	0	0	0	0	0	0	1
{b,c,d}	0	0	0	0	0	0	0	0	0	0	0	0	0	0	0	0	0	0	0	0	0	0	0	0	0	0	0	0	0	1	0	0
{a,b,c,d}	0	0	0	0	0	0	0	0	0	0	0	0	0	0	0	0	0	0	0	0	0	0	0	0	0	0	0	0	0	0	0	1

Fig. 2.8 To illustrate the concepts of existential and inverse image of a relation R

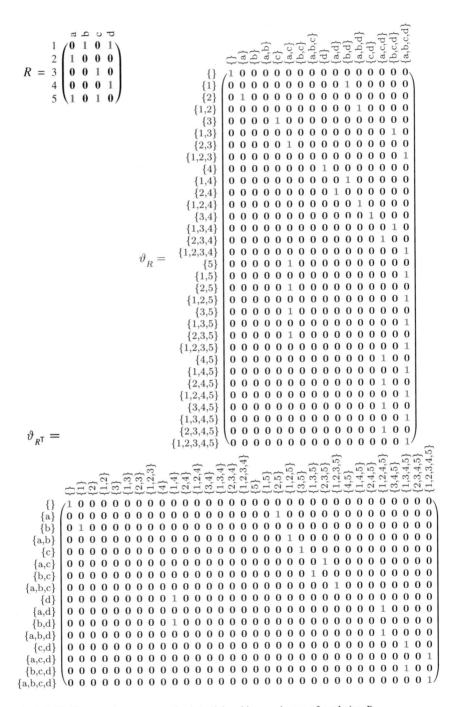

Proposition 2.2.5

i) $R^\mathsf{T}; \varepsilon = \varepsilon'; \vartheta_R^\mathsf{T}$
ii) $R; \varepsilon' = \varepsilon; \vartheta_{R^\mathsf{T}}^\mathsf{T}$
iii) $R = \overline{\varepsilon; \vartheta_R; \overline{\varepsilon'}^\mathsf{T}}$

Proof

i) $\varepsilon'; \vartheta_R^\mathsf{T} = \varepsilon'; \mathtt{syq}(\varepsilon', R^\mathsf{T}; \varepsilon) = R^\mathsf{T}; \varepsilon$
ii) is the same result as (i), but correspondingly formulated for R^T.
iii)

$$
\begin{aligned}
\overline{\varepsilon; \vartheta_R; \overline{\varepsilon'}^\mathsf{T}} &= \overline{\varepsilon; \overline{\vartheta_R; \varepsilon'^\mathsf{T}}} && \text{since } \vartheta_R \text{ is a mapping}\\
&= \overline{\varepsilon; \overline{\varepsilon^\mathsf{T}; R}} && \text{following (i)}\\
&= R && \text{membership deletion Proposition 2.2.3}
\end{aligned}
$$

□

Another rule combines the inverse image with the singleton injection.

Proposition 2.2.6 *If R is an arbitrary relation or f is a mapping, and σ, σ' are the singleton injections on source and target side, then*

i) $\sigma; \vartheta_{R^\mathsf{T}}^\mathsf{T}; \sigma'^\mathsf{T} \subseteq \varepsilon; \vartheta_{R^\mathsf{T}}^\mathsf{T}; \sigma'^\mathsf{T} = R,$
ii) $\sigma; \vartheta_f = f; \sigma'.$

Proof

i)

$$
\begin{aligned}
\sigma; \vartheta_{R^\mathsf{T}}^\mathsf{T}; \sigma'^\mathsf{T} &\subseteq \varepsilon; \vartheta_{R^\mathsf{T}}^\mathsf{T}; \sigma'^\mathsf{T}\\
&= R; \varepsilon'; \sigma'^\mathsf{T} && \text{Proposition 2.2.5.ii}\\
&= R && \text{Proposition 2.2.2.i}
\end{aligned}
$$

ii)

$$
\begin{aligned}
\sigma; \vartheta_f &= \sigma; \mathtt{syq}(f^\mathsf{T}; \varepsilon, \varepsilon') && \text{definition of } \vartheta_f\\
&= \mathtt{syq}(f^\mathsf{T}; \varepsilon; \sigma^\mathsf{T}, \varepsilon') && \text{Proposition 2.1.4.i}\\
&= \mathtt{syq}(f^\mathsf{T}, \varepsilon') && \text{Proposition 2.2.2.i}\\
&= f; \mathtt{syq}(\mathbb{I}, \varepsilon') = f; \sigma' && \text{Proposition 2.1.4 again and definition of } \sigma
\end{aligned}
$$

□

An interpretation is easy via 1-element sets; e.g. for (ii): f maps $x \mapsto f(x)$, which is with the second singleton injection sent to $\{f(x)\}$; on the other hand, singleton injection σ sends this element x to the 1-element set $\{x\}$ that is then via ϑ_f mapped to $\{f(x)\}$.

The following rules are important when the topology concept of continuity is transferred to a point-free and quantifier-free relation-algebraic version. The condition in this proposition is mainly useful for mappings, but may be formulated slightly more general for a difunctional R, i.e., a relation that satisfies: $R{:}R^{\mathsf T}{:}R = R$.

Proposition 2.2.7 *For an arbitrary relation $R : X \longrightarrow Y$ we have always*

i) $\vartheta^{\mathsf T}_{R^{\mathsf T}}{:}\vartheta_{R^{\mathsf T}}{:}\vartheta_R = \vartheta^{\mathsf T}_{R^{\mathsf T}}{:}\mathbb{T} \cap \vartheta_R$,
ii) $\vartheta^{\mathsf T}_{R^{\mathsf T}}{:}\vartheta^{\mathsf T}_R{:}\vartheta_R = \vartheta^{\mathsf T}_{R^{\mathsf T}} \cap \mathbb{T}{:}\vartheta_R$,
iii) $\vartheta^{\mathsf T}_{R^{\mathsf T}} \cap \mathbb{T}{:}\vartheta_R \subseteq \vartheta_R$ *if $R{:}R^{\mathsf T}{:}R = R$, in particular when R is a map,*
iv) $\vartheta^{\mathsf T}_{R^{\mathsf T}}{:}\mathbb{T} \cap \vartheta_R \subseteq \vartheta^{\mathsf T}_{R^{\mathsf T}}$ *if $R{:}R^{\mathsf T}{:}R = R$, in particular when R is a map.*

Proof We prove only (i,iii), since (ii,iv) are the same results, however, applied to $R^{\mathsf T}$ and then transposed.

i) The chain of containments implies equality in between:

$$
\begin{aligned}
&\vartheta^{\mathsf T}_{R^{\mathsf T}}{:}\mathbb{T} \cap \vartheta_R \\
&\subseteq (\vartheta^{\mathsf T}_{R^{\mathsf T}} \cap \vartheta_R{:}\mathbb{T}){:}(\mathbb{T} \cap \vartheta_{R^{\mathsf T}}{:}\vartheta_R) && \text{Dedekind rule}\\
&\subseteq \vartheta^{\mathsf T}_{R^{\mathsf T}}{:}\vartheta_{R^{\mathsf T}}{:}\vartheta_R \\
&\subseteq \vartheta^{\mathsf T}_{R^{\mathsf T}}{:}\mathbb{T} \cap \vartheta_R && \text{since existential images are univalent}
\end{aligned}
$$

iii)

$$
\begin{aligned}
&\vartheta^{\mathsf T}_R{:}(\vartheta^{\mathsf T}_{R^{\mathsf T}} \cap \mathbb{T}{:}\vartheta_R) \\
&= \vartheta^{\mathsf T}_R{:}\vartheta^{\mathsf T}_{R^{\mathsf T}}{:}\vartheta^{\mathsf T}_R{:}\vartheta_R && \text{due to (ii)}\\
&= \vartheta^{\mathsf T}_{R{:}R^{\mathsf T}{:}R}{:}\vartheta_R && \text{existential images act multiplicatively}\\
&= \vartheta^{\mathsf T}_R{:}\vartheta_R \subseteq \mathbb{I} && \text{by assumption } R{:}R^{\mathsf T}{:}R = R
\end{aligned}
$$

Now shunting gives the result. □

Remarkable algebraic properties may be proved when an injective, respectively surjective, mapping is lifted to its existential image.

Proposition 2.2.8 *Let f be a mapping. According to its being surjective or injective, it will in both cases have four consequences:*

i) f *surjective* $\Longrightarrow$ $\vartheta^{\mathsf T}_{f^{\mathsf T}} \subseteq \vartheta_f$ $\quad$ ϑ_f *surjective* $\quad$ $\vartheta_{f^{\mathsf T}}$ *injective*
$\vartheta^{\mathsf T}_{f^{\mathsf T}} = \vartheta_f \cap \vartheta^{\mathsf T}_{f^{\mathsf T}}{:}\mathbb{T}$

ii) f *injective* $\Longrightarrow$ $\vartheta^{\mathsf T}_{f^{\mathsf T}} \supseteq \vartheta_f$ $\quad$ ϑ_f *injective* $\quad$ $\vartheta_{f^{\mathsf T}}$ *surjective*
$\vartheta_f = \vartheta^{\mathsf T}_{f^{\mathsf T}} \cap \mathbb{T}{:}\vartheta_f$

Proof

i) Setting $X := \varepsilon$ and $Y := \varepsilon'$, the first result is simply Proposition 2.1.5.i. This in turn allows us to reason

$$\vartheta^{\mathsf T}_{f^{\mathsf T}} \subseteq \vartheta_f \quad\Longrightarrow\quad \mathbb{T} = \mathbb{T}{:}\vartheta^{\mathsf T}_{f^{\mathsf T}} \subseteq \mathbb{T}{:}\vartheta_f,$$

so that ϑ_f is surjective. And now

$$\vartheta_{f^\mathsf{T}} ; \vartheta_{f^\mathsf{T}}^\mathsf{T} \subseteq \vartheta_{f^\mathsf{T}} ; \vartheta_f = \vartheta_{f^\mathsf{T};f} = \vartheta_{\mathbb{I}} = \mathbb{I},$$

making ϑ_{f^T} injective, because the map f has been assumed to be surjective, i.e. $f^\mathsf{T};f = \mathbb{I}$. The last formula is then a consequence of Proposition 2.2.7.iv.

ii) We apply Proposition 2.1.5.iii and get the first result in a similar way. Then

$$\vartheta_{f^\mathsf{T}}^\mathsf{T} \supseteq \vartheta_f \quad\Longrightarrow\quad \vartheta_f ; \vartheta_f^\mathsf{T} \subseteq \vartheta_{f^\mathsf{T}}^\mathsf{T} ; \vartheta_{f^\mathsf{T}} \subseteq \mathbb{I}$$

provides injectivity. Concerning surjectivity, we estimate correspondingly

$$\vartheta_{f^\mathsf{T}}^\mathsf{T} ; \vartheta_{f^\mathsf{T}} \supseteq \vartheta_f ; \vartheta_{f^\mathsf{T}} = \vartheta_{f;f^\mathsf{T}} = \vartheta_{\mathbb{I}} = \mathbb{I}.$$

The last formula is then a consequence of Proposition 2.2.7.iii. □

We illustrate the sometimes astonishing behaviour of existential and inverse images with Figs. 2.9 and 2.10.

The following concerns possibly empty columns of R, resp. f.

Proposition 2.2.9 *Assume a relation and a map* $R, f : X \longrightarrow Y$.

i) $\varepsilon' ; \vartheta_{R^\mathsf{T}} \cap R^\mathsf{T} ; \mathbb{T} \subseteq R^\mathsf{T} ; \varepsilon = \varepsilon' ; \vartheta_R^\mathsf{T}$

ii) $\varepsilon' ; \vartheta_{f^\mathsf{T}} \cap f^\mathsf{T} ; \mathbb{T} = f^\mathsf{T} ; \varepsilon = \varepsilon' ; \vartheta_f^\mathsf{T}$ *when f is an injective map*

Proof

i)

$$\begin{aligned} R^\mathsf{T};\mathbb{T} \cap \varepsilon' ; \vartheta_{R^\mathsf{T}} &\subseteq (R^\mathsf{T} \cap \varepsilon' ; \vartheta_{R^\mathsf{T}} ; \mathbb{T}) ; (\mathbb{T} \cap R ; \varepsilon' ; \vartheta_{R^\mathsf{T}}) && \text{Dedekind rule} \\ &\subseteq R^\mathsf{T} ; R ; \varepsilon' ; \vartheta_{R^\mathsf{T}} \\ &\subseteq R^\mathsf{T} ; \varepsilon ; \vartheta_{R^\mathsf{T}}^\mathsf{T} ; \vartheta_{R^\mathsf{T}} && \text{Proposition 2.2.5.ii} \\ &\subseteq R^\mathsf{T} ; \varepsilon && \text{since inverse images are mappings and thus univalent} \\ &= \varepsilon' ; \vartheta_R^\mathsf{T} && \text{Proposition 2.2.5.i} \end{aligned}$$

ii) "$\subseteq$" has already been shown with (i). It remains to prove

$f^\mathsf{T} ; \varepsilon \subseteq \varepsilon' ; \vartheta_{f^\mathsf{T}} \iff \varepsilon \subseteq f ; \varepsilon' ; \vartheta_{f^\mathsf{T}} = \varepsilon ; \vartheta_{f^\mathsf{T}}^\mathsf{T} ; \vartheta_{f^\mathsf{T}}$ shunting, Proposition 2.2.5.ii

Now f injective implies ϑ_{f^T} surjective due to Proposition 2.2.8.ii. □

The existential image and the inverse image also satisfy several homomorphism-like formulae with respect to the powerset orderings. Not least is ϑ_{f^T} monotonic.

$$f = \begin{array}{c} \\ 1 \\ 2 \\ 3 \\ 4 \\ 5 \end{array}\begin{array}{c} \text{a b c d} \\ \begin{pmatrix} 0&1&0&0 \\ 1&0&0&0 \\ 0&0&1&0 \\ 0&0&0&1 \\ 0&0&1&0 \end{pmatrix} \end{array}$$

$\vartheta_f =$

	{}	{a}	{b}	{a,b}	{c}	{a,c}	{b,c}	{a,b,c}	{d}	{a,d}	{b,d}	{a,b,d}	{c,d}	{a,c,d}	{b,c,d}	{a,b,c,d}
{}	1	0	0	0	0	0	0	0	0	0	0	0	0	0	0	0
{1}	0	0	1	0	0	0	0	0	0	0	0	0	0	0	0	0
{2}	0	1	0	0	0	0	0	0	0	0	0	0	0	0	0	0
{1,2}	0	0	0	1	0	0	0	0	0	0	0	0	0	0	0	0
{3}	0	0	0	0	1	0	0	0	0	0	0	0	0	0	0	0
{1,3}	0	0	0	0	0	0	1	0	0	0	0	0	0	0	0	0
{2,3}	0	0	0	0	0	1	0	0	0	0	0	0	0	0	0	0
{1,2,3}	0	0	0	0	0	0	0	1	0	0	0	0	0	0	0	0
{4}	0	0	0	0	0	0	0	0	1	0	0	0	0	0	0	0
{1,4}	0	0	0	0	0	0	0	0	0	0	1	0	0	0	0	0
{2,4}	0	0	0	0	0	0	0	0	0	1	0	0	0	0	0	0
{1,2,4}	0	0	0	0	0	0	0	0	0	0	0	1	0	0	0	0
{3,4}	0	0	0	0	0	0	0	0	0	0	0	0	1	0	0	0
{1,3,4}	0	0	0	0	0	0	0	0	0	0	0	0	0	0	1	0
{2,3,4}	0	0	0	0	0	0	0	0	0	0	0	0	0	1	0	0
{1,2,3,4}	0	0	0	0	0	0	0	0	0	0	0	0	0	0	0	1
{5}	0	0	0	0	1	0	0	0	0	0	0	0	0	0	0	0
{1,5}	0	0	0	0	0	0	1	0	0	0	0	0	0	0	0	0
{2,5}	0	0	0	0	0	1	0	0	0	0	0	0	0	0	0	0
{1,2,5}	0	0	0	0	0	0	0	1	0	0	0	0	0	0	0	0
{3,5}	0	0	0	0	1	0	0	0	0	0	0	0	0	0	0	0
{1,3,5}	0	0	0	0	0	0	1	0	0	0	0	0	0	0	0	0
{2,3,5}	0	0	0	0	0	1	0	0	0	0	0	0	0	0	0	0
{1,2,3,5}	0	0	0	0	0	0	0	1	0	0	0	0	0	0	0	0
{4,5}	0	0	0	0	0	0	0	0	0	0	0	0	1	0	0	0
{1,4,5}	0	0	0	0	0	0	0	0	0	0	0	0	0	0	1	0
{2,4,5}	0	0	0	0	0	0	0	0	0	0	0	0	0	1	0	0
{1,2,4,5}	0	0	0	0	0	0	0	0	0	0	0	0	0	0	0	1
{3,4,5}	0	0	0	0	0	0	0	0	0	0	0	0	1	0	0	0
{1,3,4,5}	0	0	0	0	0	0	0	0	0	0	0	0	0	0	1	0
{2,3,4,5}	0	0	0	0	0	0	0	0	0	0	0	0	0	1	0	0
{1,2,3,4,5}	0	0	0	0	0	0	0	0	0	0	0	0	0	0	0	1

$\vartheta^{\mathsf{T}}_{f^{\mathsf{T}}} =$

	{}	{a}	{b}	{a,b}	{c}	{a,c}	{b,c}	{a,b,c}	{d}	{a,d}	{b,d}	{a,b,d}	{c,d}	{a,c,d}	{b,c,d}	{a,b,c,d}
{}	1	0	0	0	0	0	0	0	0	0	0	0	0	0	0	0
{1}	0	0	1	0	0	0	0	0	0	0	0	0	0	0	0	0
{2}	0	1	0	0	0	0	0	0	0	0	0	0	0	0	0	0
{1,2}	0	0	0	1	0	0	0	0	0	0	0	0	0	0	0	0
{3}	0	0	0	0	0	0	0	0	0	0	0	0	0	0	0	0
{1,3}	0	0	0	0	0	0	0	0	0	0	0	0	0	0	0	0
{2,3}	0	0	0	0	0	0	0	0	0	0	0	0	0	0	0	0
{1,2,3}	0	0	0	0	0	0	0	0	0	0	0	0	0	0	0	0
{4}	0	0	0	0	0	0	0	0	1	0	0	0	0	0	0	0
{1,4}	0	0	0	0	0	0	0	0	0	0	1	0	0	0	0	0
{2,4}	0	0	0	0	0	0	0	0	0	1	0	0	0	0	0	0
{1,2,4}	0	0	0	0	0	0	0	0	0	0	0	1	0	0	0	0
{3,4}	0	0	0	0	0	0	0	0	0	0	0	0	0	0	0	0
{1,3,4}	0	0	0	0	0	0	0	0	0	0	0	0	0	0	0	0
{2,3,4}	0	0	0	0	0	0	0	0	0	0	0	0	0	0	0	0
{1,2,3,4}	0	0	0	0	0	0	0	0	0	0	0	0	0	0	0	0
{5}	0	0	0	0	0	0	0	0	0	0	0	0	0	0	0	0
{1,5}	0	0	0	0	0	0	0	0	0	0	0	0	0	0	0	0
{2,5}	0	0	0	0	0	0	0	0	0	0	0	0	0	0	0	0
{1,2,5}	0	0	0	0	0	0	0	0	0	0	0	0	0	0	0	0
{3,5}	0	0	0	0	1	0	0	0	0	0	0	0	0	0	0	0
{1,3,5}	0	0	0	0	0	0	1	0	0	0	0	0	0	0	0	0
{2,3,5}	0	0	0	0	0	1	0	0	0	0	0	0	0	0	0	0
{1,2,3,5}	0	0	0	0	0	0	0	1	0	0	0	0	0	0	0	0
{4,5}	0	0	0	0	0	0	0	0	0	0	0	0	0	0	0	0
{1,4,5}	0	0	0	0	0	0	0	0	0	0	0	0	0	0	0	0
{2,4,5}	0	0	0	0	0	0	0	0	0	0	0	0	0	0	0	0
{1,2,4,5}	0	0	0	0	0	0	0	0	0	0	0	0	0	0	0	0
{3,4,5}	0	0	0	0	0	0	0	0	0	0	0	0	1	0	0	0
{1,3,4,5}	0	0	0	0	0	0	0	0	0	0	0	0	0	0	1	0
{2,3,4,5}	0	0	0	0	0	0	0	0	0	0	0	0	0	1	0	0
{1,2,3,4,5}	0	0	0	0	0	0	0	0	0	0	0	0	0	0	0	1

Fig. 2.9 Existential and inverse image for a surjective and non-injective mapping

$$f = \begin{array}{c} \\ 1 \\ 2 \\ 3 \end{array}\begin{array}{c} \text{a b c d} \\ \begin{pmatrix} 0&0&1&0 \\ 0&1&0&0 \\ 1&0&0&0 \end{pmatrix} \end{array}$$

$\vartheta_f =$

	{}	{a}	{b}	{a,b}	{c}	{a,c}	{b,c}	{a,b,c}	{d}	{a,d}	{b,d}	{a,b,d}	{c,d}	{a,c,d}	{b,c,d}	{a,b,c,d}
{}	1	0	0	0	0	0	0	0	0	0	0	0	0	0	0	0
{1}	0	0	0	0	1	0	0	0	0	0	0	0	0	0	0	0
{2}	0	0	1	0	0	0	0	0	0	0	0	0	0	0	0	0
{1,2}	0	0	0	0	0	0	1	0	0	0	0	0	0	0	0	0
{3}	0	1	0	0	0	0	0	0	0	0	0	0	0	0	0	0
{1,3}	0	0	0	0	0	1	0	0	0	0	0	0	0	0	0	0
{2,3}	0	0	0	1	0	0	0	0	0	0	0	0	0	0	0	0
{1,2,3}	0	0	0	0	0	0	0	1	0	0	0	0	0	0	0	0

$\vartheta^{\mathsf{T}}_{f^{\mathsf{T}}} =$

	{}	{a}	{b}	{a,b}	{c}	{a,c}	{b,c}	{a,b,c}	{d}	{a,d}	{b,d}	{a,b,d}	{c,d}	{a,c,d}	{b,c,d}	{a,b,c,d}
{}	1	0	0	0	0	0	0	0	1	0	0	0	0	0	0	0
{1}	0	0	0	0	1	0	0	0	0	0	0	0	1	0	0	0
{2}	0	0	1	0	0	0	0	0	0	0	1	0	0	0	0	0
{1,2}	0	0	0	0	0	0	1	0	0	0	0	0	0	0	1	0
{3}	0	1	0	0	0	0	0	0	0	1	0	0	0	0	0	0
{1,3}	0	0	0	0	0	1	0	0	0	0	0	0	0	1	0	0
{2,3}	0	0	0	1	0	0	0	0	0	0	0	1	0	0	0	0
{1,2,3}	0	0	0	0	0	0	0	1	0	0	0	0	0	0	0	1

Fig. 2.10 Existential and inverse image for an injective, but non-surjective, mapping

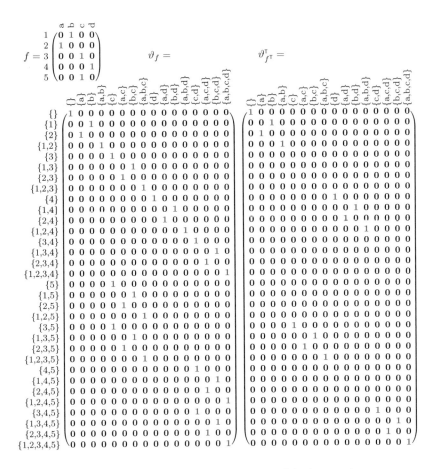

Proposition 2.2.10 *If $f : X \longrightarrow X'$ is a mapping, we have*

i) $\Omega' ; \vartheta_{f^\top} \subseteq \vartheta_{f^\top} ; \Omega$,

ii) $\Omega ; \vartheta^\top_{f^\top} = \vartheta_f ; \Omega'$.

Proof

i) The result is obtainable via shunting from

$$\Omega' = \overline{\varepsilon'^\top ; \overline{\varepsilon'}} \subseteq \overline{\varepsilon'^\top ; f^\top ; f ; \overline{\varepsilon'}} = \overline{\varepsilon'^\top ; f^\top ; \overline{f ; \varepsilon'}} = \overline{\vartheta_{f^\top} ; \varepsilon^\top ; \overline{\varepsilon ; \vartheta^\top_{f^\top}}} = \vartheta_{f^\top} ; \overline{\varepsilon^\top ; \overline{\varepsilon}} ; \vartheta^\top_{f^\top}.$$

ii)

$$\begin{aligned}\Omega ; \vartheta^\top_{f^\top} &= \overline{\varepsilon^\top ; \overline{\varepsilon}} ; \vartheta^\top_{f^\top} = \overline{\varepsilon^\top ; \overline{\varepsilon ; \vartheta^\top_{f^\top}}} = \overline{\varepsilon^\top ; \overline{f ; \varepsilon'}} = \overline{\varepsilon^\top ; f ; \overline{\varepsilon'}} \\ &= \overline{\vartheta_f ; \varepsilon'^\top ; \overline{\varepsilon'}} = \vartheta_f ; \overline{\varepsilon'^\top ; \overline{\varepsilon'}} = \vartheta_f ; \Omega'\end{aligned}$$

□

Later on, when looking at Proposition 4.3.6, one will find out that even equality holds in (i) when f is injective. The proof, however, considerably exceeds the prerequisites developed so far.

In addition, we consider the powerset negation in combination with the inverse image of a mapping $f : X \longrightarrow Y$:

Proposition 2.2.11 *In the following,* $\mathcal{N}_X := \mathtt{syq}(\varepsilon_X, \overline{\varepsilon_X}) : 2^X \longrightarrow 2^X$ *is the powerset negation of the respective set:*

$$\mathcal{N}_X ; \vartheta^\top_{f^\top} = \vartheta^\top_{f^\top} ; \mathcal{N}_Y \qquad\qquad \vartheta_{f^\top} ; \mathcal{N}_X = \mathcal{N}_Y ; \vartheta_{f^\top}$$

Proof

$$\begin{aligned}\mathcal{N}_X ; \vartheta^\top_{f^\top} &= \mathcal{N}_X ; \mathtt{syq}(\varepsilon_X, f ; \varepsilon_Y) = \mathtt{syq}(\varepsilon_X ; \mathcal{N}^\top_X, f ; \varepsilon_Y) \\ &= \mathtt{syq}(\varepsilon_X ; \mathcal{N}_X, f ; \varepsilon_Y) = \mathtt{syq}(\overline{\varepsilon_X}, f ; \varepsilon_Y) \\ &= \mathtt{syq}(\varepsilon_X, \overline{f ; \varepsilon_Y}) = \mathtt{syq}(\varepsilon_X, f ; \overline{\varepsilon_Y}) \\ &= \mathtt{syq}(\varepsilon_X, f ; \varepsilon_Y ; \mathcal{N}_Y) = \mathtt{syq}(\varepsilon_X, f ; \varepsilon_Y) ; \mathcal{N}_Y = \vartheta^\top_{f^\top} ; \mathcal{N}_Y\end{aligned}$$

The second property follows then from symmetry of $\mathcal{N}_X$ and $\mathcal{N}_Y$. □

This may be understood considering the right relation of Fig. 2.11: Multiplying $\mathcal{N}_X$ from the left means turning upside down, while $\mathcal{N}_Y$ composed from the right side flips left/right.

We identify here disjointness $\overline{\varepsilon^\top ; \varepsilon}$ which is shown in Fig. 2.11. It looks as if the powerset ordering Ω of Fig. 4.11 were rotated by an angle of $-90°$, which may

	{}	{a}	{b}	{a,b}	{c}	{a,c}	{b,c}	{a,b,c}	{d}	{a,d}	{b,d}	{a,b,d}	{c,d}	{a,c,d}	{b,c,d}	{a,b,c,d}
{}	0	0	0	0	0	0	0	0	0	0	0	0	0	0	0	1
{a}	0	0	0	0	0	0	0	0	0	0	0	0	0	0	1	0
{b}	0	0	0	0	0	0	0	0	0	0	0	0	0	1	0	0
{a,b}	0	0	0	0	0	0	0	0	0	0	0	0	1	0	0	0
{c}	0	0	0	0	0	0	0	0	0	0	0	1	0	0	0	0
{a,c}	0	0	0	0	0	0	0	0	0	0	1	0	0	0	0	0
{b,c}	0	0	0	0	0	0	0	0	0	1	0	0	0	0	0	0
{a,b,c}	0	0	0	0	0	0	0	0	1	0	0	0	0	0	0	0
{d}	0	0	0	0	0	0	0	1	0	0	0	0	0	0	0	0
{a,d}	0	0	0	0	0	0	1	0	0	0	0	0	0	0	0	0
{b,d}	0	0	0	0	0	1	0	0	0	0	0	0	0	0	0	0
{a,b,d}	0	0	0	0	1	0	0	0	0	0	0	0	0	0	0	0
{c,d}	0	0	0	1	0	0	0	0	0	0	0	0	0	0	0	0
{a,c,d}	0	0	1	0	0	0	0	0	0	0	0	0	0	0	0	0
{b,c,d}	0	1	0	0	0	0	0	0	0	0	0	0	0	0	0	0
{a,b,c,d}	1	0	0	0	0	0	0	0	0	0	0	0	0	0	0	0

	{}	{a}	{b}	{a,b}	{c}	{a,c}	{b,c}	{a,b,c}	{d}	{a,d}	{b,d}	{a,b,d}	{c,d}	{a,c,d}	{b,c,d}	{a,b,c,d}
{}	1	1	1	1	1	1	1	1	1	1	1	1	1	1	1	1
{a}	1	0	1	0	1	0	1	0	1	0	1	0	1	0	1	0
{b}	1	1	0	0	1	1	0	0	1	1	0	0	1	1	0	0
{a,b}	1	0	0	0	1	0	0	0	1	0	0	0	1	0	0	0
{c}	1	1	1	1	0	0	0	0	1	1	1	1	0	0	0	0
{a,c}	1	0	1	0	0	0	0	0	1	0	1	0	0	0	0	0
{b,c}	1	1	0	0	0	0	0	0	1	1	0	0	0	0	0	0
{a,b,c}	1	0	0	0	0	0	0	0	1	0	0	0	0	0	0	0
{d}	1	1	1	1	1	1	1	1	0	0	0	0	0	0	0	0
{a,d}	1	0	1	0	1	0	1	0	0	0	0	0	0	0	0	0
{b,d}	1	1	0	0	1	1	0	0	0	0	0	0	0	0	0	0
{a,b,d}	1	0	0	0	1	0	0	0	0	0	0	0	0	0	0	0
{c,d}	1	1	1	1	0	0	0	0	0	0	0	0	0	0	0	0
{a,c,d}	1	0	1	0	0	0	0	0	0	0	0	0	0	0	0	0
{b,c,d}	1	1	0	0	0	0	0	0	0	0	0	0	0	0	0	0
{a,b,c,d}	1	0	0	0	0	0	0	0	0	0	0	0	0	0	0	0

Fig. 2.11 Negation $\mathcal{N}$ and disjointness $\overline{\varepsilon^{\mathsf{T}} ; \varepsilon} = \Omega ; \mathcal{N}$ in the powerset

more mathematically be expressed as $\Omega ; \mathcal{N} = \overline{\varepsilon^{\mathsf{T}} ; \varepsilon}$; this time therefore mirroring left/right.

In this chapter we recalled some fundamentals of the calculus of relations used in the remainder of this book. The first mathematical investigations into this calculus were done by George Boole, Augustus De Morgan, and Charles S. Peirce. Later, their first attempts were systematically extended by Ernst Schröder. The modern and algebraic treatment of relations has its origins in the work of Alfred Tarski. His relation algebras are concerned with relations on a single universe, i.e., they are homogeneous. A heterogeneous approach using category theory, the approach taken in this book, was proposed by multiple researchers including the authors of this book starting in the 70s of the previous century. It can be shown that only this version of the calculus of relations is capable of handling the "is element" relation ε in an algebraic fashion and, hence, makes it possible to investigate topology via relations.

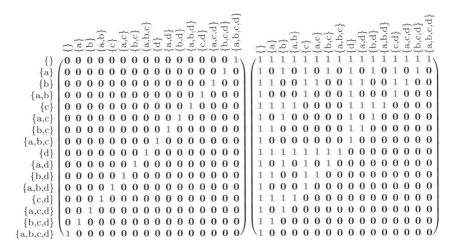

Chapter 3
Products of Relations

In Definition 2.2.1, we have introduced the direct power of a set—modelling the concept of a powerset—and shown that it is uniquely determined up to isomorphism. Even earlier, we have defined the natural projection of a set equipped with an equivalence to the set of its classes. We are now going to handle the direct product and direct sum.

Once products of sets are available, also products of relations R, S may be considered; these shall lead from pairs of arguments on the source side to e.g. pairs of results on the target side; the first component of the result determined via R and the second via S.

3.1 Products of Sets and Relations

It is everyday routine to work with pairs of elements. This is then theoretically modelled with the Cartesian product of the sets to which these elements belong. The interesting point is that there can exist just one Cartesian product of two given sets. In order to underpin this uniqueness claim, we consider the projection relations π, ρ holding between the product set and its two component sets.

If any two such heterogeneous relations π, ρ with common source are given, they are said to form a **direct product** if

$$\pi^{\mathsf{T}}; \pi = \mathbb{I}, \quad \rho^{\mathsf{T}}; \rho = \mathbb{I}, \quad \pi; \pi^{\mathsf{T}} \cap \rho; \rho^{\mathsf{T}} = \mathbb{I}, \quad \pi^{\mathsf{T}}; \rho = \mathbb{T}.$$

Interpreting these formulae, the relations are mappings, usually called **projections**. Their common source is the Cartesian product so that typing is $\pi : X \times Y \longrightarrow X$ and $\rho : X \times Y \longrightarrow Y$. In a similar way, any two heterogeneous relations ι, κ with

G. Schmidt, M. Winter, *Relational Topology*, Lecture Notes
in Mathematics 2208, https://doi.org/10.1007/978-3-319-74451-3_3

common target are said to form the left, respectively right, **injection** of a **direct sum** if

$$\iota;\iota^{\mathsf{T}} = \mathbb{I}, \quad \kappa;\kappa^{\mathsf{T}} = \mathbb{I}, \quad \iota^{\mathsf{T}};\iota \cup \kappa^{\mathsf{T}};\kappa = \mathbb{I}, \quad \iota;\kappa^{\mathsf{T}} = \mathbb{⊥}.$$

Correspondingly, the common target is a disjoint union or sum and typing should be $\iota : X \longrightarrow X + Y$ and $\kappa : Y \longrightarrow X + Y$.

Another construction has a rather similar name, but should be seen separately. It is folklore, but now treated algebraically: When an equivalence Ξ is given, one will like to proceed to quotients. This means that a **natural projection** η is conceived satisfying $\Xi = \eta;\eta^{\mathsf{T}}$ and $\eta^{\mathsf{T}};\eta = \mathbb{I}$—thus determined up to isomorphism.

Being given a subset $U \subseteq X$ of some set is an absolute standard situation. In the present environment with our strong typing discipline, however, it may be necessary to consider a copy U' of the set U as a separate entity. To this end, we introduce **extrusion** of a nonempty subset $U \subseteq X$ by a **natural injection** mapping $\iota_U : U' \longrightarrow X$, which satisfies $\iota_U;\iota_U^{\mathsf{T}} = \mathbb{I}_{U'}$ and $\iota_U^{\mathsf{T}};\mathbb{T} = U$, thus characterizing it uniquely up to isomorphism.

Starting from projections of a direct product, we take this opportunity to give an account of the Kronecker, fork, and join operator[1] with a further clarified sequence of the proofs for their properties.

Definition 3.1.1 Given any direct products by projections

$$\begin{aligned} \pi &: X \times Y \longrightarrow X, & \rho &: X \times Y \longrightarrow Y, \\ \pi' &: U \times V \longrightarrow U, & \rho' &: U \times V \longrightarrow V, \end{aligned}$$

we define as operations for relations (typed according to Fig. 3.1) the

i) **Kronecker product** of relations $A : X \longrightarrow U$ and $B : Y \longrightarrow V$ as
$(A \otimes B) := \pi;A;\pi'^{\mathsf{T}} \cap \rho;B;\rho'^{\mathsf{T}}$,

ii) **fork operator** applied to relations $C : Q \longrightarrow X$, $D : Q \longrightarrow Y$ as
$(C \otimes D) := C;\pi^{\mathsf{T}} \cap D;\rho^{\mathsf{T}}$,

iii) **join operator** applied to relations $E : U \longrightarrow P$, $F : V \longrightarrow P$ as
$(E \otimes F) := \pi';E \cap \rho';F$. □

Note how the projections of a direct product can be seen as abbreviations of particular join operations

$$\pi' := (\mathbb{I}_U \otimes \mathbb{T}), \quad \rho' := (\mathbb{T} \otimes \mathbb{I}_V).$$

[1] These operators are conceived so as to be *strict*, in contrast to what the Argentinian school around Armando Haeberer has propagated some time ago in software specification; [BHSV94].

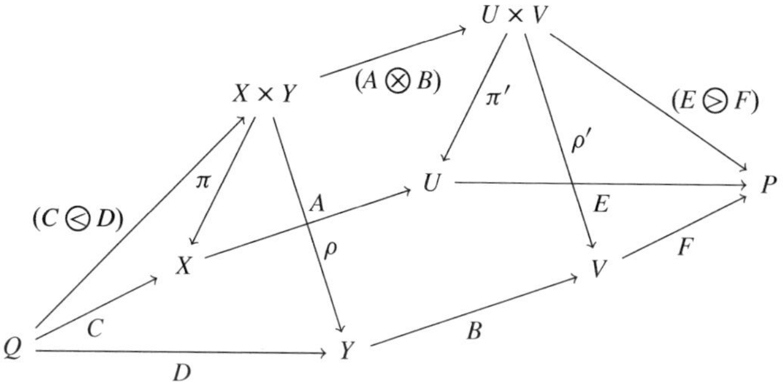

Fig. 3.1 Typing in case of the Kronecker, fork, and join operator

Immediately recognizable are the following identities:

Proposition 3.1.2

i) $(A \otimes B)^{\mathsf{T}} = (A^{\mathsf{T}} \otimes B^{\mathsf{T}})$, $\qquad (C \olessthan D)^{\mathsf{T}} = (C^{\mathsf{T}} \ogreaterthan D^{\mathsf{T}})$,
ii) $(C \cap C' \olessthan D \cap D') = (C \olessthan D) \cap (C' \olessthan D')$,
iii) $R; (C \olessthan D) = (R; C \olessthan R; D)$, *provided R is univalent.* □

Then we observe how the relational product operators behave when composed with the respective projections.

Proposition 3.1.3

i) $(A \otimes B); \pi' = \pi; A \cap \rho; B; \mathbb{T}$,
$(A \otimes B); \pi' = \pi; A$, *provided B is total*
ii) $(A \otimes B); \rho' = \rho; B \cap \pi; A; \mathbb{T}$,
$(A \otimes B); \rho' = \rho; B$, *provided A is total*

Proof i) and, analogously (ii):
$(A \otimes B); \pi' = (\pi; A; \pi'^{\mathsf{T}} \cap \rho; B; \rho'^{\mathsf{T}}); \pi'$ by definition
$= \pi; A \cap \rho; B; \rho'^{\mathsf{T}}; \pi'$ destroy and append rule for the univalent π'
$= \pi; A \cap \rho; B; \mathbb{T}$ rule for projections π', ρ'; see begin of Sect. 3.1.
Now $\rho; B; \mathbb{T} = \mathbb{T}$ when B should be total. □

We derive herefrom a corollary simply by specialization:

Corollary 3.1.4

$$(A \otimes \mathbb{I}); \pi' = \pi; A = (A \ogreaterthan \mathbb{T}), \qquad (\mathbb{I} \otimes B); \rho' = \rho; B = (\mathbb{T} \ogreaterthan B). \qquad \square$$

Trivial specialization—requiring no further proof—leads us also to the following versions for the fork operator:

Proposition 3.1.5

i) $(A \olessthan B); \pi' = A \cap B; \mathbb{T}$,
$(A \olessthan B); \pi' = A$, *provided B is total*

ii) $(A \circledless B) : \rho' = B \cap A : \mathbb{T}$,
$(A \circledless B) : \rho' = B$, *provided A is total* □

Of course, analogous formulae hold in the converse situation with $\circledgtr$; they are not reformulated here.

Now we investigate how the operators behave in connection with universal relations.

Proposition 3.1.6 *Assume relations*

$$A : X \longrightarrow U, \quad B : Y \longrightarrow V, \quad C : Q \longrightarrow X, \quad D : Q \longrightarrow Y$$

and $R : Q \longrightarrow Z$ for some Z, as well as $S : Q \longrightarrow W$ for some W, mainly as in Fig. 3.1. Then

i) $(A \otimes B) : \mathbb{T} = (A : \mathbb{T} \circledgtr B : \mathbb{T})$,
ii) $(C \circledless D) \cap R : \mathbb{T} = (C \cap R : \mathbb{T} \circledless D \cap R : \mathbb{T})$,
iii) $(A \otimes B) \cap (R : \mathbb{T} \circledgtr S : \mathbb{T}) = (A \cap R : \mathbb{T} \otimes B \cap S : \mathbb{T})$,
iv) $(A \otimes B) = (\pi : A \circledless \rho : B)$.

Proof For reasons of clarity, we mention the ever changing types of the universal relations explicitly.

i) $(A \otimes B) : \mathbb{T}_{U \times V, Z}$
$= (\pi : A : \pi'^{\mathsf{T}} \cap \rho : B : \rho'^{\mathsf{T}}) : \mathbb{T}_{U \times V, Z}$ by definition
$= (\pi : A : \pi'^{\mathsf{T}} \cap \rho : B : \rho'^{\mathsf{T}}) : \pi' : \mathbb{T}_{U,Z}$ since π' is total
$= (\pi : A \cap \rho : B : \rho'^{\mathsf{T}} : \pi') : \mathbb{T}_{U,Z}$ destroy and append rule; π' univalent
$= (\pi : A \cap \rho : B : \mathbb{T}_{V,U}) : \mathbb{T}_{U,Z}$ property of the direct product
$= \pi : A : \mathbb{T}_{U,Z} \cap \rho : B : \mathbb{T}_{V,Z}$ masking
$= (A : \mathbb{T}_{U,Z} \circledgtr B : \mathbb{T}_{V,Z})$ by definition

ii) $(C \cap R : \mathbb{T}_{Z,X} \circledless D \cap R : \mathbb{T}_{Z,Y}) = (C \cap R : \mathbb{T}_{Z,X}) : \pi^{\mathsf{T}} \cap (D \cap R : \mathbb{T}_{Z,Y}) : \rho^{\mathsf{T}}$
$= C : \pi^{\mathsf{T}} \cap R : \mathbb{T}_{Z,X} : \pi^{\mathsf{T}} \cap D : \rho^{\mathsf{T}} \cap R : \mathbb{T}_{Z,Y} : \rho^{\mathsf{T}}$
$= C : \pi^{\mathsf{T}} \cap R : \mathbb{T}_{Z,X \times Y} \cap D : \rho^{\mathsf{T}} \cap R : \mathbb{T}_{Z,X \times Y}$
$= C : \pi^{\mathsf{T}} \cap D : \rho^{\mathsf{T}} \cap R : \mathbb{T}_{Z,X \times Y}$
$= (C \circledless D) \cap R : \mathbb{T}_{Z,X \times Y}$

iii) $(A \otimes B) \cap (R : \mathbb{T}_{Z,U \times V} \circledgtr S : \mathbb{T}_{W,U \times V})$
$= \pi : A : \pi'^{\mathsf{T}} \cap \rho : B : \rho'^{\mathsf{T}} \cap \pi : R : \mathbb{T}_{Z,U \times V} \cap \rho : S : \mathbb{T}_{W,U \times V}$
$= \pi : A : \pi'^{\mathsf{T}} \cap \pi : R : \mathbb{T}_{Z,U \times V} \cap \rho : B : \rho'^{\mathsf{T}} \cap \rho : S : \mathbb{T}_{W,U \times V}$ shuffled
$= \pi : (A : \pi'^{\mathsf{T}} \cap R : \mathbb{T}_{Z,U \times V}) \cap \rho : (B : \rho'^{\mathsf{T}} \cap S : \mathbb{T}_{W,U \times V})$
$= \pi : (A \cap R : \mathbb{T}_{Z,U}) : \pi'^{\mathsf{T}} \cap \rho : (B \cap S : \mathbb{T}_{W,V}) : \rho'^{\mathsf{T}}$ masking
$= (A \cap R : \mathbb{T}_{Z,U} \otimes B \cap S : \mathbb{T}_{W,V})$

iv) trivial □

In the next propositions we investigate how the relation products behave when composed with other such products. The proofs have to be presented in some detail because of the so-called 'unsharpness' situation. In general: When some theory is defined axiomatically, there may exist no, just one, or more than one model. We anticipate that the forthcoming result Proposition 3.1.7.i holds with "=" in the

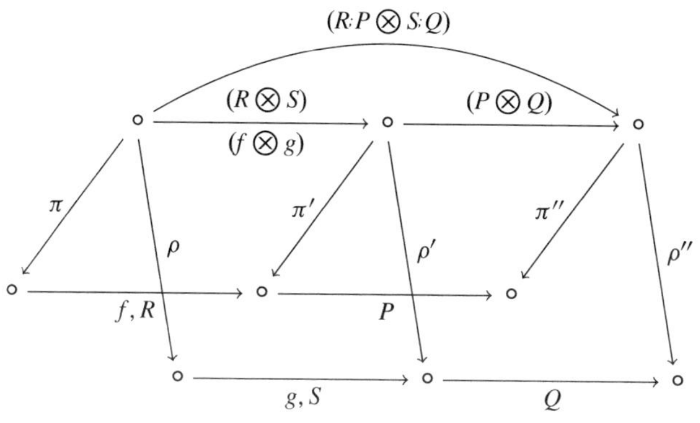

Fig. 3.2 Typing in case of unsharp composition of Kronecker products

classical interpretation of relation algebra based on the Boolean matrix model, but cannot be derived in the axiomatization we follow here. It is not simply insufficient skill that we did not succeed in proving! Indeed, there exist models[2] satisfying "$\subsetneqq$".

Of course, the question has arisen whether these models admit any reasonable interpretation. It seems possible that they allow to relationally cope with *processes* as opposed to just *programs*.[3]

Proposition 3.1.7 *Let be given the setting as in Fig. 3.2. Then*

i) $(R \otimes S){:}(P \otimes Q) \subseteq (R{:}P \otimes S{:}Q)$,
ii) If f, g *are both univalent, then so is* $(f \otimes g)$,
iii) If f, g *are both mappings, then so is* $(f \otimes g)$,
iv) $(f \otimes g){:}(P \otimes Q) = (f{:}P \otimes g{:}Q)$, *provided* f, g *are both univalent.*

Proof

i) The proof of containment "$\subseteq$" is fairly easy:
$(R \otimes S){:}(P \otimes Q) = (\pi{:}R{:}\pi'^{\mathsf{T}} \cap \rho{:}S{:}\rho'^{\mathsf{T}}){:}(\pi'{:}P{:}\pi''^{\mathsf{T}} \cap \rho'{:}Q{:}\rho''^{\mathsf{T}})$ expanded
$\subseteq \pi{:}R{:}\pi'^{\mathsf{T}}{:}\pi'{:}P{:}\pi''^{\mathsf{T}} \cap \rho{:}S{:}\rho'^{\mathsf{T}}{:}\rho'{:}Q{:}\rho''^{\mathsf{T}}$ monotony
$= \pi{:}R{:}P{:}\pi''^{\mathsf{T}} \cap \rho{:}S{:}Q{:}\rho''^{\mathsf{T}}$ since π', ρ' are univalent and surjective
$= (R{:}P \otimes S{:}Q)$ by definition

ii) $(f \otimes g)^{\mathsf{T}}{:}(f \otimes g) = (f^{\mathsf{T}} \otimes g^{\mathsf{T}}){:}(f \otimes g) \subseteq (f^{\mathsf{T}}{:}f \otimes g^{\mathsf{T}}{:}g)$ due to (i)
$\subseteq (\mathbb{I} \otimes \mathbb{I}) = \mathbb{I}$

iii) Univalency follows from (ii). It remains to prove totality:
$(f \otimes g){:}\mathbb{T} = (f \otimes g){:}\pi'{:}\mathbb{T}$ since projection π' is total
$= \pi{:}f{:}\mathbb{T}$ following Proposition 3.1.3.i since g is total
$= \pi{:}\mathbb{T} = \mathbb{T}$, since f is total

[2] See Sect. 3.2 of [KS00]: http://titurel.org/Papers/RATH-Titel.pdf.

[3] Processes may have arguments that are tuples—with the availability of its components varying over time—and in turn produce such results, in a strict or non-strict form. If tuples occur, they may just partially exist. A relational theory of partialities is already fairly developed and may be found in [Sch11a, Sch12].

iv) According to (ii), $(f \otimes g)$ is univalent, so that we may reason

$$\begin{aligned}
&(f \otimes g);(P \otimes Q) = (f \otimes g);(\pi';P;{\pi''}^{\mathsf{T}} \cap \rho';Q;{\rho''}^{\mathsf{T}}) && \text{by definition}\\
&= (f \otimes g);\pi';P;{\pi''}^{\mathsf{T}} \cap (f \otimes g);\rho';Q;{\rho''}^{\mathsf{T}} && \text{following (ii)}\\
&= (\pi;f \cap \rho;g;\mathbb{T});P;{\pi''}^{\mathsf{T}} \cap (\pi;f;\mathbb{T} \cap \rho;g);Q;{\rho''}^{\mathsf{T}} && \text{Proposition 3.1.3.i,ii}\\
&= \pi;f;P;{\pi''}^{\mathsf{T}} \cap \rho;g;\mathbb{T} \;\cap\; \pi;f;\mathbb{T} \cap \rho;g;Q;{\rho''}^{\mathsf{T}} && \text{masking}\\
&= \pi;f;P;{\pi''}^{\mathsf{T}} \cap \rho;g;Q;{\rho''}^{\mathsf{T}} && \text{trivial since, e.g., } \pi;f;P;{\pi''}^{\mathsf{T}} \subseteq \pi;f;\mathbb{T}\\
&= (f;P \otimes g;Q) && \text{by definition}
\end{aligned}$$

□

Of course, also the converse variants as well as those with fork or join instead of the Kronecker operator are satisfied, correspondingly. For reference purposes, we add these results without their proofs:

Proposition 3.1.8

i) $(R \otimes S);(P \mathbin{⧁} Q) \subseteq (R;P \mathbin{⧁} S;Q)$,
$(R \mathbin{⧀} S);(P \otimes Q) \subseteq (R;P \mathbin{⧀} S;Q)$,
$(R \mathbin{⧀} S);(P \mathbin{⧁} Q) \subseteq R;P \cap S;Q$,

ii) $(f \otimes g);(A \mathbin{⧁} B) = (f;A \mathbin{⧁} g;B)$ *when f, g are both univalent*
$(f \mathbin{⧀} g);(A \otimes B) = (f;A \mathbin{⧀} g;B)$ *when f, g are both univalent*
$(f \mathbin{⧀} g);(A \mathbin{⧁} B) = f;A \cap g;B$ *when f, g are both univalent* □

In some sense residuation distributes over Kronecker, fork, or join operators. In (i), this holds in its purest form when A and B are both surjective.

Proposition 3.1.9

i) $(A \otimes B)\backslash(C \otimes D) = (A\backslash C \otimes B\backslash D) \cup \pi;\overline{A^{\mathsf{T}};\mathbb{T}} \cup \rho;\overline{B^{\mathsf{T}};\mathbb{T}}$
ii) $(A\backslash C \mathbin{⧁} B\backslash C) \subseteq (A \mathbin{⧀} B)\backslash C$
iii) $(A/C \mathbin{⧁} B/C) = (A \mathbin{⧁} B)/C$

Proof

i)

$$\begin{aligned}
&(A \otimes B)\backslash(C \otimes D) = \overline{(A \otimes B)^{\mathsf{T}};\overline{(C \otimes D)}} && \text{expanded}\\
&= \overline{(A \otimes B)^{\mathsf{T}};[\pi_1;\overline{C};\pi_2^{\mathsf{T}} \cup \rho_1;\overline{D};\rho_2^{\mathsf{T}}]}\\
&= \overline{(A^{\mathsf{T}} \otimes B^{\mathsf{T}});\pi_1;\overline{C};\pi_2^{\mathsf{T}} \cup (A^{\mathsf{T}} \otimes B^{\mathsf{T}});\rho_1;\overline{D};\rho_2^{\mathsf{T}}}\\
&= \overline{[\pi;A^{\mathsf{T}} \cap \rho;B^{\mathsf{T}};\mathbb{T}];\overline{C};\pi_2^{\mathsf{T}} \cup [\pi;A^{\mathsf{T}};\mathbb{T} \cap \rho;B^{\mathsf{T}}];\overline{D};\rho_2^{\mathsf{T}}} && \text{Proposition 3.1.3}\\
&= \overline{[\pi;A^{\mathsf{T}};\overline{C};\pi_2^{\mathsf{T}} \cap \rho;B^{\mathsf{T}};\mathbb{T}] \cup [\rho;B^{\mathsf{T}};\overline{D};\rho_2^{\mathsf{T}} \cap \pi;A^{\mathsf{T}};\mathbb{T}]} && \text{masking}\\
&= [\pi;\overline{A^{\mathsf{T}};\overline{C}};\pi_2^{\mathsf{T}} \cup \rho;\overline{B^{\mathsf{T}};\mathbb{T}}] \cap [\rho;\overline{B^{\mathsf{T}};\overline{D}};\rho_2^{\mathsf{T}} \cup \pi;\overline{A^{\mathsf{T}};\mathbb{T}}]\\
&= [\pi;\overline{A^{\mathsf{T}};\overline{C}};\pi_2^{\mathsf{T}} \cap \rho;\overline{B^{\mathsf{T}};\overline{D}};\rho_2^{\mathsf{T}}] \cup [\pi;\overline{A^{\mathsf{T}};\overline{C}};\pi_2^{\mathsf{T}} \cap \pi;\overline{A^{\mathsf{T}};\mathbb{T}}] && \cup\\
&\qquad [\rho;\overline{B^{\mathsf{T}};\mathbb{T}} \cap \rho;\overline{B^{\mathsf{T}};\overline{D}};\rho_2^{\mathsf{T}}] \cup [\rho;\overline{B^{\mathsf{T}};\mathbb{T}} \cap \pi;\overline{A^{\mathsf{T}};\mathbb{T}}]\\
&= (\overline{A^{\mathsf{T}};\overline{C}} \otimes \overline{B^{\mathsf{T}};\overline{D}}) \cup \pi;\overline{A^{\mathsf{T}};\mathbb{T}} \cup \rho;\overline{B^{\mathsf{T}};\mathbb{T}}\\
&= (A\backslash C \otimes B\backslash D) \cup \pi;\overline{A^{\mathsf{T}};\mathbb{T}} \cup \rho;\overline{B^{\mathsf{T}};\mathbb{T}}
\end{aligned}$$

ii)

$$\begin{aligned}(A\backslash C \ogreaterthan B\backslash C) &= \pi \mathbin{;} \overline{A^{\mathsf{T}} \mathbin{;} \overline{C}} \cap \rho \mathbin{;} \overline{B^{\mathsf{T}} \mathbin{;} \overline{C}} = \overline{\pi \mathbin{;} A^{\mathsf{T}} \mathbin{;} \overline{C} \cup \rho \mathbin{;} B^{\mathsf{T}} \mathbin{;} \overline{C}} \\ &= \overline{(\pi \mathbin{;} A^{\mathsf{T}} \cup \rho \mathbin{;} B^{\mathsf{T}}) \mathbin{;} \overline{C}} \subseteq \overline{(\pi \mathbin{;} A^{\mathsf{T}} \cap \rho \mathbin{;} B^{\mathsf{T}}) \mathbin{;} \overline{C}} \\ &= \overline{(A \olessthan B)^{\mathsf{T}} \mathbin{;} \overline{C}} = (A \olessthan B)\backslash C\end{aligned}$$

iii)

$$\begin{aligned}(A/C \ogreaterthan B/C) &= \pi \mathbin{;} \overline{\overline{A} \mathbin{;} C^{\mathsf{T}}} \cap \rho \mathbin{;} \overline{\overline{B} \mathbin{;} C^{\mathsf{T}}} \\ &= \overline{\pi \mathbin{;} \overline{A} \mathbin{;} C^{\mathsf{T}} \cup \rho \mathbin{;} \overline{B} \mathbin{;} C^{\mathsf{T}}} = \overline{(\pi \mathbin{;} \overline{A} \cup \rho \mathbin{;} \overline{B}) \mathbin{;} C^{\mathsf{T}}} \\ &= \overline{\overline{(\pi \mathbin{;} A \cap \rho \mathbin{;} B)} \mathbin{;} C^{\mathsf{T}}} = \overline{\overline{(A \ogreaterthan B)} \mathbin{;} C^{\mathsf{T}}} = (A \ogreaterthan B)/C\end{aligned}$$

□

3.2 Sharp Factorizations

Our approach in the present book is to base reasoning not simply on first-order predicate logic, but on point-free relation algebra that avoids quantifiers. This may be seen as a shorthand notation; replacing, e.g., the lengthy

$$\forall x : \forall y : (\exists z : A_{xz} \wedge B_{zy}) \rightarrow C_{xy}$$

by the more concise

$$A \mathbin{;} B \subseteq C,$$

offering access to much simpler (namely quantifier-free) algebraic rules.

But now we face the situation that only $(R \otimes S) \mathbin{;} (A \otimes B) \subseteq (R \mathbin{;} A \otimes S \mathbin{;} B)$ can be proved in the style just exhibited. As already announced, Proposition 3.1.7.i as well as Proposition 3.1.8.i deserve, thus, some discussion. In contrast to the fact that in addition to "$\subseteq$" what has been proved, Boolean matrices—the only model of relation algebra uninitiated people usually think about—always satisfy "$=$".

This turns out to be a model problem that one may compare with the appearance non-Euclidian geometry: "Obviously", through every point outside a straight line precisely one parallel will exist to that line. But in the early nineteenth century, Bolyai and Lobachevsky provided a model for the geometric axioms where this does not hold.

We find ourselves in a similar situation here, since small finite counterexamples to equality are presented in [KS00]; the main one is originally due to Roger Maddux.

Therefore, any study of relational methods has to be extremely careful with regard to this so-called *unsharpness* problem. In particular, we must never indiscriminately apply rules for the Kronecker, the fork, and the join operator without checking the axiomatic basis. Should A, B, R, S be relations for which equality holds, we will say that the right side is **sharply factorized** as

$$(R \otimes S) ; (A \otimes B) = (R ; A \otimes S ; B).$$

The possibility to factorize certainly deserves further study. With the following discussion, we try to approach comprehending it. Sets between which relations are assumed to hold contain *elements*. Also first-order predicate logic, omnipresent in mathematics as well as in informatics, works with elements. In the present text, it is crucial how to conceive of an element relation-algebraically. The element has as its counterpart in relation algebra the concept of a *point* introduced earlier. We recall that a relation p is a point when it is *row-constant* ($p = p ; \mathbb{T}$), injective ($p ; p^{\mathsf{T}} \subseteq \mathbb{I}$), and surjective ($\mathbb{T} ; p = \mathbb{T}$). The so-called *point axiom* then demands that

$$R \neq \mathbb{L} \quad \Longrightarrow \quad \text{There exist points} \quad x, y \quad \text{such that} \quad x ; y^{\mathsf{T}} \subseteq R.$$

So x, y are "some" elements between which the relation R holds, which brings us back to first-order predicate logic. Very roughly we may state that relation algebra with point axiom is as powerful as first-order predicate logic.

$$\pi = \begin{array}{c} \\ (1,\spadesuit) \\ (1,\heartsuit) \\ (1,\diamondsuit) \\ (1,\clubsuit) \\ (2,\spadesuit) \\ (2,\heartsuit) \\ (2,\diamondsuit) \\ (2,\clubsuit) \\ (3,\spadesuit) \\ (3,\heartsuit) \\ (3,\diamondsuit) \\ (3,\clubsuit) \end{array} \begin{array}{c} \begin{array}{ccc} 1 & 2 & 3 \end{array} \\ \begin{pmatrix} 1&0&0\\ 1&0&0\\ 1&0&0\\ 1&0&0\\ 0&1&0\\ 0&1&0\\ 0&1&0\\ 0&1&0\\ 0&0&1\\ 0&0&1\\ 0&0&1\\ 0&0&1 \end{pmatrix} \end{array} \quad \begin{array}{c} \begin{array}{cccc} \spadesuit & \heartsuit & \diamondsuit & \clubsuit \end{array} \\ \begin{pmatrix} 1&0&0&0\\ 0&1&0&0\\ 0&0&1&0\\ 0&0&0&1\\ 1&0&0&0\\ 0&1&0&0\\ 0&0&1&0\\ 0&0&0&1\\ 1&0&0&0\\ 0&1&0&0\\ 0&0&1&0\\ 0&0&0&1 \end{pmatrix} \end{array} = \rho \qquad \text{element } (2,\clubsuit) = \begin{array}{c} (1,\spadesuit) \\ (1,\heartsuit) \\ (1,\diamondsuit) \\ (1,\clubsuit) \\ (2,\spadesuit) \\ (2,\heartsuit) \\ (2,\diamondsuit) \\ (2,\clubsuit) \\ (3,\spadesuit) \\ (3,\heartsuit) \\ (3,\diamondsuit) \\ (3,\clubsuit) \end{array} \begin{pmatrix} 0\\0\\0\\0\\0\\0\\0\\1\\0\\0\\0\\0 \end{pmatrix}$$

When given a set in the way here often presented in examples, a point is a vector with precisely one entry 1. This works fine for the sets discussed so far. A point may, however, also stand for a *pair* of elements of different sets as in the example above. At a first glance not much has changed; the entry 1 will simply mark the pair $(2, \clubsuit)$.

However, when considered in this way, we assume the pair somehow to be available "instantaneously", although we have to observe two items. This is where we may encounter a problem when observing a dynamic process: Either one of the two may be available or not.

With Fig. 3.3, we discuss the difference between the static and the dynamic case. Let the pair be located in different rooms with windows to opposite sides of the house.

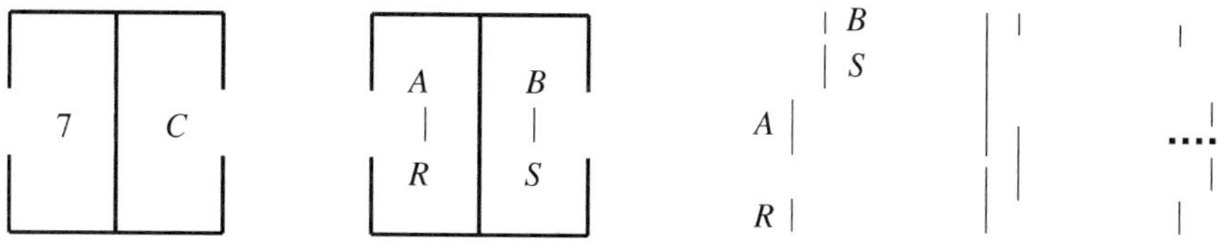

Fig. 3.3 Observing a pair, statical vs. dynamical; the latter case with 3 conceivable execution time scales

When thinking of Fig. 3.3 on the very left, it is just a minor inconvenience that one has to run to the other side of the house and finally observe the pair $(7, C)$. The left situation is what has earlier been announced as the static case.

But now assume dynamic processes to be observed in Fig. 3.3 in the middle. Running around the house does not really help; and even St. Peter assumed to be looking from above with X-raying eyes may hardly ever spot an intermediate pair, because of possibly differing speeds of the processes. In Fig. 3.3 on the very right, we show conceivable execution time intervals of the processes. An intermediate pair between $(R \otimes S)$ and $(A \otimes B)$ will co-exist only in the third case for a rather short period of time: R, S finished and A, B not yet started. (Of course, there might exist no problem when processes are synchronized somehow, but this should be considered an overly restricted scenario.)

Let us consider yet another situation exhibiting in particular that the Kronecker operator is a very general construct; this may be seen assuming for example

- $R =$ relation of car owners to cars in Christchurch, New Zealand, in 2015
- $A =$ relation of cars in Christchurch to number of accidents in 2015,
- $S =$ relation of male to female dancers on the Vienna opera ball 2016,
- $B =$ relation of corresponding female dancers to the Booleans $\mathbb{B}$, expressing whether they had been accompanied by their father.

Should one have been given the respective information, it would not be a problem to evaluate $R{:}A$ in Christchurch, resp. $S{:}A$ in Vienna, and thus obtain $(R{:}A \otimes S{:}B)$. But now imagine that the latter has been given and one is asked to find out whether $(R{:}A \otimes S{:}B) \subseteq (R \otimes S) : (A \otimes B)$. The two strands $R : A$ and $S : B$ are "heavily unrelated" and one would have severe problems to find the necessary intermediate (car, female dancer) pairs when looking at

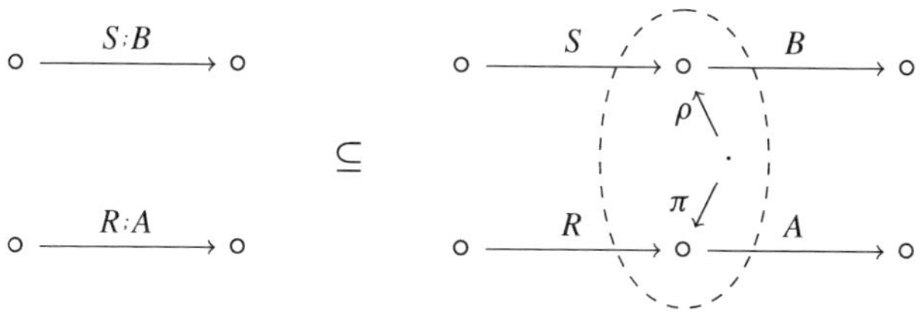

This is what happens even if one does not consider processes. Our traditional matrix computation as well as first-order predicate logic fail to model or mirror this adequately—but relation algebra does!

On the other side, one is often interested to have a sharp factorization to facilitate reasoning. So people kept working on conditions guaranteeing it, most notably Zierer [Zie88, Zie91] and Desharnais, [Des99].

This is the reason for the following proposition and its corollaries which typically assume that additional—be it far-fetched—products or relations be available, which means some sort of an "improved observability" for the pairs. Sharp factorization is sometimes possible when such additional relational connections exist. The proofs then, however, turn out to be quite difficult.

This is a rather general situation in which sharp factorization is possible:

Proposition 3.2.1 *Let be given the typing configuration of Fig. 3.4, i.e.,*

$$\begin{array}{lll} R : W_1 \longrightarrow X, & f : W_1 \longrightarrow W_2, & Q, S : W_2 \longrightarrow Y \\ A : X \longrightarrow Z, & B : Y \longrightarrow Z, & \end{array}$$

and postulate in addition

$$f \text{ univalent}, \quad f{:}Q \text{ total}, \quad \text{and} \quad Q \text{ injective}$$

(so that f is in fact a map). Then one may sharply factorize the left side so as to obtain the product of a fork- and a join-operator in

$$R{:}A \cap f{:}S{:}B = (R \otimes f{:}S){:}(A \otimes B).$$

Proof $R{:}A \cap f{:}S{:}B = (R \cap f{:}Q{:}\mathbb{T}){:}A \cap f{:}S{:}B$ since $f{:}Q$ is total
$= (R \cap f{:}Q{:}\rho^\mathsf{T}{:}\pi){:}A \cap f{:}S{:}B$ property of the direct product π, ρ
$= (R{:}\pi^\mathsf{T} \cap f{:}Q{:}\rho^\mathsf{T}){:}\pi{:}A \cap f{:}S{:}B$ destroy and append rule
$\subseteq \left[(R{:}\pi^\mathsf{T} \cap f{:}Q{:}\rho^\mathsf{T}) \cap \ldots\right]{:}\left[\pi{:}A \cap (R{:}\pi^\mathsf{T} \cap f{:}Q{:}\rho^\mathsf{T})^\mathsf{T}{:}f{:}S{:}B\right]$ Dedekind
$\subseteq (R \otimes f{:}Q){:}\left[\pi{:}A \cap \rho{:}Q^\mathsf{T}{:}f^\mathsf{T}{:}f{:}S{:}B\right]$
$\subseteq (R \otimes f{:}Q){:}\left[\pi{:}A \cap \rho{:}Q^\mathsf{T}{:}S{:}B\right]$ because f is univalent

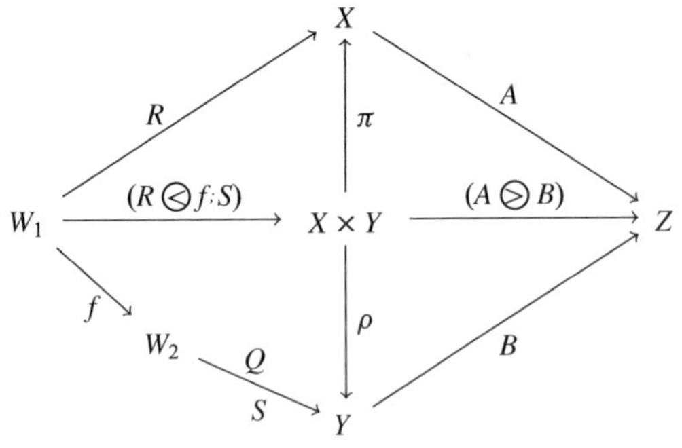

Fig. 3.4 Typing of a sharp fork-join factorization

$= (R \otimes f ; Q) ; [\pi ; A \cap (\mathbb{I} \otimes Q^{\mathsf{T}} ; S) ; \rho ; B]$ Proposition 3.1.4
$\subseteq (R \otimes f ; Q) ; [(\mathbb{I} \otimes Q^{\mathsf{T}} ; S) \cap \dots] ; [\rho ; B \cap (\mathbb{I} \otimes Q^{\mathsf{T}} ; S)^{\mathsf{T}} ; \pi ; A]$ Dedekind
$\subseteq (R \otimes f ; Q) ; (\mathbb{I} \otimes Q^{\mathsf{T}} ; S) ; [\rho ; B \cap \pi ; \mathbb{I} ; A]$ Proposition 3.1.3.i
$\subseteq (R \otimes f ; Q) ; (\mathbb{I} \otimes Q^{\mathsf{T}} ; S) ; (A \otimes B)$ by definition
$\subseteq (R \otimes f ; Q ; Q^{\mathsf{T}} ; S) ; (A \otimes B)$ Proposition 3.1.8.i
$\subseteq (R \otimes f ; S) ; (A \otimes B)$ since Q is injective
$\subseteq R ; A \cap f ; S ; B$ Proposition 3.1.8.i □

Observe, that Q doesn't show up in the formula that has been proved! Its sheer existence—in combination with f— as some sort of a catalyst, however, suffices to spot the "intermediate pair" sufficiently.

Now, that this has been proved, two independent variations in diverging directions are possible, resulting in four corollaries. One idea to proceed is to contract W_1, W_2 into W, thus making f the identity.

In Fig. 3.5, we first contract W_1, W_2 to W.

We can go to an even more special case and introduce for the relation Q of Fig. 3.5 the (injective) singleton injection $\sigma := \mathtt{syq}(\mathbb{I}, \varepsilon) : W \longrightarrow \mathbf{2}^W$, while maintaining $f := \mathbb{I}$.

Corollary 3.2.2

i) *Typed as on the left of Fig. 3.5, i.e. as*

$$R : W \longrightarrow X, \quad Q, S : W \longrightarrow Y$$
$$A : X \longrightarrow Z, \quad B : Y \longrightarrow Z,$$

with Q total and injective; i.e. a transposed map, we obtain the factorization

$$R ; A \cap S ; B = (R \otimes S) ; (A \otimes B).$$

ii) *Typed as on the right of Fig. 3.5, i.e. as*

$$R : W \longrightarrow X, \quad \sigma, S : W \longrightarrow \mathbf{2}^W$$
$$A : X \longrightarrow Z, \quad B : \mathbf{2}^W \longrightarrow Z,$$

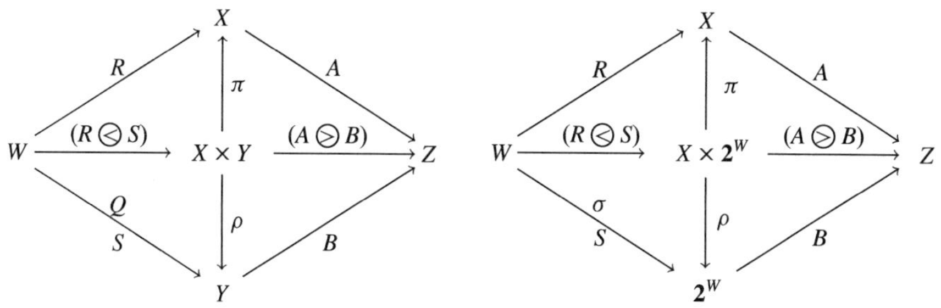

Fig. 3.5 Typing of sharp fork-join factorizations

with σ the singleton injection, we obtain the factorization

$$R{:}A \cap S{:}B = (R \circledless S){:}(A \circledgtr B).$$

Proof Both fully subsume to Proposition 3.2.1 when taking $f := \mathbb{I}$. □

We may specialize Corollary 3.2.2.ii even further, identifying X with $\mathbf{2}^W$ and admitting a direct product $U \times V$ for Z.

Corollary 3.2.3 *Given any relations $R, S : W \longrightarrow \mathbf{2}^W$ typed like membership relation and singleton injection $\varepsilon, \sigma : W \longrightarrow \mathbf{2}^W$, together with the two relations $A : \mathbf{2}^W \longrightarrow U$ and $B : \mathbf{2}^W \longrightarrow V$, one may sharply factorize*

$$(R{:}A \circledless S{:}B) = (R \circledless S){:}(A \otimes B).$$

□

The result of Proposition 3.2.1 allows two other corollaries when one proceeds to the direct product $W_1 \times W_2$.

Corollary 3.2.4

i) Typed as on the left of Fig. 3.6, i.e. as

$$\begin{array}{ll} P : W_1 \longrightarrow X, & Q, S : W_2 \longrightarrow Y \\ A : X \longrightarrow Z, & B : Y \longrightarrow Z, \end{array}$$

with Q total and injective, one may factorize

$$(P{:}A \circledgtr S{:}B) = (P \otimes S){:}(A \circledgtr B).$$

ii) Typed as on the right of Fig. 3.6, i.e. as

$$\begin{array}{ll} P : W_1 \longrightarrow X, & \sigma, S : W_2 \longrightarrow \mathbf{2}^{W_2} \\ A : X \longrightarrow Z, & B : \mathbf{2}^{W_2} \longrightarrow Z, \end{array}$$

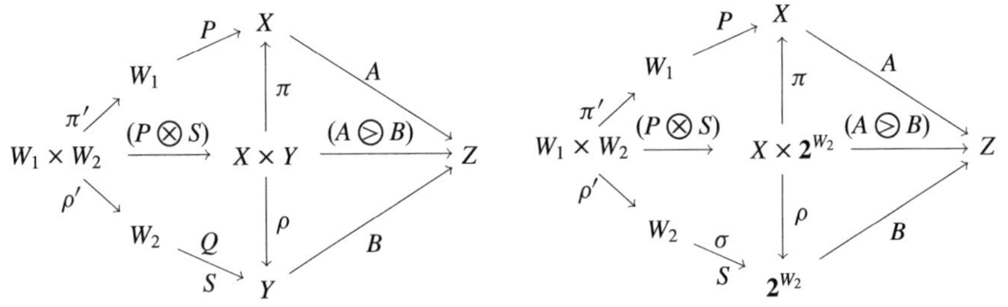

Fig. 3.6 Typing of sharp Kronecker-join factorizations

with σ the singleton injection, always

$$(P{:}A \circledgt S{:}B) = (P \otimes S){:}(A \circledgt B).$$

Proof

i) subsumes to Proposition 3.2.1 when setting $R := \pi'{:}P$ and $f := \rho'$.
ii) in turn subsumes to (i) when taking as Q the singleton injection map $\sigma :=$ $\mathtt{syq}(\mathbb{I}, \varepsilon)$ and $f := \mathbb{I}$. □

Again, neither Q nor σ show up in the final factorization statements!

When formulating aspects that—be it just loosely—belong together, such difficulty may not arise, and this was the case in the preceding corollaries. The proofs where relations "did not belong to different continents" but were "coherent with one another in some sense" turned out to be really difficult when executed formally. How is it possible, in a relation-algebraic manner, to formulate such coherence? Always some relations had been available besides the A, B, R, S scenario that could be made use of in the proof. But in the end, they were not visible in the result of the respective proposition—a situation hardly ever being met in mathematics.

3.3 Binary Mappings in General

Binary mappings are most generally typed $f : X \times Y \longrightarrow Z$. The following investigations are concerned with their standard properties such as commutativity, etc. First, we define what it means algebraically to exchange components of a pair without effect on the result.

Definition 3.3.1 We consider a binary mapping $f : X \times X \longrightarrow Z$ as well as the corresponding projections $\pi, \rho : X \times X \longrightarrow X$ and define as follows:

i) $\mathfrak{P} := \pi{:}\rho^{\mathsf{T}} \cap \rho{:}\pi^{\mathsf{T}}$ **commutativity flip**
ii) f **commutative** $:\Longleftrightarrow \quad \mathfrak{P}{:}f = f$ □

Being commutative means to flip the arguments without changing the result.

Lemma 3.3.2 *The following identities hold for $\mathfrak{P}$ together with relations*

$$R : X \longrightarrow U \quad \text{and} \quad S : X \longrightarrow U, \qquad A : X \longrightarrow Z \quad \text{and} \quad B : X \longrightarrow Z$$

with additional projections

$$\pi' : U \times U \longrightarrow U, \qquad \rho' : U \times U \longrightarrow U :$$

i) $\mathfrak{P}$ *is a bijective mapping.*
ii) $\mathfrak{P}^{\mathsf{T}} = \mathfrak{P} \qquad \mathfrak{P}{:}\pi = \rho \qquad \mathfrak{P}{:}\rho = \pi$
iii) $\mathfrak{P}{:}(R \otimes S) = (S \otimes R){:}\mathfrak{P}'$
iv) $\mathfrak{P}{:}(A \circledgt B) = (B \circledgt A)$

Proof
i) and the first identity of (ii) are trivial.
$\mathfrak{P}{:}\pi = (\pi{:}\rho^{\mathsf{T}} \cap \rho{:}\pi^{\mathsf{T}}){:}\pi = \pi{:}\rho^{\mathsf{T}}{:}\pi \cap \rho$ destroy and append
$= \pi{:}\mathbb{T} \cap \rho = \mathbb{T} \cap \rho = \rho$
iii) $\mathfrak{P}{:}(R \otimes S) = \mathfrak{P}{:}\,[\pi{:}R{:}\pi'^{\mathsf{T}} \cap \rho{:}S{:}\rho'^{\mathsf{T}}]$ by definition
$= \mathfrak{P}{:}\pi{:}R{:}\pi'^{\mathsf{T}} \cap \mathfrak{P}{:}\rho{:}S{:}\rho'^{\mathsf{T}}$
$= \rho{:}R{:}\pi'^{\mathsf{T}} \cap \pi{:}S{:}\rho'^{\mathsf{T}}$ following (ii)
$= \ldots = (S \otimes R){:}\,\mathfrak{P}'$ analogously in reverse direction
iv) $\mathfrak{P}{:}(A \oslash B) = \mathfrak{P}{:}(\pi{:}A \cap \rho{:}B)$ by definition
$= \mathfrak{P}{:}\pi{:}A \cap \mathfrak{P}{:}\rho{:}B$ since $\mathfrak{P}$ is univalent
$= \rho{:}A \cap \pi{:}B = (B \oslash A)$ by definition □

Next to commutativity, associativity is discussed. The shuffling executed by the associative law is meant to catch up with

$$\forall a,\ \forall b,\ \forall c : (a + b) + c = a + (b + c).$$

Definition 3.3.3 We consider a binary mapping $f : X \times X \longrightarrow X$ together with all the respective projections according to Fig. 3.7 and define as follows:

i) $\mathfrak{T} := \pi'{:}\pi{:}\pi_1^{\mathsf{T}} \cap \pi'{:}\rho{:}\pi^{\mathsf{T}}{:}\rho_1^{\mathsf{T}} \cap \rho'{:}\rho^{\mathsf{T}}{:}\rho_1^{\mathsf{T}}$ **associativity shuffle**,
or, differently grouped, when so required:
$= \pi'{:}\pi{:}\pi_1^{\mathsf{T}} \cap (\pi'{:}\rho{:}\pi^{\mathsf{T}} \cap \rho'{:}\rho^{\mathsf{T}}){:}\rho_1^{\mathsf{T}} = (\pi'{:}\pi \oslash (\rho \otimes \mathbb{I}))$
$= \pi'{:}(\pi{:}\pi_1^{\mathsf{T}} \cap \rho{:}\pi^{\mathsf{T}}{:}\rho_1^{\mathsf{T}}) \cap \rho'{:}\rho^{\mathsf{T}}{:}\rho_1^{\mathsf{T}} = ((\mathbb{I} \otimes \pi^{\mathsf{T}}) \oslash \rho^{\mathsf{T}}{:}\rho_1^{\mathsf{T}})$
ii) f **associative** $:\Longleftrightarrow \quad (f \otimes \mathbb{I}){:}f = \mathfrak{T}{:}(\mathbb{I} \otimes f){:}f$ □

The appearance of so many projections may seem terrifying. However, one need not introduce them all when checking a binary map for being associative. With an advanced proof assistant the projections will correctly be derived via unification procedures when asking whether f is associative.

Lemma 3.3.4 *The following identities hold for* $\mathfrak{T}$*:*

i) $\mathfrak{T}$ *is a bijective mapping.*
ii) $\mathfrak{T}{:}\pi_1 = \pi'{:}\pi \qquad \mathfrak{T}{:}\rho_1 = (\rho \otimes \mathbb{I})$
iii) $((Q \otimes R) \otimes S){:}\ \mathfrak{T} = \mathfrak{T}{:}(Q \otimes (R \otimes S))$
iv) $((A \oslash B) \oslash C){:}\ \mathfrak{T} = (A \oslash (B \oslash C))$.

Proof The proof of (i) is omitted; it is trivial—but tedious.
ii) $\mathfrak{T}{:}\pi_1 = (\pi'{:}\pi \oslash (\rho \otimes \mathbb{I})){:}\ \pi_1$ Definition 3.3.3.i
$= \pi'{:}\pi \cap (\rho \otimes \mathbb{I}){:}\ \mathbb{T}$ Proposition 3.1.5.i
$= \pi'{:}\pi \cap \mathbb{T} = \pi'{:}\pi$

$\mathfrak{T}{:}\rho_1 = (\pi'{:}\pi \oslash (\rho \otimes \mathbb{I})){:}\ \rho_1$
$= \pi'{:}\pi{:}\mathbb{T} \cap (\rho \otimes \mathbb{I})$ Proposition 3.1.5.ii
$= \mathbb{T} \cap (\rho \otimes \mathbb{I}) = (\rho \otimes \mathbb{I})$

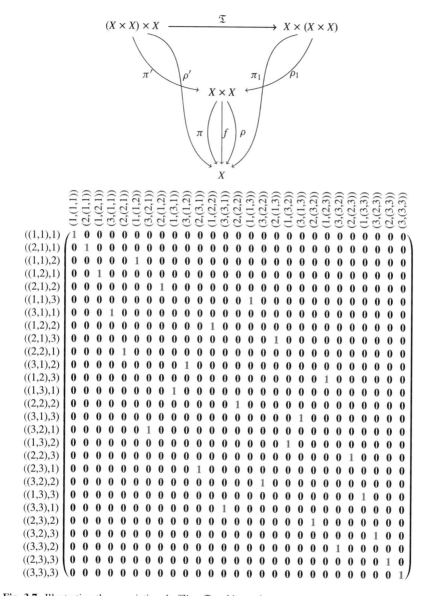

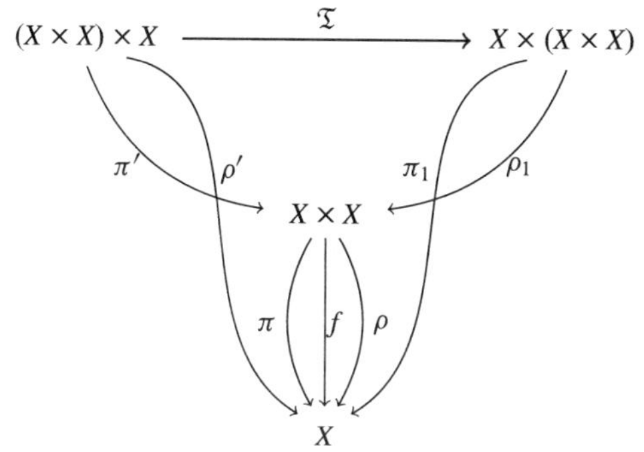

	(1,(1,1))	(2,(1,1))	(1,(2,1))	(3,(1,1))	(2,(2,1))	(1,(1,2))	(3,(2,1))	(2,(1,2))	(1,(3,1))	(3,(1,2))	(2,(3,1))	(1,(2,2))	(3,(3,1))	(2,(2,2))	(1,(1,3))	(3,(2,2))	(2,(1,3))	(1,(3,2))	(3,(1,3))	(2,(3,2))	(1,(2,3))	(3,(3,2))	(2,(2,3))	(1,(3,3))	(3,(2,3))	(2,(3,3))	(3,(3,3))
((1,1),1)	1	0	0	0	0	0	0	0	0	0	0	0	0	0	0	0	0	0	0	0	0	0	0	0	0	0	0
((2,1),1)	0	1	0	0	0	0	0	0	0	0	0	0	0	0	0	0	0	0	0	0	0	0	0	0	0	0	0
((1,1),2)	0	0	0	0	0	1	0	0	0	0	0	0	0	0	0	0	0	0	0	0	0	0	0	0	0	0	0
((1,2),1)	0	0	1	0	0	0	0	0	0	0	0	0	0	0	0	0	0	0	0	0	0	0	0	0	0	0	0
((2,1),2)	0	0	0	0	0	0	0	1	0	0	0	0	0	0	0	0	0	0	0	0	0	0	0	0	0	0	0
((1,1),3)	0	0	0	0	0	0	0	0	0	0	0	0	0	0	1	0	0	0	0	0	0	0	0	0	0	0	0
((3,1),1)	0	0	0	1	0	0	0	0	0	0	0	0	0	0	0	0	0	0	0	0	0	0	0	0	0	0	0
((1,2),2)	0	0	0	0	0	0	0	0	0	0	0	1	0	0	0	0	0	0	0	0	0	0	0	0	0	0	0
((2,1),3)	0	0	0	0	0	0	0	0	0	0	0	0	0	0	0	0	1	0	0	0	0	0	0	0	0	0	0
((2,2),1)	0	0	0	0	1	0	0	0	0	0	0	0	0	0	0	0	0	0	0	0	0	0	0	0	0	0	0
((3,1),2)	0	0	0	0	0	0	0	0	0	1	0	0	0	0	0	0	0	0	0	0	0	0	0	0	0	0	0
((1,2),3)	0	0	0	0	0	0	0	0	0	0	0	0	0	0	0	0	0	0	0	0	1	0	0	0	0	0	0
((1,3),1)	0	0	0	0	0	0	0	0	1	0	0	0	0	0	0	0	0	0	0	0	0	0	0	0	0	0	0
((2,2),2)	0	0	0	0	0	0	0	0	0	0	0	0	0	1	0	0	0	0	0	0	0	0	0	0	0	0	0
((3,1),3)	0	0	0	0	0	0	0	0	0	0	0	0	0	0	0	0	0	0	1	0	0	0	0	0	0	0	0
((3,2),1)	0	0	0	0	0	0	1	0	0	0	0	0	0	0	0	0	0	0	0	0	0	0	0	0	0	0	0
((1,3),2)	0	0	0	0	0	0	0	0	0	0	0	0	0	0	0	0	0	1	0	0	0	0	0	0	0	0	0
((2,2),3)	0	0	0	0	0	0	0	0	0	0	0	0	0	0	0	0	0	0	0	0	0	0	1	0	0	0	0
((2,3),1)	0	0	0	0	0	0	0	0	0	0	1	0	0	0	0	0	0	0	0	0	0	0	0	0	0	0	0
((3,2),2)	0	0	0	0	0	0	0	0	0	0	0	0	0	0	0	1	0	0	0	0	0	0	0	0	0	0	0
((1,3),3)	0	0	0	0	0	0	0	0	0	0	0	0	0	0	0	0	0	0	0	0	0	0	0	1	0	0	0
((3,3),1)	0	0	0	0	0	0	0	0	0	0	0	0	1	0	0	0	0	0	0	0	0	0	0	0	0	0	0
((2,3),2)	0	0	0	0	0	0	0	0	0	0	0	0	0	0	0	0	0	0	0	1	0	0	0	0	0	0	0
((3,2),3)	0	0	0	0	0	0	0	0	0	0	0	0	0	0	0	0	0	0	0	0	0	0	0	0	1	0	0
((3,3),2)	0	0	0	0	0	0	0	0	0	0	0	0	0	0	0	0	0	0	0	0	0	1	0	0	0	0	0
((2,3),3)	0	0	0	0	0	0	0	0	0	0	0	0	0	0	0	0	0	0	0	0	0	0	0	0	0	1	0
((3,3),3)	0	0	0	0	0	0	0	0	0	0	0	0	0	0	0	0	0	0	0	0	0	0	0	0	0	0	1

Fig. 3.7 Illustrating the associative shuffling $\mathfrak{T}$ and its typing

iii) $\mathfrak{T};(Q\otimes(R\otimes S)) = (\pi';\pi\olessthan(\rho\otimes\mathbb{I}));(Q\otimes(R\otimes S))$
$= (\pi';\pi;Q\olessthan(\rho\otimes\mathbb{I});(R\otimes S))$ Proposition 3.1.8.ii
$= (\pi';\pi;Q\olessthan(\rho;R\otimes S))$ since $(\rho\otimes\mathbb{I})$ is a mapping
$= \pi';\pi;Q;\pi_1^{\mathsf{T}} \cap (\rho;R\otimes S);\rho_1^{\mathsf{T}}$
$= \pi';\pi;Q;\pi_1^{\mathsf{T}} \cap (\pi';\rho;R;\pi^{\mathsf{T}} \cap \rho';S;\rho^{\mathsf{T}});\rho_1^{\mathsf{T}}$
$= \pi';\pi;Q;\pi_1^{\mathsf{T}} \cap \pi';\rho;R;\pi^{\mathsf{T}};\rho_1^{\mathsf{T}} \cap \rho';S;\rho^{\mathsf{T}};\rho_1^{\mathsf{T}}$
$= \pi';(Q\otimes R;\pi^{\mathsf{T}}) \cap \rho';S;\rho^{\mathsf{T}};\rho_1^{\mathsf{T}}$
$= ((Q\otimes R;\pi^{\mathsf{T}})\ogreaterthan S;\rho^{\mathsf{T}};\rho_1^{\mathsf{T}})$
$= ((Q\otimes R);(\mathbb{I}\otimes\pi^{\mathsf{T}})\ogreaterthan S;\rho^{\mathsf{T}};\rho_1^{\mathsf{T}})$
$= ((Q\otimes R)\otimes S);((\mathbb{I}\otimes\pi^{\mathsf{T}})\ogreaterthan\rho^{\mathsf{T}};\rho_1^{\mathsf{T}})$ transposed mappings
$= ((Q\otimes R)\otimes S);\mathfrak{T}$ Definition 3.3.3.i

iv) $((A\olessthan B)\olessthan C);\mathfrak{T}$
$= ((A\olessthan B)\olessthan C);((\mathbb{I}\otimes\pi^{\mathsf{T}})\ogreaterthan\rho^{\mathsf{T}};\rho_1^{\mathsf{T}})$ by definition variant of $\mathfrak{T}$
$= (A\olessthan B);(\mathbb{I}\otimes\pi^{\mathsf{T}}) \cap C;\rho^{\mathsf{T}};\rho_1^{\mathsf{T}}$
Proposition 3.1.8.ii since $(\mathbb{I}\otimes\pi^{\mathsf{T}})$ and $\rho^{\mathsf{T}};\rho_1^{\mathsf{T}}$ are both injective
$= (A\olessthan B;\pi^{\mathsf{T}}) \cap C;\rho^{\mathsf{T}};\rho_1^{\mathsf{T}}$
$= A;\pi_1^{\mathsf{T}} \cap B;\pi^{\mathsf{T}};\rho_1^{\mathsf{T}} \cap C;\rho^{\mathsf{T}};\rho_1^{\mathsf{T}}$
$= A;\pi_1^{\mathsf{T}} \cap (B;\pi^{\mathsf{T}} \cap C;\rho^{\mathsf{T}});\rho_1^{\mathsf{T}}$
$= A;\pi_1^{\mathsf{T}} \cap (B\olessthan C);\rho_1^{\mathsf{T}}$
$= (A\olessthan(B\olessthan C))$ □

We now address the concept of distributivity. When investigating such rules, we have to consider two binary mappings. The traditional formulation for addition and multiplication of numbers is rather simple

$$a(b+c) = ab + ac,$$

or, since one definitely must not use denotation-less operations in a theoretical investigation,

$$a \cdot (b+c) = a \cdot b + a \cdot c,$$

but even this is not given in sufficient clarity: Quantifiers are not explicitly mentioned; it is supposed that the necessary quantification is known to everybody, so that one may avoid expressing distributivity more precisely (Fig. 3.8)

$$\forall a, b, c: \quad a \cdot (b+c) = a \cdot b + a \cdot c.$$

We are going to make such rules quantifier-free. They may also be applied to the multiplication of a scalar to added vectors. In this case, typing would be slightly more general

$$\mathfrak{S} : S \times V \longrightarrow V \text{ and } \mathfrak{A} : V \times V \longrightarrow V.$$

	((1,W),(1,a))	((1,W),(1,b))	((1,W),(1,c))	((1,W),(2,a))	((1,W),(2,b))	((1,W),(2,c))	((1,D),(1,a))	((1,D),(1,b))	((1,D),(1,c))	((1,D),(2,a))	((1,D),(2,b))	((1,D),(2,c))	((1,L),(1,a))	((1,L),(1,b))	((1,L),(1,c))	((1,L),(2,a))	((1,L),(2,b))	((1,L),(2,c))	((2,W),(1,a))	((2,W),(1,b))	((2,W),(1,c))	((2,W),(2,a))	((2,W),(2,b))	((2,W),(2,c))	((2,D),(1,a))	((2,D),(1,b))	((2,D),(1,c))	((2,D),(2,a))	((2,D),(2,b))	((2,D),(2,c))	((2,L),(1,a))	((2,L),(1,b))	((2,L),(1,c))	((2,L),(2,a))	((2,L),(2,b))	((2,L),(2,c))
(1,(W,a))	1	0	0	0	0	0	0	0	0	0	0	0	0	0	0	0	0	0	0	0	0	0	0	0	0	0	0	0	0	0	0	0	0	0	0	0
(1,(W,b))	0	1	0	0	0	0	0	0	0	0	0	0	0	0	0	0	0	0	0	0	0	0	0	0	0	0	0	0	0	0	0	0	0	0	0	0
(1,(W,c))	0	0	1	0	0	0	0	0	0	0	0	0	0	0	0	0	0	0	0	0	0	0	0	0	0	0	0	0	0	0	0	0	0	0	0	0
(1,(D,a))	0	0	0	0	0	0	1	0	0	0	0	0	0	0	0	0	0	0	0	0	0	0	0	0	0	0	0	0	0	0	0	0	0	0	0	0
(1,(D,b))	0	0	0	0	0	0	0	1	0	0	0	0	0	0	0	0	0	0	0	0	0	0	0	0	0	0	0	0	0	0	0	0	0	0	0	0
(1,(D,c))	0	0	0	0	0	0	0	0	1	0	0	0	0	0	0	0	0	0	0	0	0	0	0	0	0	0	0	0	0	0	0	0	0	0	0	0
(1,(L,a))	0	0	0	0	0	0	0	0	0	0	0	0	1	0	0	0	0	0	0	0	0	0	0	0	0	0	0	0	0	0	0	0	0	0	0	0
(1,(L,b))	0	0	0	0	0	0	0	0	0	0	0	0	0	1	0	0	0	0	0	0	0	0	0	0	0	0	0	0	0	0	0	0	0	0	0	0
(1,(L,c))	0	0	0	0	0	0	0	0	0	0	0	0	0	0	1	0	0	0	0	0	0	0	0	0	0	0	0	0	0	0	0	0	0	0	0	0
(2,(W,a))	0	0	0	0	0	0	0	0	0	0	0	0	0	0	0	0	0	0	0	0	0	1	0	0	0	0	0	0	0	0	0	0	0	0	0	0
(2,(W,b))	0	0	0	0	0	0	0	0	0	0	0	0	0	0	0	0	0	0	0	0	0	0	1	0	0	0	0	0	0	0	0	0	0	0	0	0
(2,(W,c))	0	0	0	0	0	0	0	0	0	0	0	0	0	0	0	0	0	0	0	0	0	0	0	1	0	0	0	0	0	0	0	0	0	0	0	0
(2,(D,a))	0	0	0	0	0	0	0	0	0	0	0	0	0	0	0	0	0	0	0	0	0	0	0	0	0	0	0	1	0	0	0	0	0	0	0	0
(2,(D,b))	0	0	0	0	0	0	0	0	0	0	0	0	0	0	0	0	0	0	0	0	0	0	0	0	0	0	0	0	1	0	0	0	0	0	0	0
(2,(D,c))	0	0	0	0	0	0	0	0	0	0	0	0	0	0	0	0	0	0	0	0	0	0	0	0	0	0	0	0	0	1	0	0	0	0	0	0
(2,(L,a))	0	0	0	0	0	0	0	0	0	0	0	0	0	0	0	0	0	0	0	0	0	0	0	0	0	0	0	0	0	0	0	0	0	1	0	0
(2,(L,b))	0	0	0	0	0	0	0	0	0	0	0	0	0	0	0	0	0	0	0	0	0	0	0	0	0	0	0	0	0	0	0	0	0	0	1	0
(2,(L,c))	0	0	0	0	0	0	0	0	0	0	0	0	0	0	0	0	0	0	0	0	0	0	0	0	0	0	0	0	0	0	0	0	0	0	0	1

Fig. 3.8 An example of distributivity shuffling

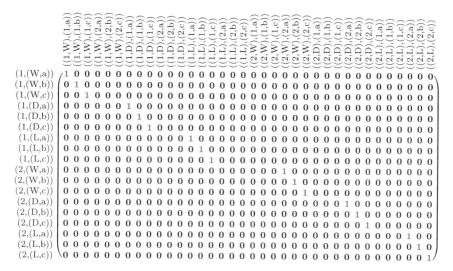

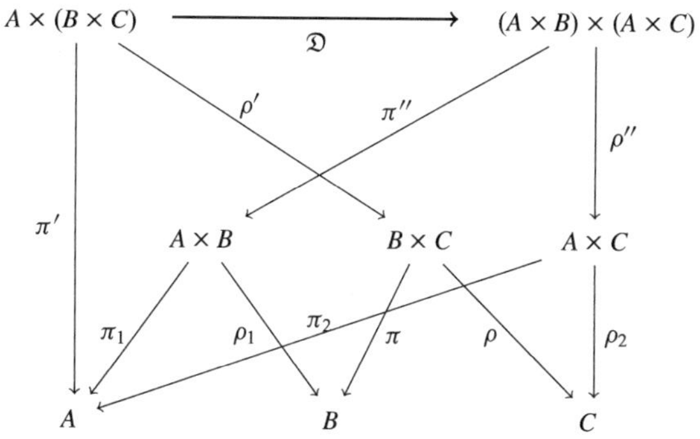

Fig. 3.9 Typing the distributivity rearrangement

The next definition prepares a rather schematic investigation of distributivity. Its typing is even more liberal than that for scalar multiplication.

The following $\mathfrak{D}$ is usually not a surjective mapping; from the left to the right side it will not map onto quadruples with first and third element different.

Definition 3.3.5 For this situation we introduce the following notation:

i) $\mathfrak{D} := ((\mathbb{I} \otimes \pi) \oslash (\mathbb{I} \otimes \rho))$ **distributivity shuffle**
or, differently grouped, when so required:
$= ((\pi_1^\mathsf{T} \oslash \pi_2^\mathsf{T}) \otimes (\rho_1^\mathsf{T} \otimes \rho_2^\mathsf{T}))$

ii) $\mathfrak{S}$ **distributes over** $\mathfrak{A}$ $:\Longleftrightarrow$ $(\mathbb{I} \otimes \mathfrak{A}) ; \mathfrak{S} = \mathfrak{D} ; (\mathfrak{S} \otimes \mathfrak{S}) ; \mathfrak{A}$ □

The transition between the two variants is obtained expanding the respective constituents and re-shuffling. Looking at Fig. 3.9, the respective typings may directly be identified.

These are elementary results concerning distributivity:

Proposition 3.3.6

i) $\mathfrak{D}$ *is an injective mapping.*
ii) $(R \otimes (S \otimes T)) ; \mathfrak{D} \subseteq \mathfrak{D}_0 ; ((R \otimes S) \otimes (R \otimes T))$
$(R \otimes (S \otimes T)) ; \mathfrak{D} = \mathfrak{D}_0 ; ((R \otimes S) \otimes (R \otimes T))$ *when R is univalent*

Proof

i) We need mainly to show

$$\begin{aligned}
\mathfrak{D} ; \mathfrak{D}^{\mathsf{T}} &= ((\mathbb{I} \otimes \pi) \mathbin{\textcircled{<}} (\mathbb{I} \otimes \rho)) ; ((\mathbb{I} \otimes \pi) \mathbin{\textcircled{<}} (\mathbb{I} \otimes \rho))^{\mathsf{T}} \\
&= (\mathbb{I} \otimes \pi) ; (\mathbb{I} \otimes \pi^{\mathsf{T}}) \cap (\mathbb{I} \otimes \rho) ; (\mathbb{I} \otimes \rho^{\mathsf{T}}) \\
&= (\mathbb{I} \otimes \pi ; \pi^{\mathsf{T}}) \cap (\mathbb{I} \otimes \rho ; \rho^{\mathsf{T}}) \\
&= \pi' ; \pi'^{\mathsf{T}} \cap \rho' ; \pi ; \pi^{\mathsf{T}} ; \rho'^{\mathsf{T}} \cap \pi' ; \pi'^{\mathsf{T}} \cap \rho' ; \rho ; \rho^{\mathsf{T}} ; \rho'^{\mathsf{T}} \\
&= \pi' ; \pi'^{\mathsf{T}} \cap \rho' ; (\pi ; \pi^{\mathsf{T}} \cap \rho ; \rho^{\mathsf{T}}) ; \rho'^{\mathsf{T}} \\
&= \pi' ; \pi'^{\mathsf{T}} \cap \rho' ; \rho'^{\mathsf{T}} = \mathbb{I}
\end{aligned}$$

ii) We use the second variant of the definition of $\mathfrak{D}$ to have

$$\pi'^{\mathsf{T}} ; \mathfrak{D} = (\pi_1^{\mathsf{T}} \mathbin{\textcircled{<}} \pi_2^{\mathsf{T}}) \quad \text{and} \quad \rho'^{\mathsf{T}} ; \mathfrak{D} = (\rho_1^{\mathsf{T}} \otimes \rho_2^{\mathsf{T}})$$

and continue with

$$\begin{aligned}
&(R \otimes (S \otimes T)) ; \mathfrak{D} = (R ; \pi'^{\mathsf{T}} \mathbin{\textcircled{>}} (S \otimes T) ; \rho'^{\mathsf{T}}) ; \mathfrak{D} \\
&= (R ; \pi'^{\mathsf{T}} ; \mathfrak{D} \mathbin{\textcircled{>}} (S \otimes T) ; \rho'^{\mathsf{T}} ; \mathfrak{D}) \quad \text{injectivity of } \mathfrak{D} \\
&= (R ; (\pi_1^{\mathsf{T}} \mathbin{\textcircled{<}} \pi_2^{\mathsf{T}}) \mathbin{\textcircled{>}} (S \otimes T) ; (\rho_1^{\mathsf{T}} \otimes \rho_2^{\mathsf{T}})) \quad \text{see above} \\
&= (R ; (\pi_1^{\mathsf{T}} \mathbin{\textcircled{<}} \pi_2^{\mathsf{T}}) \mathbin{\textcircled{>}} (S ; \rho_1^{\mathsf{T}} \otimes T ; \rho_2^{\mathsf{T}})) \\
&= ((R ; \pi_1^{\mathsf{T}} \mathbin{\textcircled{<}} R ; \pi_2^{\mathsf{T}}) \mathbin{\textcircled{>}} (S ; \rho_1^{\mathsf{T}} \otimes T ; \rho_2^{\mathsf{T}})) \quad \text{if } R \text{ is univalent, otherwise“ } \subseteq'' \\
&= \pi_0' ; (R ; \pi_1^{\mathsf{T}} \mathbin{\textcircled{<}} R ; \pi_2^{\mathsf{T}}) \cap \rho_0' ; (S ; \rho_1^{\mathsf{T}} \otimes T ; \rho_2^{\mathsf{T}}) \\
&= \pi_0' ; R ; \pi_1^{\mathsf{T}} ; \pi''^{\mathsf{T}} \cap \pi_0' ; R ; \pi_2^{\mathsf{T}} ; \rho''^{\mathsf{T}} \cap \rho_0' ; \pi_0 ; S ; \rho_1^{\mathsf{T}} ; \pi''^{\mathsf{T}} \cap \rho_0' ; \rho_0 ; T ; \rho_2^{\mathsf{T}} ; \rho''^{\mathsf{T}} \\
&= \pi_0' ; R ; \pi_1^{\mathsf{T}} ; \pi''^{\mathsf{T}} \cap \rho_0' ; \pi_0 ; S ; \rho_1^{\mathsf{T}} ; \pi''^{\mathsf{T}} \cap \pi_0' ; R ; \pi_2^{\mathsf{T}} ; \rho''^{\mathsf{T}} \cap \rho_0' ; \rho_0 ; T ; \rho_2^{\mathsf{T}} ; \rho''^{\mathsf{T}} \\
&= (\pi_0' ; R ; \pi_1^{\mathsf{T}} \cap \rho_0' ; \pi_0 ; S ; \rho_1^{\mathsf{T}} \mathbin{\textcircled{<}} \pi_0' ; R ; \pi_2^{\mathsf{T}} \cap \rho_0' ; \rho_0 ; T ; \rho_2^{\mathsf{T}}) \\
&= ((R \otimes \pi_0 ; S) \mathbin{\textcircled{<}} (R \otimes \rho_0 ; T)) \\
&= ((\mathbb{I} \otimes \pi_0) ; (R \otimes S) \mathbin{\textcircled{<}} (\mathbb{I} \otimes \rho_0) ; (R \otimes T)) \\
&= ((\mathbb{I} \otimes \pi_0) \mathbin{\textcircled{<}} (\mathbb{I} \otimes \rho_0)) ; ((R \otimes S) \otimes (R \otimes T)) \\
&= \mathfrak{D}_0 ; ((R \otimes S) \otimes (R \otimes T))
\end{aligned}$$

□

Later we will need yet another result of this type, namely the Kronecker-fork shuffling. The typing scheme is as follows. One might say in view of the following result that $\otimes$ and $\mathbin{\textcircled{<}}$ together are **bi-commutative** (Fig. 3.10).

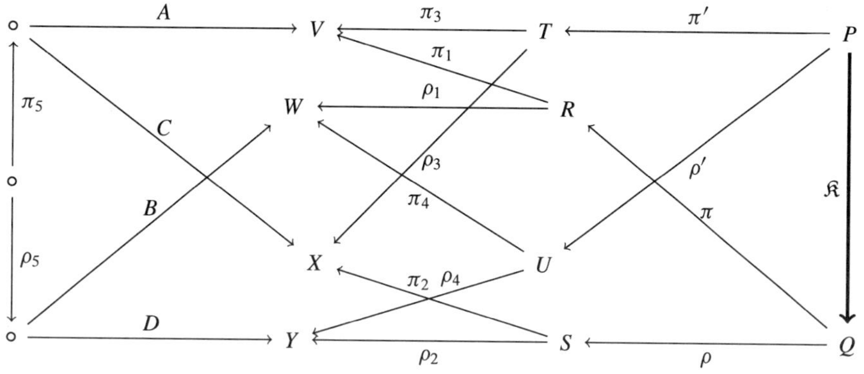

Fig. 3.10 Typing the Kronecker-fork shuffle

Proposition 3.3.7 *We define the following Kronecker-fork shuffling as a relation* $\mathfrak{K} : (V \times X) \times (W \times Y) \longrightarrow (V \times W) \times (X \times Y)$

$$\begin{aligned}
\mathfrak{K} :=\ & \pi' ; \pi_3 ; \pi_1^{\mathsf{T}} ; \pi^{\mathsf{T}} \cap \rho' ; \pi_4 ; \rho_1^{\mathsf{T}} ; \pi^{\mathsf{T}} \cap \pi' ; \rho_3 ; \pi_2^{\mathsf{T}} ; \rho^{\mathsf{T}} \cap \rho' ; \rho_4 ; \rho_2^{\mathsf{T}} ; \rho^{\mathsf{T}} \\
& \qquad\qquad\qquad\qquad \textit{or, differently grouped, when so required:} \\
=\ & \pi' ; (\pi_3 ; \pi_1^{\mathsf{T}} ; \pi^{\mathsf{T}} \cap \rho_3 ; \pi_2^{\mathsf{T}} ; \rho^{\mathsf{T}}) \cap \rho' ; (\pi_4 ; \rho_1^{\mathsf{T}} ; \pi^{\mathsf{T}} \cap \rho_4 ; \rho_2^{\mathsf{T}} ; \rho^{\mathsf{T}}) \\
=\ & ((\pi_1 \otimes \pi_2) \textcircled{<} (\rho_1 \otimes \rho_2))^{\mathsf{T}} \\
=\ & (\pi' ; \pi_3 ; \pi_1^{\mathsf{T}} \cap \rho' ; \pi_4 ; \rho_1^{\mathsf{T}}) ; \pi^{\mathsf{T}} \cap (\pi' ; \rho_3 ; \pi_2^{\mathsf{T}} \cap \rho' ; \rho_4 ; \rho_2^{\mathsf{T}}) ; \rho^{\mathsf{T}} \\
=\ & ((\pi_3 \otimes \pi_4) \textcircled{<} (\rho_3 \otimes \rho_4)).
\end{aligned}$$

Then the following holds

$$((A \textcircled{<} C) \otimes (B \textcircled{<} D)) ; \mathfrak{K} = ((A \otimes B) \textcircled{<} (C \otimes D)).$$

Proof We omit the lengthy proof that $\mathfrak{K}$ is a bijective mapping and focus on proving that it satisfies useful formulae such as

$$\begin{aligned}
& \mathfrak{K} ; (\pi_1 \otimes \pi_2) = ((\pi_3 \otimes \pi_4) \textcircled{<} (\rho_3 \otimes \rho_4)) ; (\pi_1 \otimes \pi_2) \quad \text{expanded} \\
& = ((\pi_3 \otimes \pi_4) ; \pi_1 \textcircled{<} (\rho_3 \otimes \rho_4) ; \pi_2) \\
& = (\pi' ; \pi_3 \textcircled{<} \pi' ; \rho_3) = \pi' ; (\pi_3 \textcircled{<} \rho_3) = \pi' ; (\pi_3 ; \pi_3^{\mathsf{T}} \cap \rho_3 ; \rho_3^{\mathsf{T}}) = \pi' ; \mathbb{I} = \pi' \\
& \mathfrak{K} ; (\rho_1 \otimes \rho_2) = \rho' \\
& \pi'^{\mathsf{T}} ; \mathfrak{K} = (\pi_1 \otimes \pi_2)^{\mathsf{T}} \qquad \rho'^{\mathsf{T}} ; \mathfrak{K} = (\rho_1 \otimes \rho_2)^{\mathsf{T}}
\end{aligned}$$

Using this $\mathfrak{K}$, we prove

$$\begin{aligned}
& ((A \textcircled{<} C) \otimes (B \textcircled{<} D)) ; \mathfrak{K} = (\pi_5 ; (A \textcircled{<} C) ; \pi'^{\mathsf{T}} \cap \rho_5 ; (B \textcircled{<} D) ; \rho'^{\mathsf{T}}) ; \mathfrak{K} \\
& = \pi_5 ; (A \textcircled{<} C) ; (\pi_1 \otimes \pi_2)^{\mathsf{T}} \cap \rho_5 ; (B \textcircled{<} D) ; (\rho_1 \otimes \rho_2)^{\mathsf{T}} \\
& = \pi_5 ; (A ; \pi_1^{\mathsf{T}} \textcircled{<} C ; \pi_2^{\mathsf{T}}) \cap \rho_5 ; (B ; \rho_1^{\mathsf{T}} \textcircled{<} D ; \rho_2^{\mathsf{T}}) \\
& = \pi_5 ; A ; \pi_1^{\mathsf{T}} ; \pi^{\mathsf{T}} \cap \pi_5 ; C ; \pi_2^{\mathsf{T}} ; \rho^{\mathsf{T}} \cap \rho_5 ; B ; \rho_1^{\mathsf{T}} ; \pi^{\mathsf{T}} \cap \rho_5 ; D ; \rho_2^{\mathsf{T}} ; \rho^{\mathsf{T}} \\
& = \left[\pi_5 ; A ; \pi_1^{\mathsf{T}} \cap \rho_5 ; B ; \rho_1^{\mathsf{T}}\right] ; \pi^{\mathsf{T}} \cap \left[\pi_5 ; C ; \pi_2^{\mathsf{T}} \cap \rho_5 ; D ; \rho_2^{\mathsf{T}}\right] ; \rho^{\mathsf{T}} \\
& = ((A \otimes B) \textcircled{<} (C \otimes D))
\end{aligned}$$

□

Looking back, this chapter started introducing product operators on relations, based on the direct product of sets (Cartesian product). This was the starting point for the initially unexpected and intricate discussion of sharp factorization. Using the techniques thus developed, a very general study of binary mappings took place.

In the last propositions, we have managed to lift several standard constructions such as commutativity, associativity, distributivity, e.g., that require quantifiers when being defined, to the algebraic level. These steps, we are traditionally used to execute element-wise, may now be handled on the upper level. One may complain that the proofs necessary to establish these results were sometimes fairly difficult, in particular the shuffling results. However, looking at the final resulting formulae, one will agree that these purely algebraic formulae are concise and open to intuitive comprehension.

We are now prepared to treat binary mappings algebraically. This offers the opportunity to so investigate meet and join forming.

Exercises

Exercise 3.1 Prove that whenever one has two direct products leading to common targets, i.e., $\pi_i : P_i \longrightarrow X$ and $\rho_i : P_i \longrightarrow Y$ with $i = 1, 2$, these will be isomorphic, meaning that one can define a mapping $\varphi : P_1 \longrightarrow P_2$ that satisfies $\pi_1 = \varphi ; \pi_2$ and $\rho_1 = \varphi ; \rho_2$.

Exercise 3.2 Prove that whenever one has two direct sums starting from common sources, i.e., $\iota_i : X \longrightarrow S_i$ and $\kappa_i : Y \longrightarrow S_i$ with $i = 1, 2$, these will be isomorphic, meaning that one can define a mapping $\psi : S_1 \longrightarrow S_2$ that satisfies $\iota_1 ; \psi = \iota_2$ and $\kappa_1 ; \psi = \kappa_2$.

Chapter 4
Meet and Join as Relations

When, in the preceding chapter, we had pairs (and so iterated also tuples), we immediately proceeded to handling binary mappings with relational means. This concerned the very general concepts such as being commutative, distributive, or associative. A more specific law concerns absorption mainly occurring in one traditional environment, namely for binary meets and joins. They will be handled here accordingly when the following cone mappings are available.

4.1 Cone Mappings

Powerset operations together with the symmetric quotient allow, for instance, to define the lower cone mappings for any relation (and in particular for an ordering on a set which gave the name for the operations) as presented in the following definitions. We distinguish lower cones $\bigwedge_1$ for a single element, lower cones $\bigwedge_2$ for a pair (and so iterated, any finite set) of elements and lower cones $\bigwedge$ for an arbitrary (i.e., not least possibly infinite) subset of elements.

Definition 4.1.1 We consider any relation $R : X \longrightarrow Y$, embedded in the typing environment with projections and memberships as provided by Fig. 4.1. Then we define the mappings to obtain the

lower cone of an element	$\bigwedge_1 = \mathtt{syq}(R, \varepsilon_X) : Y \longrightarrow \mathbf{2}^X$
lower cone for two elements	$\bigwedge_2 = \mathtt{syq}((R \otimes R), \varepsilon_X) : Y \times Y \longrightarrow \mathbf{2}^X$
lower cone of an element set	$\bigwedge = \mathtt{syq}(\mathtt{lbd}_R(\varepsilon_Y), \varepsilon_X) : \mathbf{2}^Y \longrightarrow \mathbf{2}^X$

□

For the lower R-cones, a point is assumed, then a pair of points with their lower R-cones intersected, or lastly the lower bound set of a—possibly infinite—point set. In any case, this is then compared with the subsets contained as columns in the relation ε_X (Fig. 4.2).

G. Schmidt, M. Winter, *Relational Topology*, Lecture Notes
in Mathematics 2208, https://doi.org/10.1007/978-3-319-74451-3_4

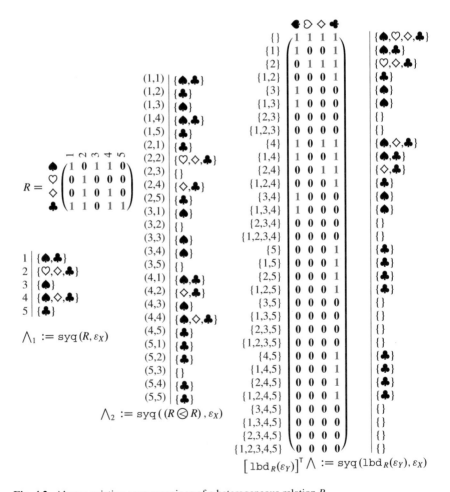

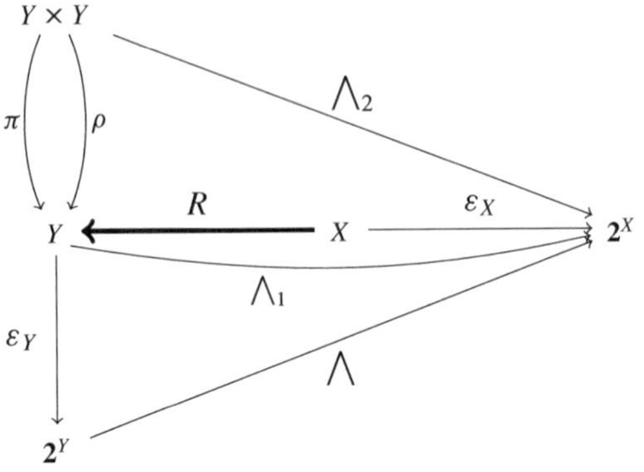

Fig. 4.1 Typing for the cone mappings of an arbitrary relation R

$R =$

	1	2	3	4	5
♠	1	0	1	1	0
♡	0	1	0	0	0
◇	0	1	0	1	0
♣	1	1	0	1	1

1	{♠,♣}
2	{♡,◇,♣}
3	{♠}
4	{♠,◇,♣}
5	{♣}

$\bigwedge_1 := \mathtt{syq}(R, \varepsilon_X)$

(1,1)	{♠,♣}
(1,2)	{♣}
(1,3)	{♠}
(1,4)	{♠,♣}
(1,5)	{♣}
(2,1)	{♣}
(2,2)	{♡,◇,♣}
(2,3)	{}
(2,4)	{◇,♣}
(2,5)	{♣}
(3,1)	{♠}
(3,2)	{}
(3,3)	{♠}
(3,4)	{♠}
(3,5)	{}
(4,1)	{♠,♣}
(4,2)	{◇,♣}
(4,3)	{♠}
(4,4)	{♠,◇,♣}
(4,5)	{♣}
(5,1)	{♣}
(5,2)	{♣}
(5,3)	{}
(5,4)	{♣}
(5,5)	{♣}

$\bigwedge_2 := \mathtt{syq}((R \oslash R), \varepsilon_X)$

	♠	♡	◇	♣	
{}	1	1	1	1	{♠,♡,◇,♣}
{1}	1	0	0	1	{♠,♣}
{2}	0	1	1	1	{♡,◇,♣}
{1,2}	0	0	0	1	{♣}
{3}	1	0	0	0	{♠}
{1,3}	1	0	0	0	{♠}
{2,3}	0	0	0	0	{}
{1,2,3}	0	0	0	0	{}
{4}	1	0	1	1	{♠,◇,♣}
{1,4}	1	0	0	1	{♠,♣}
{2,4}	0	0	1	1	{◇,♣}
{1,2,4}	0	0	0	1	{♣}
{3,4}	1	0	0	0	{♠}
{1,3,4}	1	0	0	0	{♠}
{2,3,4}	0	0	0	0	{}
{1,2,3,4}	0	0	0	0	{}
{5}	0	0	0	1	{♣}
{1,5}	0	0	0	1	{♣}
{2,5}	0	0	0	1	{♣}
{1,2,5}	0	0	0	1	{♣}
{3,5}	0	0	0	0	{}
{1,3,5}	0	0	0	0	{}
{2,3,5}	0	0	0	0	{}
{1,2,3,5}	0	0	0	0	{}
{4,5}	0	0	0	1	{♣}
{1,4,5}	0	0	0	1	{♣}
{2,4,5}	0	0	0	1	{♣}
{1,2,4,5}	0	0	0	1	{♣}
{3,4,5}	0	0	0	0	{}
{1,3,4,5}	0	0	0	0	{}
{2,3,4,5}	0	0	0	0	{}
{1,2,3,4,5}	0	0	0	0	{}

$\left[\mathtt{lbd}_R(\varepsilon_Y)\right]^{\mathsf{T}}$ $\bigwedge := \mathtt{syq}(\mathtt{lbd}_R(\varepsilon_Y), \varepsilon_X)$

Fig. 4.2 Always existing cone mappings of a heterogeneous relation R

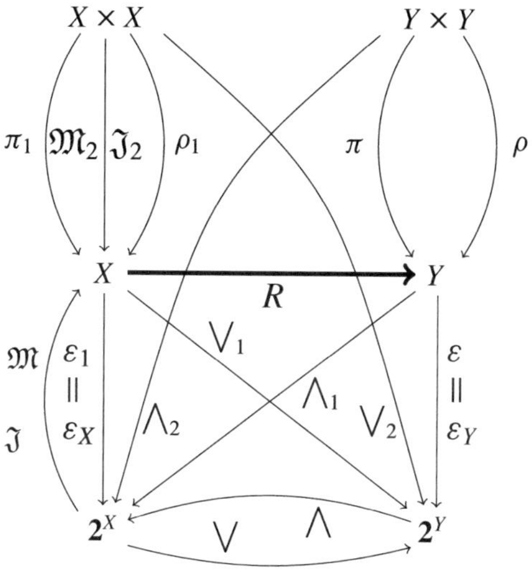

Fig. 4.3 Typing of cone mappings as well as meet and join relations for R

All these, $\bigwedge_1$, $\bigwedge_2$, $\bigwedge$, are necessarily mappings; see the remark following Definition 2.2.1.

Of course, there exist the upper counterparts to these mappings; see Definition 4.1.2. It may be helpful to recall that $\mathtt{lbd}_{R^\mathsf{T}}(C) = \mathtt{ubd}_R(C)$. (It should be mentioned that in other texts, e.g. in [Sch11], $\bigvee_1 = \mathtt{syq}(R^\mathsf{T}, \varepsilon_Y) =: \Lambda_R$ has also been called the **power transpose**.)

Definition 4.1.2 For an arbitrary relation R, considered together with the projections and membership relations according to Fig. 4.3, we define accordingly the mapping to obtain the

upper cone of an element	$\bigvee_1 = \mathtt{syq}(R^\mathsf{T}, \varepsilon_Y) : X \longrightarrow \mathbf{2}^Y$
upper cone for two elements	$\bigvee_2 = \mathtt{syq}((R^\mathsf{T} \otimes R^\mathsf{T}), \varepsilon_Y) : X \times X \longrightarrow \mathbf{2}^Y$
upper cone of an element set	$\bigvee = \mathtt{syq}(\mathtt{ubd}_R(\varepsilon_X), \varepsilon_Y) : \mathbf{2}^X \longrightarrow \mathbf{2}^Y$

□

These constructs satisfy in particular $\bigwedge ; \bigvee ; \bigwedge = \bigwedge$ and $\bigvee ; \bigwedge ; \bigvee = \bigvee$. The proof reduces to the non-lifted familiar version $\mathtt{lbd}_R(\mathtt{ubd}_R(\mathtt{lbd}_R(X))) = \mathtt{lbd}_R(X)$:

$$\begin{aligned}
\bigwedge ; \bigvee ; \bigwedge &= \bigwedge ; \bigvee ; \mathtt{syq}(\mathtt{lbd}_R(\varepsilon_Y), \varepsilon_X) = \bigwedge ; \mathtt{syq}(\mathtt{lbd}_R(\varepsilon_Y) ; \bigvee^\mathsf{T}, \varepsilon_X) \\
&= \bigwedge ; \mathtt{syq}(\mathtt{lbd}_R(\varepsilon_Y ; \bigvee^\mathsf{T}), \varepsilon_X) \\
&= \bigwedge ; \mathtt{syq}(\mathtt{lbd}_R(\varepsilon_Y ; \mathtt{syq}(\varepsilon_Y, \mathtt{ubd}_R(\varepsilon_X))), \varepsilon_X) \\
&= \bigwedge ; \mathtt{syq}(\mathtt{lbd}_R(\mathtt{ubd}_R(\varepsilon_X)), \varepsilon_X) \\
&= \mathtt{syq}(\mathtt{lbd}_R(\mathtt{ubd}_R(\varepsilon_X)) ; \bigwedge^\mathsf{T}, \varepsilon_X)
\end{aligned}$$

$$= \mathtt{syq}(\mathtt{lbd}_R(\mathtt{ubd}_R(\varepsilon_X ; \bigwedge^{\mathsf{T}})), \varepsilon_X)$$
$$= \mathtt{syq}(\mathtt{lbd}_R(\mathtt{ubd}_R(\varepsilon_X ; \mathtt{syq}(\varepsilon_X, \mathtt{lbd}_R(\varepsilon_Y)))), \varepsilon_X)$$
$$= \mathtt{syq}(\mathtt{lbd}_R(\mathtt{ubd}_R(\mathtt{lbd}_R(\varepsilon_Y))), \varepsilon_X)$$
$$= \mathtt{syq}(\mathtt{lbd}_R(\varepsilon_Y), \varepsilon_X)$$
$$= \bigwedge$$

One is tempted to say that, once lifted, affairs become increasingly simpler!

4.2 Binary and Arbitrary Meets and Joins

Next, we define the meet and join relations of an arbitrary ordering E, i.e., of a reflexive, transitive, and antisymmetric relation. The definition is based on the typing according to Fig. 4.4. It is presented on the left side of the table in Definition 4.2.1. Its right side is reserved for the special case of a powerset ordering Ω allowing slightly simpler formulations via membership deletion.

In the following definition, row groups separate the finite from the arbitrary (and possibly infinite) case. Columns distinguish the arbitrary ordering E as opposed to the special case of a powerset ordering Ω.

Definition 4.2.1 Assume an order $E : X \longrightarrow X$,—or in the more specific case $X := \mathbf{2}^W$ with a powerset order $\Omega : \mathbf{2}^W \longrightarrow \mathbf{2}^W$ and corresponding membership $\varepsilon : W \longrightarrow \mathbf{2}^W$. Then we define **meet formings** and **join formings** as

order E in general	special case of a powerset order Ω
$\mathfrak{M}_2 := \mathtt{syq}((E \otimes E), E) = \left[\mathtt{lub}_E((E \otimes E))\right]^{\mathsf{T}}$	$\mathfrak{M}_2 := \mathtt{syq}((\varepsilon \otimes \varepsilon), \varepsilon)$
$\mathfrak{J}_2 := \mathtt{syq}((E^{\mathsf{T}} \otimes E^{\mathsf{T}}), E^{\mathsf{T}}) = \left[\mathtt{glb}_E((E^{\mathsf{T}} \otimes E^{\mathsf{T}}))\right]^{\mathsf{T}}$	$\mathfrak{J}_2 := \mathtt{syq}((\overline{\varepsilon} \otimes \overline{\varepsilon}), \overline{\varepsilon})$
$\mathfrak{M} := \mathtt{syq}(\mathtt{lbd}_E(\varepsilon_1), E) = \left[\mathtt{glb}_E(\varepsilon_1)\right]^{\mathsf{T}}$	$\mathfrak{M} := \mathtt{syq}(\overline{\varepsilon} ; \varepsilon_1, \overline{\varepsilon})$
$\mathfrak{J} := \mathtt{syq}(\mathtt{ubd}_E(\varepsilon_1), E^{\mathsf{T}}) = \left[\mathtt{lub}_E(\varepsilon_1)\right]^{\mathsf{T}}$	$\mathfrak{J} := \mathtt{syq}(\varepsilon ; \varepsilon_1, \varepsilon)$

□

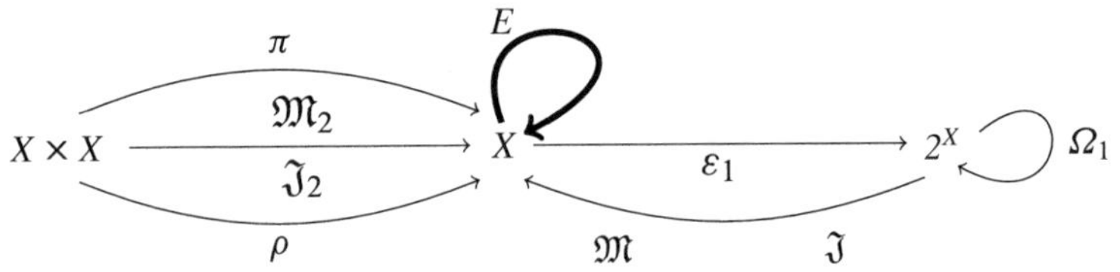

Fig. 4.4 Meet and join relations for order E; binary $\mathfrak{M}_2$, $\mathfrak{J}_2$ or arbitrary $\mathfrak{M}$, $\mathfrak{J}$

A distinction is made between the binary case $\mathfrak{M}_2$, $\mathfrak{J}_2$ and the general case of obtaining meet and join for arbitrary (possibly non-finite) sets with $\mathfrak{M}$, $\mathfrak{J}$. In the binary case of an arbitrary ordering, we have

$$(E \otimes E) = E ; \pi^\mathsf{T} \cap E ; \rho^\mathsf{T} = \overline{\overline{E} ; \pi^\mathsf{T} \cup \overline{E} ; \rho^\mathsf{T}} = \overline{\overline{E} ; (\pi^\mathsf{T} \cup \rho^\mathsf{T})} = \mathtt{lbd}_E(\pi^\mathsf{T} \cup \rho^\mathsf{T}),$$

showing that $(E \otimes E)$ has column-wise all the lower bound cones of the pairs. For $\mathfrak{M}_2$ in Definition 4.2.1, these columns are then either compared via the symmetric quotient with sets below an *element* of the ordering, or in the variant form their least upper bound is taken.

For a pair of elements as well as for any subset of elements, these relations are necessarily univalent; however, meets and joins need not exist in general, so that these relations need not be mappings for arbitrary E. On the right hand side, binary as well as arbitrary meet and join relations in powerset orderings Ω are provided, all of which are necessarily mappings.

In the left column always two versions are presented that may not immediately be recognized as being equivalent; therefore equivalence needs a proof which is presented in Proposition 4.2.2.i for $\mathfrak{M}_2$ only.

Proposition 4.2.2

i) The variants in Definition 4.2.1 are indeed equivalent, i.e.:

$$\mathfrak{M}_2 = \mathtt{syq}((E \otimes E), E) = \left[\mathtt{lub}_E((E \otimes E))\right]^\mathsf{T}$$

ii) $\mathfrak{M}_2$ *is commutative,* $\mathfrak{P} ; \mathfrak{M}_2 = \mathfrak{M}_2$.
iii) $E ; \mathfrak{M}_2^\mathsf{T} \subseteq (E \otimes E)$, $\qquad E ; \mathfrak{M}_2^\mathsf{T} = (E \otimes E)$ *when* $\mathfrak{M}_2$ *is a map*
iv) When $\mathfrak{M}_2$ *is a map, then* $\mathfrak{M}_2$ *is necessarily associative, i.e.*

$$(\mathfrak{M}_2 \otimes \mathbb{I}) ; \mathfrak{M}_2 = \mathfrak{T} ; (\mathbb{I} \otimes \mathfrak{M}_2) ; \mathfrak{M}_2 .$$

v) $\pi \cap \rho \subseteq \mathfrak{M}_2$
vi) $\mathfrak{M}_2^\mathsf{T} ; \pi = E \qquad \mathfrak{M}_2^\mathsf{T} ; \rho = E$
vii) $\mathfrak{J}_2^\mathsf{T} ; \pi = E^\mathsf{T} \qquad \mathfrak{J}_2^\mathsf{T} ; \rho = E^\mathsf{T}$

Proof

i) We use that $(E \otimes E) = \mathtt{lbd}_E(\pi^\mathsf{T} \cup \rho^\mathsf{T})$.

$$\begin{aligned}
\mathtt{lub}_E((E \otimes E)) &= \mathtt{lbd}_E(\mathtt{ubd}_E((E \otimes E))) \cap \mathtt{ubd}_E((E \otimes E)) \\
&= \mathtt{lbd}_E(\mathtt{ubd}_E(\mathtt{lbd}_E(\pi^\mathsf{T} \cup \rho^\mathsf{T}))) \cap \mathtt{ubd}_E((E \otimes E)) \\
&= \mathtt{lbd}_E(\pi^\mathsf{T} \cup \rho^\mathsf{T}) \cap \mathtt{ubd}_E((E \otimes E)) \quad \text{traditional rule} \\
&= \overline{\overline{E} ; (\pi^\mathsf{T} \cup \rho^\mathsf{T})} \cap \mathtt{ubd}_E((E \otimes E)) \\
&= \overline{E^\mathsf{T} ; \overline{E} ; (\pi^\mathsf{T} \cup \rho^\mathsf{T})} \cap \mathtt{ubd}_E((E \otimes E)) \quad \text{since } E^\mathsf{T} ; \overline{E} = \overline{E} \text{ for an ordering } E \\
&= \overline{E^\mathsf{T} ; \overline{(E \otimes E)}} \cap \overline{\overline{E}^\mathsf{T} ; (E \otimes E)} \\
&= \mathtt{syq}(E, (E \otimes E)) \\
&= \mathfrak{M}_2^\mathsf{T} \quad \text{by definition}
\end{aligned}$$

ii) We give a strictly formal proof:

$$\begin{aligned}
\mathfrak{P} ; \mathfrak{M}_2 &= \mathfrak{P} ; \mathtt{syq}((E \olessthan E), E) && \text{by definition}\\
&= \mathtt{syq}((E \olessthan E) ; \mathfrak{P}, E) && \text{since } \mathfrak{P} \text{ is a symmetric bijective map}\\
&= \mathtt{syq}(E ; \pi^{\mathsf{T}} ; \mathfrak{P} \cap E ; \rho^{\mathsf{T}} ; \mathfrak{P}, E) &&\\
&= \mathtt{syq}(E ; \rho^{\mathsf{T}} \cap E ; \pi^{\mathsf{T}}, E) && \text{Proposition 3.3.2.ii}\\
&= \mathtt{syq}((E \olessthan E), E) = \mathfrak{M}_2 &&
\end{aligned}$$

iii) Both results follow immediately with Proposition 2.1.1 when expanded to

$$E ; \mathfrak{M}_2^{\mathsf{T}} = E ; \mathtt{syq}(E, (E \olessthan E)) = (E \olessthan E) \cap \mathbb{T} ; \mathtt{syq}(E, (E \olessthan E)).$$

iv)

$$\begin{aligned}
&\mathfrak{T} ; (\mathbb{I} \otimes \mathfrak{M}_2) ; \mathfrak{M}_2 &&\\
&= \mathfrak{T} ; (\mathbb{I} \otimes \mathfrak{M}_2)\, \mathtt{syq}((E \olessthan E), E) && \text{Definition 4.2.1}\\
&= \mathfrak{T} ; \mathtt{syq}((E \olessthan E) ; (\mathbb{I} \otimes \mathfrak{M}_2^{\mathsf{T}}), E) && \text{since } (\mathbb{I} \otimes \mathfrak{M}_2) \text{ is a mapping}\\
&= \mathfrak{T} ; \mathtt{syq}((E \olessthan E ; \mathfrak{M}_2^{\mathsf{T}}), E) && \text{Proposition 3.1.7.iv}\\
&= \mathfrak{T} ; \mathtt{syq}((E \olessthan (E \olessthan E)), E) && \text{due to (iii) with } \mathfrak{M}_2 \text{ a map}\\
&= \mathtt{syq}((E \olessthan (E \olessthan E)) ; \mathfrak{T}^{\mathsf{T}}, E) && \mathfrak{T} \text{ is a map}\\
&= \mathtt{syq}(((E \olessthan E) \olessthan E), E) && \text{Lemma 3.3.4.iv}\\
&= \mathtt{syq}((E ; \mathfrak{M}_2^{\mathsf{T}} \olessthan E), E) && \text{using (iii) with } \mathfrak{M}_2 \text{ a map}\\
&= \mathtt{syq}((E \olessthan E) ; (\mathfrak{M}_2^{\mathsf{T}} \otimes \mathbb{I}), E) &&\\
&= (\mathfrak{M}_2 \otimes \mathbb{I}) ; \mathtt{syq}((E \olessthan E), E) &&\\
&= (\mathfrak{M}_2 \otimes \mathbb{I}) ; \mathfrak{M}_2 &&
\end{aligned}$$

v)

$$\begin{aligned}
\mathfrak{M}_2^{\mathsf{T}} &= \mathtt{syq}(E, (E \olessthan E)) = \overline{\overline{E}^{\mathsf{T}} ; (E \olessthan E)} \cap \overline{E^{\mathsf{T}} ; \overline{E ; \pi^{\mathsf{T}} \cap E ; \rho^{\mathsf{T}}}}\\
&= \overline{\overline{E}^{\mathsf{T}} ; (E \olessthan E)} \cap \overline{E^{\mathsf{T}} ; \overline{E ; \pi^{\mathsf{T}}}} \cap \overline{E^{\mathsf{T}} ; \overline{E ; \rho^{\mathsf{T}}}} = \overline{\overline{E}^{\mathsf{T}} ; (E \olessthan E)} \cap E ; \pi^{\mathsf{T}} \cap E ; \rho^{\mathsf{T}}\\
&\supseteq \overline{\overline{E}^{\mathsf{T}} ; E ; \rho^{\mathsf{T}}} \cap E ; \pi^{\mathsf{T}} \cap E ; \rho^{\mathsf{T}} = E^{\mathsf{T}} ; \rho^{\mathsf{T}} \cap E ; \pi^{\mathsf{T}} \cap E ; \rho^{\mathsf{T}}\\
&= E ; \pi^{\mathsf{T}} \cap (E^{\mathsf{T}} \cap E) ; \rho^{\mathsf{T}} \supseteq \pi^{\mathsf{T}} \cap \rho^{\mathsf{T}}
\end{aligned}$$

vi, vii): Having in mind $\mathfrak{M}_2^{\mathsf{T}} \subseteq E ; \pi^{\mathsf{T}}$ from the proof of (v), we prove just

$$\begin{aligned}
\mathfrak{M}_2^{\mathsf{T}} &= \mathtt{syq}(E, (E \olessthan E)) = \overline{\overline{E}^{\mathsf{T}} ; (E \olessthan E)} \cap \overline{E^{\mathsf{T}} ; \overline{(E \olessthan E)}}\\
&= \;\ldots\; \cap \overline{E^{\mathsf{T}} ; (\overline{E} ; \pi^{\mathsf{T}} \cup \overline{E} ; \rho^{\mathsf{T}})} = \;\ldots\; \cap \overline{E^{\mathsf{T}} ; \overline{E} ; \pi^{\mathsf{T}} \cup E^{\mathsf{T}} ; \overline{E} ; \rho^{\mathsf{T}}}\\
&= \ldots \cap \overline{\overline{\overline{E}} ; \pi^{\mathsf{T}} \cup \overline{\overline{E}} ; \rho^{\mathsf{T}}} = \overline{\overline{E}^{\mathsf{T}} ; (E \olessthan E)} \cap (E \olessthan E)
\end{aligned}$$

Now

$$
\begin{aligned}
\mathfrak{M}_2^{\mathsf{T}};\pi &= \left[\overline{\overline{E}^{\mathsf{T}};(E \circledS E)} \cap E;\pi^{\mathsf{T}} \cap E;\rho^{\mathsf{T}}\right];\pi && \text{expanded}\\
&= \left[E;\pi^{\mathsf{T}} \cap \{\overline{\overline{E}^{\mathsf{T}};(E;\pi^{\mathsf{T}} \cap E;\rho^{\mathsf{T}})} \cap E;\rho^{\mathsf{T}}\}\right];\pi && \text{further expanded, shuffled}\\
&\supseteq \left[E;\pi^{\mathsf{T}} \cap (\overline{\overline{E}^{\mathsf{T}};E;\rho^{\mathsf{T}}} \cap E;\rho^{\mathsf{T}})\right];\pi && \text{monotonic}\\
&= \left[E;\pi^{\mathsf{T}} \cap (\overline{\overline{E}^{\mathsf{T}};\rho^{\mathsf{T}}} \cap E;\rho^{\mathsf{T}})\right];\pi && \text{reflexive and transitive } E\\
&= \left[E;\pi^{\mathsf{T}} \cap (E^{\mathsf{T}} \cap E);\rho^{\mathsf{T}}\right];\pi\\
&= E \cap (E^{\mathsf{T}} \cap E);\rho^{\mathsf{T}};\pi && \text{destroy and append}\\
&= E \cap \mathbb{I};\mathbb{T} = E
\end{aligned}
$$

□

We see here that a pair with coinciding first and second component will have precisely this coinciding element as its meet, i.e.

$$\pi \cap \rho \subseteq \mathfrak{M}_2 \quad \text{or} \quad (\mathbb{I} \circledR \mathbb{I}) \subseteq \mathfrak{M}_2.$$

Figure 4.5 provides an example. First it contains the tiny ordering E in the upper left together with its Hasse diagram. We may then look for the binary meet or intersection of elements 5 and 1. Both rows in $\mathfrak{M}_2$ therefore, $(1,5)$ as well as $(5,1)$, only show **0** s, indicating that there is no element less than both. However, the rows $(2,4)$ and $(4,2)$ show precisely that the intersection of 2 and 4 in this ordering is 3. Correspondingly in $\mathfrak{M}$: The row for the subset $\{1,5\}$ is a row of **0** s, while the row $\{2,1,3,4\}$ points to 3.

This means in particular that $\mathfrak{M}_2$, $\mathfrak{J}_2$ are surjective; in a formal proof one has, namely,

$$
\begin{aligned}
\mathbb{T};\mathfrak{M}_2 \supseteq \mathbb{T};(\mathbb{I} \circledS \mathbb{I});\mathfrak{M}_2 &= \mathbb{T};(\mathbb{I} \circledS \mathbb{I});\mathtt{syq}((E \circledS E), E)\\
&= \mathbb{T};\mathtt{syq}((E \circledS E);(\mathbb{I} \circledR \mathbb{I}), E)\\
&= \mathbb{T};\mathtt{syq}(E \cap E, E) = \mathbb{T};\mathtt{syq}(E, E) = \mathbb{T};\mathbb{I} = \mathbb{T}
\end{aligned}
$$

The cone mappings $\bigwedge$, $\bigvee$ introduced earlier allow remarkable factorizations for meet and join relations in full generality.

Proposition 4.2.3 *The following hold for an arbitrary ordering E and its corresponding relations:*

i) $\mathfrak{M} = \bigwedge;\mathfrak{J}$,
ii) $\mathfrak{J} = \bigvee;\mathfrak{M}$.
iii) $\mathfrak{J}$ *is a mapping if and only if* $\mathfrak{M}$ *is a mapping.*
iv) In any case, $\mathfrak{M}$ *and* $\mathfrak{J}$ *are surjective relations.*

	1	2	3	4	5
1	1	0	0	1	0
2	0	1	0	0	0
3	1	1	1	1	0
4	0	0	0	1	0
5	0	1	0	0	1

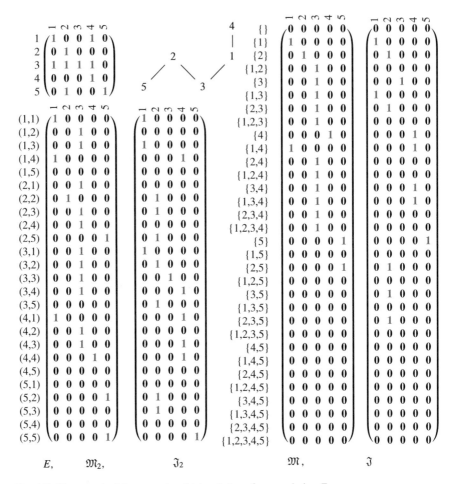

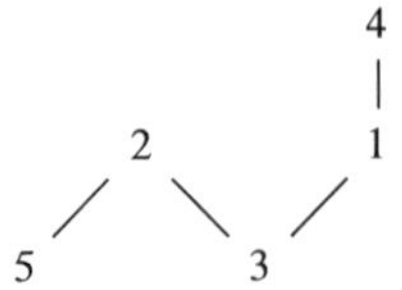

	1	2	3	4	5	1	2	3	4	5
(1,1)	1	0	0	0	0	1	0	0	0	0
(1,2)	0	0	1	0	0	0	0	0	0	0
(1,3)	0	0	1	0	0	1	0	0	0	0
(1,4)	1	0	0	0	0	0	0	0	1	0
(1,5)	0	0	0	0	0	0	0	0	0	0
(2,1)	0	0	1	0	0	0	0	0	0	0
(2,2)	0	1	0	0	0	0	1	0	0	0
(2,3)	0	0	1	0	0	0	1	0	0	0
(2,4)	0	0	1	0	0	0	0	0	0	0
(2,5)	0	0	0	0	1	0	1	0	0	0
(3,1)	0	0	1	0	0	1	0	0	0	0
(3,2)	0	0	1	0	0	0	1	0	0	0
(3,3)	0	0	1	0	0	0	0	1	0	0
(3,4)	0	0	1	0	0	0	0	0	1	0
(3,5)	0	0	0	0	0	0	1	0	0	0
(4,1)	1	0	0	0	0	0	0	0	1	0
(4,2)	0	0	1	0	0	0	0	0	0	0
(4,3)	0	0	1	0	0	0	0	0	1	0
(4,4)	0	0	0	1	0	0	0	0	1	0
(4,5)	0	0	0	0	0	0	0	0	0	0
(5,1)	0	0	0	0	0	0	0	0	0	0
(5,2)	0	0	0	0	1	0	1	0	0	0
(5,3)	0	0	0	0	0	0	1	0	0	0
(5,4)	0	0	0	0	0	0	0	0	0	0
(5,5)	0	0	0	0	1	0	0	0	0	1

	1	2	3	4	5	1	2	3	4	5
{}	0	0	0	0	0	0	0	0	0	0
{1}	1	0	0	0	0	1	0	0	0	0
{2}	0	1	0	0	0	0	1	0	0	0
{1,2}	0	0	1	0	0	0	0	0	0	0
{3}	0	0	1	0	0	0	0	1	0	0
{1,3}	0	0	1	0	0	1	0	0	0	0
{2,3}	0	0	1	0	0	0	1	0	0	0
{1,2,3}	0	0	1	0	0	0	0	0	0	0
{4}	0	0	0	1	0	0	0	0	1	0
{1,4}	1	0	0	0	0	0	0	0	1	0
{2,4}	0	0	1	0	0	0	0	0	0	0
{1,2,4}	0	0	1	0	0	0	0	0	0	0
{3,4}	0	0	1	0	0	0	0	0	1	0
{1,3,4}	0	0	1	0	0	0	0	0	1	0
{2,3,4}	0	0	1	0	0	0	0	0	0	0
{1,2,3,4}	0	0	1	0	0	0	0	0	0	0
{5}	0	0	0	0	1	0	0	0	0	1
{1,5}	0	0	0	0	0	0	0	0	0	0
{2,5}	0	0	0	0	1	0	1	0	0	0
{1,2,5}	0	0	0	0	0	0	0	0	0	0
{3,5}	0	0	0	0	0	0	1	0	0	0
{1,3,5}	0	0	0	0	0	0	0	0	0	0
{2,3,5}	0	0	0	0	0	0	1	0	0	0
{1,2,3,5}	0	0	0	0	0	0	0	0	0	0
{4,5}	0	0	0	0	0	0	0	0	0	0
{1,4,5}	0	0	0	0	0	0	0	0	0	0
{2,4,5}	0	0	0	0	0	0	0	0	0	0
{1,2,4,5}	0	0	0	0	0	0	0	0	0	0
{3,4,5}	0	0	0	0	0	0	0	0	0	0
{1,3,4,5}	0	0	0	0	0	0	0	0	0	0
{2,3,4,5}	0	0	0	0	0	0	0	0	0	0
{1,2,3,4,5}	0	0	0	0	0	0	0	0	0	0

E, $\mathfrak{M}_2$, $\mathfrak{J}_2$ $\mathfrak{M}$, $\mathfrak{J}$

Fig. 4.5 Binary and arbitrary meet and join relations for an ordering E

Proof From (i,ii), only (i) is proved. Due to the typing with the homogeneous relation E, now $\varepsilon_1 = \varepsilon$. We use that $\bigwedge$ is a mapping.

$$\begin{aligned}\bigwedge;\mathfrak{J} &= \bigwedge;[\mathtt{lub}_E(\varepsilon_1)]^{\mathsf{T}} \\ &= \left[\mathtt{lub}_E(\varepsilon_1);\bigwedge^{\mathsf{T}}\right]^{\mathsf{T}} = \left[\mathtt{lub}_E(\varepsilon_1;\bigwedge^{\mathsf{T}})\right]^{\mathsf{T}} = \left[\mathtt{lub}_E(\varepsilon_1;\mathtt{syq}(\varepsilon_1,\mathtt{lbd}_E(\varepsilon)))\right]^{\mathsf{T}} \\ &= \left[\mathtt{lub}_E(\mathtt{lbd}_E(\varepsilon))\right]^{\mathsf{T}} = \left[\mathtt{glb}_E(\varepsilon)\right]^{\mathsf{T}} = \mathfrak{M}\end{aligned}$$

Clearly, the least upper of all the lower bounds equals the greatest lower bound.

iii) Using (i,ii), this is trivial, since a product of mappings will again be a mapping.

iv) We prove surjectivity of $\mathfrak{M}$:

$$\begin{aligned}\mathfrak{M}^{\mathsf{T}} &= \mathtt{syq}(E, \mathtt{lbd}_E(\varepsilon_1)) = \overline{\overline{E}^{\mathsf{T}} ; \mathtt{lbd}_E(\varepsilon_1)} \cap \overline{E^{\mathsf{T}} ; \overline{\mathtt{lbd}_E(\varepsilon_1)}} \\ &= \overline{\overline{E}^{\mathsf{T}} ; \overline{\overline{E} ; \varepsilon_1}} \cap \overline{E^{\mathsf{T}} ; \overline{E} ; \varepsilon_1} = \overline{\overline{E}^{\mathsf{T}} ; \overline{\overline{E} ; \varepsilon_1}} \cap \overline{\overline{E} ; \varepsilon_1}\end{aligned}$$

Now we introduce the singleton injection map $\sigma_1 \subseteq \varepsilon_1$:

$$\begin{aligned}\mathfrak{M}^{\mathsf{T}};\mathbb{T} \supseteq \mathfrak{M}^{\mathsf{T}};\sigma_1^{\mathsf{T}};\mathbb{T} &= \left[\overline{\overline{E}^{\mathsf{T}};\overline{\overline{E};\varepsilon_1}} \cap \overline{\overline{E};\varepsilon_1}\right];\sigma_1^{\mathsf{T}};\mathbb{T} \\ &= \left[\overline{\overline{E}^{\mathsf{T}};\overline{\overline{E};\varepsilon_1;\sigma_1^{\mathsf{T}}}} \cap \overline{\overline{E};\varepsilon_1;\sigma_1^{\mathsf{T}}}\right];\mathbb{T} = \left[\overline{\overline{E}^{\mathsf{T}};\overline{\overline{E};\mathbb{I}}} \cap \overline{\overline{E};\mathbb{I}}\right];\mathbb{T} \\ &= \left[E^{\mathsf{T}} \cap E\right];\mathbb{T} = \mathbb{T}\end{aligned}$$ □

Properties (i,ii) hold in general, i.e. also when E does not describe a lattice. Of the four $\bigwedge$, $\bigvee$, $\mathfrak{M}$, $\mathfrak{J}$, only $\bigwedge$, $\bigvee$ are necessarily mappings.

Also in the binary case, such factorizations for meet and join via the bound mappings $\bigwedge_2$, $\bigvee_2$ are possible.

Proposition 4.2.4 *The following hold for an arbitrary ordering E and its corresponding relations:*

i) $\mathfrak{M}_2 = \bigwedge_2 ; \mathfrak{J}$,
ii) $\mathfrak{J}_2 = \bigvee_2 ; \mathfrak{M}$.
iii) Should $\mathfrak{J}$ be a mapping, so will be $\mathfrak{M}_2$.
iv) Should $\mathfrak{M}$ be a mapping, so will be $\mathfrak{J}_2$.
v) In any case, $\mathfrak{M}_2$ and $\mathfrak{J}_2$ are surjective relations.

Proof The proof—here restricted to (i)—rests on $\bigwedge_2$ being a mapping.

$$\begin{aligned}\textstyle\bigwedge_2 ; \mathfrak{J} &= \textstyle\bigwedge_2 ; \mathtt{syq}(\mathtt{ubd}_E(\varepsilon_1), E^{\mathsf{T}}) = \mathtt{syq}(\mathtt{ubd}_E(\varepsilon_1) ; \bigwedge_2^{\mathsf{T}}, E^{\mathsf{T}}) \\ &= \textstyle\mathtt{syq}(\mathtt{ubd}_E(\varepsilon_1 ; \bigwedge_2^{\mathsf{T}}), E^{\mathsf{T}}) \\ &= \mathtt{syq}(\mathtt{ubd}_E(\varepsilon_1 ; \mathtt{syq}(\varepsilon_1, (E \otimes E))), E^{\mathsf{T}}) \\ &= \mathtt{syq}(\mathtt{ubd}_E((E \otimes E)), E^{\mathsf{T}}) \\ &= \mathtt{syq}((E \otimes E), E) \quad \text{since } \mathtt{lbd}_E(\mathtt{ubd}_E((E \otimes E))) = (E \otimes E) \\ &= \left[\mathtt{lub}_E((E \otimes E))\right]^{\mathsf{T}} \quad \text{Definition 4.2.1} \\ &= \mathfrak{M}_2 \quad \text{Proposition 4.2.2}\end{aligned}$$

v) We use Proposition 4.2.2.v and show

$$\mathfrak{M}_2^{\mathsf{T}} ; \mathbb{T} \supseteq (\pi^{\mathsf{T}} \cap \rho^{\mathsf{T}}) ; \mathbb{T} = (\pi^{\mathsf{T}} \cap \rho^{\mathsf{T}}) ; \pi ; \mathbb{T} = (\mathbb{I} \cap \rho^{\mathsf{T}} ; \pi) ; \mathbb{T} = (\mathbb{I} \cap \mathbb{T}) ; \mathbb{T} = \mathbb{T}.$$ □

4.3 Join and Meet in a Powerset

In case of a powerset ordering Ω instead of E, one has in addition slightly more convenient possibilities to define these concepts—already shown in Definition 4.2.1. The typing is recollected in Fig. 4.6.

Powerset operations together with the symmetric quotient allow to lift many concepts to a point-free as well as quantifier-free level. Via lifting one may often replace predicate logic proofs with relation-algebraic ones. These proofs address in particular those formulae that the Kronecker-, fork-, and join-operators satisfy in specific situations.

Focussing in a powerset on binary meet and join as well as negation, we consider

$$\forall a, b : \overline{a} \cap \overline{b} = \overline{a \cup b},$$

the well-known fact that a binary join may be obtained by negating the arguments, forming the meet and negating the result afterwards. The lifted counterpart in (ii), expressing this fact in algebraic form, might be termed the *point-free* De Morgan rule (Figs. 4.7 and 4.8). The powerset negation $\mathcal{N}$ has been introduced in Definition 2.2.1.

Proposition 4.3.1

i)

$$(\mathcal{N} \otimes \mathcal{N}) ; \pi = \pi ; \mathcal{N}, \qquad (\mathcal{N} \otimes \mathcal{N}) ; \rho = \rho ; \mathcal{N},$$

ii)

$$(\mathcal{N} \otimes \mathcal{N}) ; \mathfrak{M}_2 = \mathfrak{J}_2 ; \mathcal{N}, \qquad (\mathcal{N} \otimes \mathcal{N}) ; \mathfrak{J}_2 = \mathfrak{M}_2 ; \mathcal{N}.$$

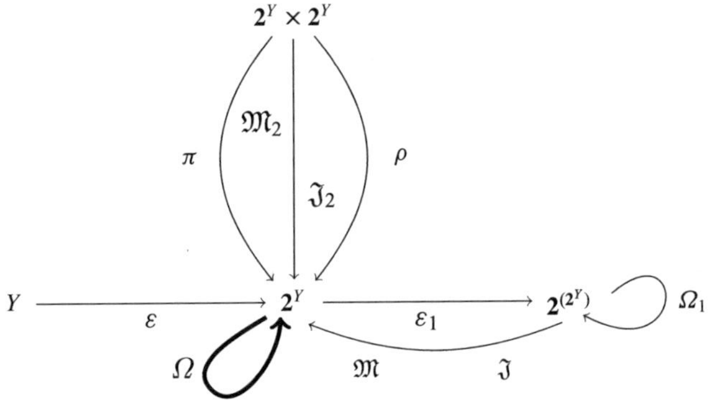

Fig. 4.6 Typing meet and join in case of a powerset order $\Omega = \overline{\varepsilon^\mathsf{T} ; \overline{\varepsilon}}$

	{}	{a}	{b}	{a,b}	{c}	{a,c}	{b,c}	{a,b,c}	{d}	{a,d}	{b,d}	{a,b,d}	{c,d}	{a,c,d}	{b,c,d}	{a,b,c,d}
({},{})	1	0	0	0	0	0	0	0	0	0	0	0	0	0	0	0
({a},{})	0	1	0	0	0	0	0	0	0	0	0	0	0	0	0	0
({},{a})	0	1	0	0	0	0	0	0	0	0	0	0	0	0	0	0
({b},{})	0	0	1	0	0	0	0	0	0	0	0	0	0	0	0	0
({a},{a})	0	1	0	0	0	0	0	0	0	0	0	0	0	0	0	0
({},{b})	0	0	1	0	0	0	0	0	0	0	0	0	0	0	0	0
({a,b},{})	0	0	0	1	0	0	0	0	0	0	0	0	0	0	0	0
({b},{a})	0	0	0	1	0	0	0	0	0	0	0	0	0	0	0	0
({a},{b})	0	0	0	1	0	0	0	0	0	0	0	0	0	0	0	0
({},{a,b})	0	0	0	1	0	0	0	0	0	0	0	0	0	0	0	0
({c},{})	0	0	0	0	1	0	0	0	0	0	0	0	0	0	0	0
({a,b},{a})	0	0	0	1	0	0	0	0	0	0	0	0	0	0	0	0
({b},{b})	0	0	1	0	0	0	0	0	0	0	0	0	0	0	0	0
({a},{a,b})	0	0	0	1	0	0	0	0	0	0	0	0	0	0	0	0
({},{c})	0	0	0	0	1	0	0	0	0	0	0	0	0	0	0	0
({a,c},{})	0	0	0	0	0	1	0	0	0	0	0	0	0	0	0	0
({c},{a})	0	0	0	0	0	1	0	0	0	0	0	0	0	0	0	0
({a,b},{b})	0	0	0	1	0	0	0	0	0	0	0	0	0	0	0	0
({b},{a,b})	0	0	0	1	0	0	0	0	0	0	0	0	0	0	0	0
({a},{c})	0	0	0	0	0	1	0	0	0	0	0	0	0	0	0	0
({},{a,c})	0	0	0	0	0	1	0	0	0	0	0	0	0	0	0	0
({b,c},{})	0	0	0	0	0	0	1	0	0	0	0	0	0	0	0	0
({a,c},{a})	0	0	0	0	0	1	0	0	0	0	0	0	0	0	0	0
({c},{b})	0	0	0	0	0	0	1	0	0	0	0	0	0	0	0	0
({a,b},{a,b})	0	0	0	1	0	0	0	0	0	0	0	0	0	0	0	0
({b},{c})	0	0	0	0	0	0	1	0	0	0	0	0	0	0	0	0
({a},{a,c})	0	0	0	0	0	1	0	0	0	0	0	0	0	0	0	0
({},{b,c})	0	0	0	0	0	0	1	0	0	0	0	0	0	0	0	0
({a,b,c},{})	0	0	0	0	0	0	0	1	0	0	0	0	0	0	0	0
({b,c},{a})	0	0	0	0	0	0	0	1	0	0	0	0	0	0	0	0
({a,c},{b})	0	0	0	0	0	0	0	1	0	0	0	0	0	0	0	0
({c},{a,b})	0	0	0	0	0	0	0	1	0	0	0	0	0	0	0	0
({a,b},{c})	0	0	0	0	0	0	0	1	0	0	0	0	0	0	0	0
({b},{a,c})	0	0	0	0	0	0	0	1	0	0	0	0	0	0	0	0
({a},{b,c})	0	0	0	0	0	0	0	1	0	0	0	0	0	0	0	0
({},{a,b,c})	0	0	0	0	0	0	0	1	0	0	0	0	0	0	0	0
({d},{})	0	0	0	0	0	0	0	0	1	0	0	0	0	0	0	0
({a,b,c},{a})	0	0	0	0	0	0	0	1	0	0	0	0	0	0	0	0
({b,c},{b})	0	0	0	0	0	0	1	0	0	0	0	0	0	0	0	0
({a,c},{a,b})	0	0	0	0	0	0	0	1	0	0	0	0	0	0	0	0
({c},{c})	0	0	0	0	1	0	0	0	0	0	0	0	0	0	0	0
({a,b},{a,c})	0	0	0	0	0	0	0	1	0	0	0	0	0	0	0	0
({b},{b,c})	0	0	0	0	0	0	1	0	0	0	0	0	0	0	0	0
({a},{a,b,c})	0	0	0	0	0	0	0	1	0	0	0	0	0	0	0	0
({},{d})	0	0	0	0	0	0	0	0	1	0	0	0	0	0	0	0
({a,d},{})	0	0	0	0	0	0	0	0	0	1	0	0	0	0	0	0
({d},{a})	0	0	0	0	0	0	0	0	0	1	0	0	0	0	0	0
({a,b,c},{b})	0	0	0	0	0	0	0	1	0	0	0	0	0	0	0	0
({b,c},{a,b})	0	0	0	0	0	0	0	1	0	0	0	0	0	0	0	0
({a,c},{c})	0	0	0	0	0	1	0	0	0	0	0	0	0	0	0	0
({c},{a,c})	0	0	0	0	0	1	0	0	0	0	0	0	0	0	0	0
({a,b},{b,c})	0	0	0	0	0	0	0	1	0	0	0	0	0	0	0	0
({b},{a,b,c})	0	0	0	0	0	0	0	1	0	0	0	0	0	0	0	0
({a},{d})	0	0	0	0	0	0	0	0	0	1	0	0	0	0	0	0
({},{a,d})	0	0	0	0	0	0	0	0	0	1	0	0	0	0	0	0

	{}	{a}	{b}	{a,b}	{c}	{a,c}	{b,c}	{a,b,c}	{d}	{a,d}	{b,d}	{a,b,d}	{c,d}	{a,c,d}	{b,c,d}	{a,b,c,d}
({},{})	1	0	0	0	0	0	0	0	0	0	0	0	0	0	0	0
({a},{})	1	0	0	0	0	0	0	0	0	0	0	0	0	0	0	0
({},{a})	1	0	0	0	0	0	0	0	0	0	0	0	0	0	0	0
({b},{})	1	0	0	0	0	0	0	0	0	0	0	0	0	0	0	0
({a},{a})	0	1	0	0	0	0	0	0	0	0	0	0	0	0	0	0
({},{b})	1	0	0	0	0	0	0	0	0	0	0	0	0	0	0	0
({a,b},{})	1	0	0	0	0	0	0	0	0	0	0	0	0	0	0	0
({b},{a})	1	0	0	0	0	0	0	0	0	0	0	0	0	0	0	0
({a},{b})	1	0	0	0	0	0	0	0	0	0	0	0	0	0	0	0
({},{a,b})	1	0	0	0	0	0	0	0	0	0	0	0	0	0	0	0
({c},{})	1	0	0	0	0	0	0	0	0	0	0	0	0	0	0	0
({a,b},{a})	0	1	0	0	0	0	0	0	0	0	0	0	0	0	0	0
({b},{b})	0	0	1	0	0	0	0	0	0	0	0	0	0	0	0	0
({a},{a,b})	0	1	0	0	0	0	0	0	0	0	0	0	0	0	0	0
({},{c})	1	0	0	0	0	0	0	0	0	0	0	0	0	0	0	0
({a,c},{})	1	0	0	0	0	0	0	0	0	0	0	0	0	0	0	0
({c},{a})	1	0	0	0	0	0	0	0	0	0	0	0	0	0	0	0
({a,b},{b})	0	0	1	0	0	0	0	0	0	0	0	0	0	0	0	0
({b},{a,b})	0	0	1	0	0	0	0	0	0	0	0	0	0	0	0	0
({a},{c})	1	0	0	0	0	0	0	0	0	0	0	0	0	0	0	0
({},{a,c})	1	0	0	0	0	0	0	0	0	0	0	0	0	0	0	0
({b,c},{})	1	0	0	0	0	0	0	0	0	0	0	0	0	0	0	0
({a,c},{a})	0	1	0	0	0	0	0	0	0	0	0	0	0	0	0	0
({c},{b})	1	0	0	0	0	0	0	0	0	0	0	0	0	0	0	0
({a,b},{a,b})	0	0	0	1	0	0	0	0	0	0	0	0	0	0	0	0
({b},{c})	1	0	0	0	0	0	0	0	0	0	0	0	0	0	0	0
({a},{a,c})	0	1	0	0	0	0	0	0	0	0	0	0	0	0	0	0
({},{b,c})	1	0	0	0	0	0	0	0	0	0	0	0	0	0	0	0
({a,b,c},{})	1	0	0	0	0	0	0	0	0	0	0	0	0	0	0	0
({b,c},{a})	1	0	0	0	0	0	0	0	0	0	0	0	0	0	0	0
({a,c},{b})	1	0	0	0	0	0	0	0	0	0	0	0	0	0	0	0
({c},{a,b})	1	0	0	0	0	0	0	0	0	0	0	0	0	0	0	0
({a,b},{c})	1	0	0	0	0	0	0	0	0	0	0	0	0	0	0	0
({b},{a,c})	1	0	0	0	0	0	0	0	0	0	0	0	0	0	0	0
({a},{b,c})	1	0	0	0	0	0	0	0	0	0	0	0	0	0	0	0
({},{a,b,c})	1	0	0	0	0	0	0	0	0	0	0	0	0	0	0	0
({d},{})	1	0	0	0	0	0	0	0	0	0	0	0	0	0	0	0
({a,b,c},{a})	0	1	0	0	0	0	0	0	0	0	0	0	0	0	0	0
({b,c},{b})	0	0	1	0	0	0	0	0	0	0	0	0	0	0	0	0
({a,c},{a,b})	0	1	0	0	0	0	0	0	0	0	0	0	0	0	0	0
({c},{c})	0	0	0	0	1	0	0	0	0	0	0	0	0	0	0	0
({a,b},{a,c})	0	1	0	0	0	0	0	0	0	0	0	0	0	0	0	0
({b},{b,c})	0	0	1	0	0	0	0	0	0	0	0	0	0	0	0	0
({a},{a,b,c})	0	1	0	0	0	0	0	0	0	0	0	0	0	0	0	0
({},{d})	1	0	0	0	0	0	0	0	0	0	0	0	0	0	0	0
({a,d},{})	1	0	0	0	0	0	0	0	0	0	0	0	0	0	0	0
({d},{a})	1	0	0	0	0	0	0	0	0	0	0	0	0	0	0	0
({a,b,c},{b})	0	0	1	0	0	0	0	0	0	0	0	0	0	0	0	0
({b,c},{a,b})	0	0	1	0	0	0	0	0	0	0	0	0	0	0	0	0
({a,c},{c})	0	0	0	0	1	0	0	0	0	0	0	0	0	0	0	0
({c},{a,c})	0	0	0	0	1	0	0	0	0	0	0	0	0	0	0	0
({a,b},{b,c})	0	0	1	0	0	0	0	0	0	0	0	0	0	0	0	0
({b},{a,b,c})	0	0	1	0	0	0	0	0	0	0	0	0	0	0	0	0
({a},{d})	1	0	0	0	0	0	0	0	0	0	0	0	0	0	0	0
({},{a,d})	1	0	0	0	0	0	0	0	0	0	0	0	0	0	0	0

Fig. 4.7 The initial ones of 256 rows of the mappings $\mathfrak{J}_2, \mathfrak{M}_2 : \mathbf{2}^X \times \mathbf{2}^X \longrightarrow \mathbf{2}^X$

	{}	{a}	{b}	{a,b}	{c}	{a,c}	{b,c}	{abc}	{d}	{a,d}	{b,d}	{a,b,d}	{c,d}	{a,c,d}	{b,c,d}	{a,b,c,d}
{}	{}	a	b	ab	c	ac	bc	abc	d	ad	bd	abd	cd	acd	bcd	all
{a}	a	a	ab	ab	ac	ac	abc	abc	ad	ad	abd	abd	acd	acd	all	all
{b}	b	ab	b	ab	bc	abc	bc	abc	bd	abd	bd	abd	bcd	all	bcd	all
{a,b}	ab	ab	ab	ab	abc	abc	abc	abc	abd	abd	abd	abd	all	all	all	all
{c}	c	ac	bc	abc	c	ac	bc	abc	cd	acd	bcd	all	cd	acd	bcd	all
{a,c}	ac	ac	abc	abc	ac	ac	abc	abc	acd	acd	abcd	abcd	acd	acd	abcd	abcd
{b,c}	bc	abc	bc	abc	bc	abc	bc	abc	bcd	abcd	bcd	abcd	bcd	abcd	bcd	abcd
{a,b,c}	abc	abc	abc	abc	abc	abc	abc	abc	abcd	abcd	abcd	abcd	abcd	abcd	abcd	abcd
{d}	d	ad	bd	abd	cd	acd	bcd	abcd	d	ad	bd	abd	cd	acd	bcd	abcd
{a,d}	ad	ad	abd	abd	acd	acd	abcd	abcd	ad	ad	abd	abd	acd	acd	abcd	abcd
{b,d}	bd	abd	bd	abd	bcd	abcd	bcd	abcd	bd	abd	bd	abd	bcd	abcd	bcd	abcd
{a,b,d}	abd	abd	abd	abd	abcd	abcd	abcd	abcd	abd	abd	abd	acd	abcd	abcd	abcd	abcd
{c,d}	cd	acd	bcd	abcd	cd	acd	bcd	abcd	cd	acd	bcd	abcd	cd	acd	bcd	abcd
{a,c,d}	acd	acd	abcd	abcd	acd	acd	abcd	abcd	acd	acd	abcd	abcd	acd	acd	abcd	abcd
{b,c,d}	bcd	abcd	bcd	abcd	bcd	abcd	bcd	abcd	bcd	abcd	bcd	abcd	bcd	abcd	bcd	abcd
{a,b,c,d}	abcd	abcd	abcd	abcd	abcd	abcd	abcd	abcd	abcd	abcd	abcd	abcd	abcd	abcd	abcd	abcd

Fig. 4.8 $\mathfrak{J}_2$ as function table $\mathfrak{J}_2 \in [\mathbf{2}^X]^{2^X \times 2^X}$; abbreviated notation for sets as table entries

Proof

i) Since $\mathcal{N}$, $(\mathcal{N} \otimes \mathcal{N})$ are mappings, we may apply Proposition 3.1.3.ii.

ii)

$$\begin{aligned}
(\mathcal{N} \otimes \mathcal{N}) ; \mathfrak{M}_2 &= (\mathcal{N} \otimes \mathcal{N}) ; \mathtt{syq}((\varepsilon \otimes \varepsilon), \varepsilon) && \text{by definition} \\
&= \mathtt{syq}((\varepsilon \otimes \varepsilon) ; (\mathcal{N} \otimes \mathcal{N})^{\mathsf{T}}, \varepsilon) && \text{Proposition 2.1.4.i} \\
&= \mathtt{syq}((\varepsilon \otimes \varepsilon) ; (\mathcal{N} \otimes \mathcal{N}), \varepsilon) && \mathcal{N} \text{ is symmetric} \\
&= \mathtt{syq}((\varepsilon ; \mathcal{N} \otimes \varepsilon ; \mathcal{N}), \overline{\varepsilon} ; \mathcal{N}) \\
&= \mathtt{syq}((\overline{\varepsilon} \otimes \overline{\varepsilon}), \overline{\varepsilon}) ; \mathcal{N} = \mathfrak{J}_2 ; \mathcal{N} && \text{by Definition 4.2.1}
\end{aligned}$$

□

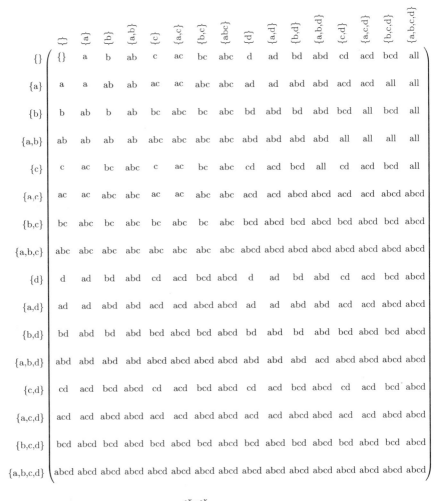

We now present statements concerning the strict fork with respect to membership and powerset ordering. Proposition 4.3.2.iii gives yet another example of a sharp factorization.

Proposition 4.3.2 *We consider membership ε and powerset ordering Ω.*

i) $\varepsilon ; \mathfrak{M}_2^\mathsf{T} = (\varepsilon \circledless \varepsilon)$ $\qquad (\varepsilon \circledless \varepsilon) ; \mathfrak{M}_2 = \varepsilon$
$\overline{\varepsilon} ; \mathfrak{J}_2^\mathsf{T} = (\overline{\varepsilon} \circledless \overline{\varepsilon})$ $\qquad (\overline{\varepsilon} \circledless \overline{\varepsilon}) ; \mathfrak{J}_2 = \overline{\varepsilon}$

ii) $\varepsilon ; (\Omega \circledless \Omega) = (\varepsilon \circledless \varepsilon)$ $\qquad (\Omega \circledless \Omega) = \overline{\varepsilon^\mathsf{T} ; \overline{(\varepsilon \circledless \varepsilon)}}$

iii) $(\varepsilon \circledless \varepsilon) = (\varepsilon \circledless \varepsilon) ; (\Omega \otimes \Omega)$

Proof

i)

$$\begin{aligned}
&\varepsilon ; \mathfrak{M}_2^\mathsf{T} = \varepsilon ; \mathtt{syq}(\varepsilon, (\varepsilon \circledless \varepsilon)) = (\varepsilon \circledless \varepsilon)\\
&(\varepsilon \circledless \varepsilon) ; \mathfrak{M}_2 = \varepsilon ; \mathfrak{M}_2^\mathsf{T} ; \mathfrak{M}_2 = \varepsilon \quad \text{initial result, } \mathfrak{M}_2 \text{ is a surjective map}\\
&\overline{\varepsilon} ; \mathfrak{J}_2^\mathsf{T} = \varepsilon ; \mathcal{N} ; \mathfrak{J}_2^\mathsf{T} = \varepsilon ; \mathfrak{M}_2^\mathsf{T} ; (\mathcal{N} \otimes \mathcal{N}) \quad \text{point-free De Morgan rule}\\
&= (\varepsilon \circledless \varepsilon) ; (\mathcal{N} \otimes \mathcal{N}) \quad \text{using (i)}\\
&= (\varepsilon ; \mathcal{N} \circledless \varepsilon ; \mathcal{N}) = (\overline{\varepsilon} \circledless \overline{\varepsilon})
\end{aligned}$$

ii)

$$\varepsilon ; (\Omega \circledless \Omega) \subseteq (\varepsilon ; \Omega \circledless \varepsilon ; \Omega) = (\varepsilon \circledless \varepsilon)$$

For the reverse direction, we show an even stronger result

$$(\varepsilon \circledless \varepsilon) \subseteq \sigma ; (\Omega \circledless \Omega) \quad \iff \quad \sigma^\mathsf{T} ; (\varepsilon \circledless \varepsilon) \subseteq (\Omega \circledless \Omega) \quad \text{via shunting}$$

and

$$\sigma^\mathsf{T} ; (\varepsilon \circledless \varepsilon) = (\sigma^\mathsf{T} ; \varepsilon \circledless \sigma^\mathsf{T} ; \varepsilon) \subseteq (\Omega \circledless \Omega) \quad \sigma \text{ is injective, Proposition 2.2.2.i}$$

iii) Finally $(\varepsilon \circledless \varepsilon) ; (\Omega \otimes \Omega) \subseteq (\varepsilon ; \Omega \circledless \varepsilon ; \Omega) = (\varepsilon \circledless \varepsilon)$ and

$$\mathbb{I} = (\mathbb{I} \otimes \mathbb{I}) \subseteq (\Omega \otimes \Omega)$$

□

The following results are easily interpreted. When one in (i) traces back how the meet is formed from two constituents, one will, of course, find only subsets at least as big as the original one.

Proposition 4.3.3

i) $(\Omega \circledless \Omega) = \Omega ; \mathfrak{M}_2^\mathsf{T}$ $\qquad (\Omega \circledgreater \Omega) = \mathfrak{J}_2 ; \Omega$

ii) $(\Omega \circledgreater \mathbb{I}) \subseteq \mathfrak{J}_2$

Proof

i)

$$\begin{aligned}\Omega ; \mathfrak{M}_2^\mathsf{T} &= \overline{\varepsilon^\mathsf{T} ; \overline{\varepsilon}} ; \mathfrak{M}_2^\mathsf{T} = \overline{\varepsilon^\mathsf{T} ; \overline{\varepsilon ; \mathfrak{M}_2^\mathsf{T}}} \\ &= \overline{\varepsilon^\mathsf{T} ; \overline{(\varepsilon \circledless \varepsilon)}} = (\Omega \circledless \Omega) \quad \text{using Proposition 4.3.2.ii,iii}\end{aligned}$$

ii)

$$\begin{aligned}(\Omega \circledgtr \mathbb{I}) &= \pi ; \Omega \cap \rho ; \mathbb{I} = \pi ; \Omega \cap \rho ; (\Omega \cap \Omega^\mathsf{T}) = (\Omega \circledgtr \Omega) \cap \rho ; \Omega^\mathsf{T} \\ &= \mathfrak{J}_2 ; \Omega \cap \rho ; \Omega^\mathsf{T} \quad \text{due to (i)} \\ &\subseteq \mathfrak{J}_2 ; \Omega \cap \mathfrak{J}_2 ; \Omega^\mathsf{T} = \mathfrak{J}_2 ; (\Omega \cap \Omega^\mathsf{T}) = \mathfrak{J}_2 \quad \text{using that} \\ &\quad \text{with Proposition 4.2.2.vii } \mathfrak{J}_2^\mathsf{T} ; \rho ; \Omega^\mathsf{T} = \Omega^\mathsf{T} ; \Omega^\mathsf{T} \subseteq \Omega^\mathsf{T} \text{ and shunting } \mathfrak{J}_2.\end{aligned}$$

□

One will understand Proposition 4.3.3.i when interpreting it with cone intersection: The lower cone of a meet coincides with the intersection of the lower cones of the constituent pairs. Similarly: The upper cone of a join of a pair is the intersection of the upper cones of the component elements of that pair.

Concerning Definition 4.2.1, the transition from the general version with E to the special version with powerset ordering Ω has to be shown. We start with Ω instead of E and can be sure that $\mathfrak{M}_2$, $\mathfrak{M}$ are mappings. The binary case is proved with

$$\begin{aligned}\mathtt{syq}((\Omega \circledless \Omega), \Omega) &= \mathtt{syq}((\overline{\varepsilon^\mathsf{T} ; \overline{\varepsilon}} \circledless \overline{\varepsilon^\mathsf{T} ; \overline{\varepsilon}}), \overline{\varepsilon^\mathsf{T} ; \overline{\varepsilon}}) && \text{expanded} \\ &= \mathtt{syq}(\varepsilon^\mathsf{T} ; \overline{\varepsilon} ; \pi^\mathsf{T} \cup \varepsilon^\mathsf{T} ; \overline{\varepsilon} ; \rho^\mathsf{T}, \varepsilon^\mathsf{T} ; \overline{\varepsilon}) && \text{both sides negated} \\ &= \mathtt{syq}(\overline{\varepsilon} ; \pi^\mathsf{T} \cup \overline{\varepsilon} ; \rho^\mathsf{T}, \overline{\varepsilon}) && \text{membership deletion} \\ &= \mathtt{syq}((\varepsilon \circledless \varepsilon), \varepsilon) = \mathfrak{M}_2 && \text{both sides negated}\end{aligned}$$

while the arbitrary case follows from

$$\begin{aligned}\mathfrak{M} &= \mathtt{syq}(\mathtt{lbd}_\Omega(\varepsilon_1), \Omega) = \mathtt{syq}(\overline{\overline{\Omega} ; \varepsilon_1}, \Omega) \\ &= \mathtt{syq}(\overline{\Omega} ; \varepsilon_1, \overline{\Omega}) = \mathtt{syq}(\varepsilon^\mathsf{T} ; \overline{\varepsilon} ; \varepsilon_1, \varepsilon^\mathsf{T} ; \overline{\varepsilon}) \\ &= \mathtt{syq}(\overline{\varepsilon} ; \varepsilon_1, \overline{\varepsilon}) \quad \text{Proposition 2.2.3.ii, i.e. membership deletion}\end{aligned}$$

The section proceeds by proving a homomorphism (i.e. monotony) condition for powerset join and meet with respect to the *inverse* image mapping. Since it is considered in the powerset, it works on Ω and not just on an arbitrary ordering E.

Proposition 4.3.4 *Assume a mapping $f : X \longrightarrow X'$ of a set into another one. Then the binary meet-, resp. join-, operations in the respective powersets satisfy some sort of a homomorphism property with regard to the existential and also to the inverse image:*

i) $\mathfrak{M}'_2 ; \vartheta_{f^\mathsf{T}} = (\vartheta_{f^\mathsf{T}} \otimes \vartheta_{f^\mathsf{T}}) ; \mathfrak{M}_2 \qquad \mathfrak{J}'_2 ; \vartheta_{f^\mathsf{T}} = (\vartheta_{f^\mathsf{T}} \otimes \vartheta_{f^\mathsf{T}}) ; \mathfrak{J}_2$

ii) When f is injective, also

$$\mathfrak{M}_2 ; \vartheta_f = (\vartheta_f \otimes \vartheta_f) ; \mathfrak{M}'_2 \qquad \mathfrak{J}_2 ; \vartheta_f = (\vartheta_f \otimes \vartheta_f) ; \mathfrak{J}'_2$$

Proof

i) We provide two different methods of proof for these two formulae:

$$\begin{aligned}
\mathfrak{M}'_2 ; \vartheta_{f^\mathsf{T}} &= \mathfrak{M}'_2 ; \operatorname{syq}(f ; \varepsilon', \varepsilon) && \text{by definition}\\
&= \operatorname{syq}(f ; \varepsilon' ; \mathfrak{M}'^{\mathsf{T}}_2, \varepsilon) && \text{Proposition 2.1.4.i}\\
&= \operatorname{syq}(f ; (\varepsilon' \otimes \varepsilon'), \varepsilon') && \text{Proposition 4.3.2.i}\\
&= \operatorname{syq}((f ; \varepsilon' \otimes f ; \varepsilon'), \varepsilon) && \text{since } f \text{ is univalent}\\
&= \operatorname{syq}((\varepsilon ; \vartheta^{\mathsf{T}}_{f^\mathsf{T}} \otimes \varepsilon ; \vartheta^{\mathsf{T}}_{f^\mathsf{T}}), \varepsilon) && \text{Proposition 2.2.5.ii}\\
&= \operatorname{syq}((\varepsilon \otimes \varepsilon) ; (\vartheta^{\mathsf{T}}_{f^\mathsf{T}} \otimes \vartheta^{\mathsf{T}}_{f^\mathsf{T}}), \varepsilon) && \text{since } (\vartheta_{f^\mathsf{T}} \otimes \vartheta_{f^\mathsf{T}}) \text{ is univalent}\\
&= (\vartheta_{f^\mathsf{T}} \otimes \vartheta_{f^\mathsf{T}}) ; \operatorname{syq}((\varepsilon \otimes \varepsilon), \varepsilon) && \text{Proposition 2.1.4.i}\\
&= (\vartheta_{f^\mathsf{T}} \otimes \vartheta_{f^\mathsf{T}}) ; \mathfrak{M}_2 && \text{by definition}
\end{aligned}$$

Now we use negations $\mathcal{N}, \mathcal{N}'$ to invert order direction and reuse (i):

$$\begin{aligned}
\mathfrak{J}'_2 ; \vartheta_{f^\mathsf{T}} &= (\mathcal{N}' \otimes \mathcal{N}') ; \mathfrak{M}'_2 ; \mathcal{N}' ; \vartheta_{f^\mathsf{T}} && \text{Proposition 4.3.1.ii}\\
&= (\mathcal{N}' \otimes \mathcal{N}') ; \mathfrak{M}'_2 ; \vartheta_{f^\mathsf{T}} ; \mathcal{N} && \text{Proposition 2.2.11}\\
&= (\mathcal{N}' \otimes \mathcal{N}') ; (\vartheta_{f^\mathsf{T}} \otimes \vartheta_{f^\mathsf{T}}) ; \mathfrak{M}_2 ; \mathcal{N} && \text{the first result of this proposition}\\
&= (\mathcal{N}' ; \vartheta_{f^\mathsf{T}} \otimes \mathcal{N}' ; \vartheta_{f^\mathsf{T}}) ; \mathfrak{M}_2 ; \mathcal{N} && \text{since } \mathcal{N}' \text{ is a bijection}\\
&= (\vartheta_{f^\mathsf{T}} ; \mathcal{N} \otimes \vartheta_{f^\mathsf{T}} ; \mathcal{N}) ; \mathfrak{M}_2 ; \mathcal{N} && \text{Proposition 2.2.11}\\
&= (\vartheta_{f^\mathsf{T}} \otimes \vartheta_{f^\mathsf{T}}) ; (\mathcal{N} \otimes \mathcal{N}) ; \mathfrak{M}_2 ; \mathcal{N} && \text{because } \mathcal{N} \text{ is a bijection}\\
&= (\vartheta_{f^\mathsf{T}} \otimes \vartheta_{f^\mathsf{T}}) ; \mathfrak{J}_2 && \text{Proposition 4.3.1.ii again}
\end{aligned}$$

ii)

$$\begin{aligned}
(\vartheta_f \otimes \vartheta_f) ; \mathfrak{M}'_2 &= \operatorname{syq}((\varepsilon' \otimes \varepsilon') ; (\vartheta_f \otimes \vartheta_f)^{\mathsf{T}}, \varepsilon')\\
&= \operatorname{syq}((\varepsilon' ; \vartheta^{\mathsf{T}}_f \otimes \varepsilon' ; \vartheta^{\mathsf{T}}_f), \varepsilon')\\
&= \operatorname{syq}((f^{\mathsf{T}} ; \varepsilon \otimes f^{\mathsf{T}} ; \varepsilon), \varepsilon') && \text{using Proposition 2.2.5.i}\\
&= \operatorname{syq}(f^{\mathsf{T}} ; (\varepsilon \otimes \varepsilon), \varepsilon') && \text{since } f \text{ is injective}\\
&= \operatorname{syq}(f^{\mathsf{T}} ; \varepsilon ; \mathfrak{M}^{\mathsf{T}}_2, \varepsilon') && \text{Proposition 4.3.2.i}\\
&= \mathfrak{M}_2 ; \operatorname{syq}(f^{\mathsf{T}} ; \varepsilon, \varepsilon')\\
&= \mathfrak{M}_2 ; \vartheta_f
\end{aligned}$$

□

Of course, also the traditional reasoning with orderings, e.g.,

$$a \leq c, a \leq d \implies a \leq c \cap d,$$

now assumes a different shape, namely $(\Omega \otimes \Omega) = \Omega ; \mathfrak{M}^{\mathsf{T}}_2$ of Proposition 4.3.3.i.

Proposition 4.3.5

i) For points a, c, d we have

$$\begin{array}{l} a \subseteq \Omega ; c \\ a \subseteq \Omega ; d \end{array} \implies a \subseteq \Omega ; \mathfrak{M}^{\mathsf{T}}_2 ; (c \otimes d) = (\Omega \otimes \Omega) ; (c \otimes d)$$

ii) For points b, c, d we have

$$\begin{matrix} b \subseteq \Omega^{\mathsf{T}};c \\ b \subseteq \Omega^{\mathsf{T}};d \end{matrix} \quad \Longrightarrow \quad b \subseteq \Omega^{\mathsf{T}};\mathfrak{J}_2^{\mathsf{T}};(c \otimes d) = (\Omega^{\mathsf{T}} \otimes \Omega^{\mathsf{T}});(c \otimes d)$$

Proof

i) $\Omega;\mathfrak{M}_2^{\mathsf{T}};(c \otimes d) = (\Omega \otimes \Omega);(c \otimes d) = \Omega;c \cap \Omega;d$ Proposition 3.1.8.ii
ii) is proved similarly. □

Only now, with the join operation available, can we prove the following result, that one might at first sight consider belonging to Proposition 2.2.10 (Fig. 4.9).

Proposition 4.3.6 *If $f : X \longrightarrow X'$ is an injective mapping and Ω, Ω' are the respective powerset orderings, we have*

$$\Omega';\vartheta_{f^{\mathsf{T}}} = \vartheta_{f^{\mathsf{T}}};\Omega.$$

Proof One half of the statement is simply Proposition 2.2.10.i. The other needs again a proof via sharp factorization of a product, made possible by a catalyst relation the statement proper does not mention in the first place, namely $\mathfrak{J}'_2$. Once we have shown as an intermediate result that

$$(\vartheta_{f^{\mathsf{T}}};\Omega \otimes \mathbb{I}) \subseteq (\mathbb{I} \otimes \vartheta_f);\mathfrak{J}'_2;\vartheta_{f^{\mathsf{T}}},$$

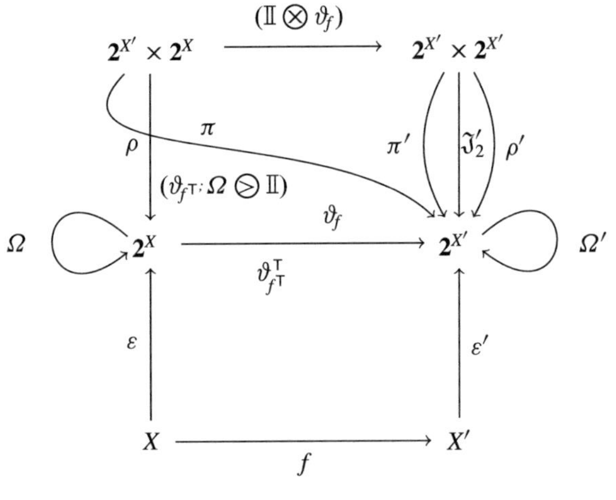

Fig. 4.9 Typing the existential and the inverse image when injective

this proof may be completed as follows:

$$\begin{aligned}
\vartheta_{f^\mathsf{T}} ; \Omega &= \pi^\mathsf{T} ; (\pi ; \vartheta_{f^\mathsf{T}} ; \Omega \cap \rho) && \text{destroy and append with } \pi^\mathsf{T} ; \rho = \mathbb{T} \\
&= \pi^\mathsf{T} ; (\vartheta_{f^\mathsf{T}} ; \Omega \oslash \mathbb{I}) && \text{definition of } \oslash \\
&\subseteq \pi^\mathsf{T} ; (\mathbb{I} \otimes \vartheta_f) ; \mathfrak{J}'_2 ; \vartheta_{f^\mathsf{T}} && \text{yet to be shown as announced} \\
&\subseteq \pi'^\mathsf{T} ; \mathfrak{J}'_2 ; \vartheta_{f^\mathsf{T}} && \text{Proposition 3.1.3.i} \\
&= \Omega' ; \vartheta_{f^\mathsf{T}} && \text{Proposition 4.2.2.vii}
\end{aligned}$$

For the estimation anticipated above we modify the upper part:

$$\begin{aligned}
&(\mathbb{I} \otimes \vartheta_f) ; \mathfrak{J}'_2 ; \vartheta_{f^\mathsf{T}} \\
&= (\mathbb{I} \otimes \vartheta_f) ; \mathfrak{J}'_2 ; \mathtt{syq}(f ; \varepsilon', \varepsilon) && \text{expanding } \vartheta_{f^\mathsf{T}} \\
&= \mathtt{syq}(f ; \varepsilon' ; \mathfrak{J}'^\mathsf{T}_2 ; (\mathbb{I} \otimes \vartheta^\mathsf{T}_f), \varepsilon) && \text{Proposition 2.1.4.i} \\
&= \mathtt{syq}(f ; \overline{\overline{\varepsilon'} ; \mathfrak{J}'^\mathsf{T}_2} ; (\mathbb{I} \otimes \vartheta^\mathsf{T}_f), \varepsilon) && \text{map } \mathfrak{J}'_2 \text{ slips below negation} \\
&= \mathtt{syq}(f ; (\overline{\overline{\varepsilon'} \oslash \overline{\varepsilon'}}) ; (\mathbb{I} \otimes \vartheta^\mathsf{T}_f), \varepsilon) && \text{Proposition 4.3.2.i} \\
&= \mathtt{syq}(f ; (\varepsilon' ; \pi'^\mathsf{T} \cup \varepsilon' ; \rho'^\mathsf{T}) ; (\mathbb{I} \otimes \vartheta^\mathsf{T}_f), \varepsilon) \\
&= \mathtt{syq}(f ; \varepsilon' ; \pi'^\mathsf{T} ; (\mathbb{I} \otimes \vartheta^\mathsf{T}_f) \cup f ; \varepsilon' ; \rho'^\mathsf{T} ; (\mathbb{I} \otimes \vartheta^\mathsf{T}_f), \varepsilon) \\
&= \mathtt{syq}(f ; \varepsilon' ; \pi^\mathsf{T} \cup f ; \varepsilon' ; \vartheta^\mathsf{T}_f ; \rho^\mathsf{T}, \varepsilon) && \text{Proposition 3.1.3.i with } \vartheta_f \text{ total} \\
&= \mathtt{syq}(f ; \varepsilon' ; \pi^\mathsf{T} \cup f ; f^\mathsf{T} ; \varepsilon ; \rho^\mathsf{T}, \varepsilon) && \text{Proposition 2.2.5.i} \\
&= \mathtt{syq}(f ; \varepsilon' ; \pi^\mathsf{T} \cup \varepsilon ; \rho^\mathsf{T}, \varepsilon) && \text{since } f \text{ is an injective map}
\end{aligned}$$

Now the estimation may be separated into the two parts

$$\begin{aligned}
&(\vartheta_{f^\mathsf{T}} ; \Omega \oslash \mathbb{I}) \subseteq \overline{\overline{f ; \varepsilon' ; \pi^\mathsf{T} \cup \varepsilon ; \rho^\mathsf{T}}^\mathsf{T} ; \varepsilon} \\
&\iff \quad (\vartheta_{f^\mathsf{T}} ; \Omega \oslash \mathbb{I}) ; \varepsilon^\mathsf{T} \subseteq \rho ; \varepsilon^\mathsf{T} \subseteq \pi ; \varepsilon'^\mathsf{T} ; f^\mathsf{T} \cup \rho ; \varepsilon^\mathsf{T}
\end{aligned}$$

and

$$\begin{aligned}
&(\vartheta_{f^\mathsf{T}} ; \Omega \oslash \mathbb{I}) \subseteq \overline{(f ; \varepsilon' ; \pi^\mathsf{T} \cup \varepsilon ; \rho^\mathsf{T})^\mathsf{T} ; \overline{\varepsilon}} \\
&\iff \quad (f ; \varepsilon' ; \pi^\mathsf{T} \cup \varepsilon ; \rho^\mathsf{T}) ; (\vartheta_{f^\mathsf{T}} ; \Omega \oslash \mathbb{I}) \\
&\qquad = f ; \varepsilon' ; \pi^\mathsf{T} ; (\vartheta_{f^\mathsf{T}} ; \Omega \oslash \mathbb{I}) \cup \varepsilon ; \rho^\mathsf{T} ; (\vartheta_{f^\mathsf{T}} ; \Omega \oslash \mathbb{I}) \\
&\qquad = f ; \varepsilon' ; \vartheta_{f^\mathsf{T}} ; \Omega \cup \varepsilon \quad \text{with } \vartheta_{f^\mathsf{T}} \text{ surjective due to Proposition 2.2.8.ii} \\
&\qquad \subseteq \varepsilon ; \Omega \cup \varepsilon = \varepsilon
\end{aligned}$$

□

In contrast to commutativity etc. the rules for absorption are fully restricted to lattices and therefore handled only now. The interpretation of the forthcoming result is immediate when looking at the traditional way

$$\forall a, b : \ a \vee (a \wedge b) = a$$

of expressing absorption: When starting from the pair (a, b), of which the first component is maintained via π, while the pair is processed via meet forming $\mathfrak{M}_2$, their join $\mathfrak{J}_2$ will reproduce the first component.

Proposition 4.3.7 *The binary meet and join $\mathfrak{J}_2$ satisfy the absorption rules*

$$(\pi \otimes \mathfrak{M}_2) ; \mathfrak{J}_2 = \pi \qquad (\pi \otimes \mathfrak{J}_2) ; \mathfrak{M}_2 = \pi.$$

Proof

$$\begin{aligned}
(\pi \otimes \mathfrak{M}_2) ; \mathfrak{J}_2 &= (\pi \otimes \mathfrak{M}_2) ; \mathtt{syq}((\overline{\varepsilon} \otimes \overline{\varepsilon}), \overline{\varepsilon}) \quad \text{by definition} \\
&= \mathtt{syq}((\overline{\varepsilon} \otimes \overline{\varepsilon}) ; (\pi \otimes \mathfrak{M}_2)^{\mathsf{T}}, \overline{\varepsilon}) \quad \text{Proposition 2.1.4.i} \\
&= \mathtt{syq}(\overline{\varepsilon} ; \pi^{\mathsf{T}} \cap \overline{\varepsilon} ; \mathfrak{M}_2^{\mathsf{T}}, \overline{\varepsilon}) \quad \text{since } \pi,\ \mathfrak{M}_2 \text{ are mappings} \\
&= \mathtt{syq}(\varepsilon ; \pi^{\mathsf{T}} \cup \varepsilon ; \mathfrak{M}_2^{\mathsf{T}}, \varepsilon) \quad \text{negated; } \pi,\ \mathfrak{M}_2 \text{ are mappings} \\
&= \mathtt{syq}(\varepsilon ; \pi^{\mathsf{T}} \cup (\varepsilon \otimes \varepsilon), \varepsilon) \quad \text{Proposition 4.3.2.i} \\
&= \mathtt{syq}(\varepsilon ; \pi^{\mathsf{T}}, \varepsilon) \quad \text{since } (\varepsilon \otimes \varepsilon) = \varepsilon ; \pi^{\mathsf{T}} \cap \varepsilon ; \rho^{\mathsf{T}} \\
&= \pi ; \mathtt{syq}(\varepsilon, \varepsilon) \quad \text{Proposition 2.1.4.i} \\
&= \pi \quad \text{standard property of the membership relation } \varepsilon
\end{aligned}$$

□

Recalling the general concept of distributivity from Definition 3.3.5, we have now to prove it for meet and join. One should compare the following results in (iii) with, e.g.,

$$a \vee (b \wedge c) = (a \vee b) \wedge (a \vee c).$$

The first item of the following proposition is intuitively clear: When both components coincide, the projection on either component leads to precisely the paired element. The intersection $p := \pi \cap \rho$ of these projections will, thus, be univalent and surjective (Fig. 4.10).

Proposition 4.3.8 *Given any direct product with projections*

$$\pi, \rho : \mathbf{2}^X \times \mathbf{2}^X \longrightarrow \mathbf{2}^X,$$

as well as meet- and join-forming $\mathfrak{M}_2$, $\mathfrak{J}_2$,

i) the construct $p := \pi \cap \rho$ is univalent and surjective,
ii) meet-forming $\mathfrak{M}_2$ and join-forming $\mathfrak{J}_2$ are surjective mappings,
iii) concerning meet- and join-forming,

$$\begin{aligned}
&\mathfrak{J}_2 \textit{ distributes over } \mathfrak{M}_2, \qquad (\mathbb{I} \otimes \mathfrak{J}_2) ; \mathfrak{M}_2 = \mathfrak{D} ; (\mathfrak{M}_2 \otimes \mathfrak{M}_2) ; \mathfrak{J}_2, \\
&\mathfrak{M}_2 \textit{ distributes over } \mathfrak{J}_2, \qquad (\mathbb{I} \otimes \mathfrak{M}_2) ; \mathfrak{J}_2 = \mathfrak{D} ; (\mathfrak{J}_2 \otimes \mathfrak{J}_2) ; \mathfrak{M}_2,
\end{aligned}$$

iv) meet-forming $\mathfrak{M}_2$ is a homomorphism (i.e. monotonic) and, even stronger,

$$(\Omega \otimes \Omega) ; \mathfrak{M}_2 = \mathfrak{M}_2 ; \Omega.$$

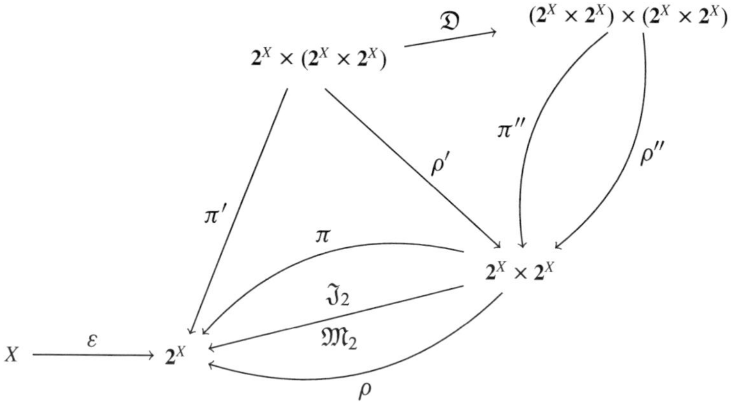

Fig. 4.10 Typing distributivity of meet and join mappings $\mathfrak{M}_2$, $\mathfrak{J}_2$,

Proof

i) We use that the direct product encompasses every pair and that projections are surjective before applying the Dedekind formula in

$$\mathbb{I} = \mathbb{T} \cap \mathbb{I} = \pi^\mathsf{T};\rho \cap \rho^\mathsf{T};\rho \subseteq (\pi^\mathsf{T} \cap \rho^\mathsf{T};\rho;\rho^\mathsf{T});(\rho \cap \pi;\rho^\mathsf{T};\rho) = (\pi^\mathsf{T} \cap \rho^\mathsf{T});(\rho \cap \pi).$$

ii) This has already been shown with Proposition 4.2.4.v.

iii)

$$\begin{aligned}
\mathfrak{D};(\mathfrak{J}_2 \otimes \mathfrak{J}_2);\mathfrak{M}_2 &= \mathfrak{D};(\mathfrak{J}_2 \otimes \mathfrak{J}_2);\mathrm{syq}((\varepsilon \textcircled{$<$} \varepsilon), \varepsilon) \quad \text{expanded} \\
&= \mathrm{syq}((\varepsilon;\mathfrak{J}_2^\mathsf{T} \textcircled{$<$} \varepsilon;\mathfrak{J}_2^\mathsf{T});\mathfrak{D}^\mathsf{T}, \varepsilon) \\
&= \mathrm{syq}((\varepsilon;\mathfrak{J}_2^\mathsf{T} \textcircled{$<$} \varepsilon;\mathfrak{J}_2^\mathsf{T});((\mathbb{I} \otimes \pi^\mathsf{T}) \textcircled{$>$} (\mathbb{I} \otimes \rho^\mathsf{T})), \varepsilon) \\
&\qquad \text{Definition 3.3.5.i} \\
&= \mathrm{syq}(\varepsilon;\mathfrak{J}_2^\mathsf{T};(\mathbb{I} \otimes \pi^\mathsf{T}) \cap \varepsilon;\mathfrak{J}_2^\mathsf{T};(\mathbb{I} \otimes \rho^\mathsf{T}), \varepsilon) \\
&= \mathrm{syq}(\varepsilon;\mathrm{syq}(\overline{\varepsilon}, (\overline{\varepsilon} \textcircled{$<$} \overline{\varepsilon};\pi^\mathsf{T})) \cap \varepsilon;\mathrm{syq}(\overline{\varepsilon}, (\overline{\varepsilon} \textcircled{$<$} \overline{\varepsilon};\rho^\mathsf{T})), \varepsilon) \\
&= \mathrm{syq}(\overline{(\overline{\varepsilon} \textcircled{$<$} \overline{\varepsilon};\pi^\mathsf{T})} \cap \overline{(\overline{\varepsilon} \textcircled{$<$} \overline{\varepsilon};\rho^\mathsf{T})}, \varepsilon) \\
&= \mathrm{syq}(\left[\varepsilon;\pi'^\mathsf{T} \cup \varepsilon;\pi^\mathsf{T};\rho'^\mathsf{T}\right] \cap \left[\varepsilon;\pi'^\mathsf{T} \cup \varepsilon;\rho^\mathsf{T};\rho'^\mathsf{T}\right], \varepsilon) \\
&= \mathrm{syq}(\varepsilon;\pi'^\mathsf{T} \cup (\varepsilon;\pi^\mathsf{T};\rho'^\mathsf{T} \cap \varepsilon;\rho^\mathsf{T};\rho'^\mathsf{T}), \varepsilon) \\
&= \mathrm{syq}(\overline{\varepsilon};\pi'^\mathsf{T} \cap \overline{(\varepsilon \textcircled{$<$} \varepsilon)};\rho'^\mathsf{T}, \overline{\varepsilon}) \quad \text{arguments both negated} \\
&= \mathrm{syq}((\overline{\varepsilon} \textcircled{$<$} \overline{\varepsilon;\mathfrak{M}_2^\mathsf{T}}), \overline{\varepsilon}) \\
&= \mathrm{syq}((\overline{\varepsilon} \textcircled{$<$} \overline{\varepsilon});(\mathbb{I} \otimes \mathfrak{M}_2)^\mathsf{T}, \overline{\varepsilon}) \\
&= (\mathbb{I} \otimes \mathfrak{M}_2);\mathrm{syq}((\overline{\varepsilon} \textcircled{$<$} \overline{\varepsilon}), \overline{\varepsilon}) \\
&= (\mathbb{I} \otimes \mathfrak{M}_2);\mathfrak{J}_2
\end{aligned}$$

iv)

$$
\begin{aligned}
&\text{“}\subseteq\text{” follows with shunting}\\
&(\Omega \otimes \Omega) ; \mathfrak{M}_2 \subseteq \mathfrak{M}_2 ; \Omega \iff (\Omega \otimes \Omega) \subseteq \mathfrak{M}_2 ; \Omega ; \mathfrak{M}_2^{\mathsf{T}} \text{ from}\\
&\mathfrak{M}_2 ; \Omega ; \mathfrak{M}_2^{\mathsf{T}} = \overline{\mathfrak{M}_2 ; \varepsilon^{\mathsf{T}} ; \overline{\varepsilon ; \mathfrak{M}_2^{\mathsf{T}}}}\\
&= \overline{(\varepsilon^{\mathsf{T}} \ogreaterthan \varepsilon^{\mathsf{T}}) ; \overline{(\varepsilon \olessthan \varepsilon)}} = \overline{(\varepsilon \olessthan \varepsilon)^{\mathsf{T}} ; \overline{(\varepsilon \olessthan \varepsilon)}} = (\varepsilon \olessthan \varepsilon) \backslash (\varepsilon \olessthan \varepsilon)\\
&\supseteq (\varepsilon \backslash \varepsilon \otimes \varepsilon \backslash \varepsilon) \quad \text{following Proposition 3.1.7.viii}\\
&= (\Omega \otimes \Omega)
\end{aligned}
$$

The other direction “$\supseteq$” applies distributivity as proved for (iii):

$$
\begin{aligned}
\mathfrak{M}_2 ; \Omega &= \mathfrak{M}_2 ; \rho^{\mathsf{T}} ; \mathfrak{J}_2 && \text{Proposition 4.2.2.vi}\\
&= \rho'^{\mathsf{T}} ; (\mathbb{I} \otimes \mathfrak{M}_2) ; \mathfrak{J}_2\\
&= \rho'^{\mathsf{T}} ; \mathfrak{D} ; (\mathfrak{J}_2 \otimes \mathfrak{J}_2) ; \mathfrak{M}_2 && \text{distributivity}\\
&= (\rho^{\mathsf{T}} \otimes \rho^{\mathsf{T}}) ; (\mathfrak{J}_2 \otimes \mathfrak{J}_2) ; \mathfrak{M}_2 && \text{intermediate result for}\\
& && \text{Proposition 3.3.6.ii}\\
&\subseteq (\rho^{\mathsf{T}} ; \mathfrak{J}_2 \otimes \rho^{\mathsf{T}} ; \mathfrak{J}_2) ; \mathfrak{M}_2\\
&= (\Omega \otimes \Omega) ; \mathfrak{M}_2 && \text{Proposition 4.2.2.vi again} \quad \square
\end{aligned}
$$

The interpretation of (iv) is evident: When we proceed from a pair of sets to a pair of possibly larger ones and form their meet, the statement is that we might also first form the meet and then increase. This again was an example that proving a sharp factorization needs a highly specific treatment—here, to go for the proof the rather long way via the initially uninvolved $\mathbf{2}^X \times (\mathbf{2}^X \times \mathbf{2}^X)$.

Distributivity allows a slightly more general result. So far, the mappings $\mathfrak{J}_2$, $\mathfrak{M}_2$ had both been binary operations, but we may also prove it when one of the operations, e.g. $\mathfrak{J} : \mathbf{2}^{(\mathbf{2}^Y)} \longrightarrow \mathbf{2}^Y$ belongs to the general case.

Definition 4.3.9 When

$$(\mathbb{I} \otimes \mathfrak{J}) ; \mathfrak{M}_2 = \mathtt{syq}(\mathfrak{M}_2^{\mathsf{T}} ; (\mathbb{I} \otimes \varepsilon_1), \varepsilon_1) ; \mathfrak{J},$$

we say that *binary* meet $\mathfrak{M}_2$ **distributes** over the *arbitrary* join $\mathfrak{J}$. □

The typing of these complex terms may not be immediate, so it is provided with $\mathbf{2}^Y \times \mathbf{2}^{(\mathbf{2}^Y)} \longrightarrow \mathbf{2}^Y$. As an exercise, we prove such distributivity for the case of a powerset ordering.

Proposition 4.3.10 *Binary meet $\mathfrak{M}_2$ distributes over arbitrary join $\mathfrak{J}$.*

Proof The abbreviation $f := \mathtt{syq}(\mathfrak{M}_2^{\mathsf{T}} ; (\mathbb{I} \otimes \varepsilon_1), \varepsilon_1)$ is used temporarily; it is a mapping since a membership relation occurs on the right side.

$$\begin{aligned}
&\mathtt{syq}(\mathfrak{M}_2^{\mathsf{T}} ; (\mathbb{I} \otimes \varepsilon_1), \varepsilon_1) ; \mathfrak{J} = f ; \mathfrak{J} = f ; \mathtt{syq}(\varepsilon ; \varepsilon_1, \varepsilon) = \mathtt{syq}(\varepsilon ; \varepsilon_1 ; f^{\mathsf{T}}, \varepsilon) \\
&\quad = \mathtt{syq}(\varepsilon ; \varepsilon_1 ; \mathtt{syq}(\varepsilon_1, \mathfrak{M}_2^{\mathsf{T}} ; (\mathbb{I} \otimes \varepsilon_1)), \varepsilon) && f \text{ expanded} \\
&\quad = \mathtt{syq}(\varepsilon ; \mathfrak{M}_2^{\mathsf{T}} ; (\mathbb{I} \otimes \varepsilon_1), \varepsilon) \\
&\quad = \mathtt{syq}((\varepsilon \otimes \varepsilon) ; (\mathbb{I} \otimes \varepsilon_1), \varepsilon) && \text{Proposition 4.3.2.i} \\
&\quad = \mathtt{syq}((\varepsilon \otimes \varepsilon ; \varepsilon_1), \varepsilon) \\
&\quad = \mathtt{syq}((\varepsilon \otimes \varepsilon ; \mathtt{syq}(\varepsilon, \varepsilon ; \varepsilon_1)), \varepsilon) \\
&\quad = \mathtt{syq}((\varepsilon \otimes \varepsilon ; \mathfrak{J}^{\mathsf{T}}), \varepsilon) && \text{by definition} \\
&\quad = \mathtt{syq}((\varepsilon \otimes \varepsilon) ; (\mathbb{I} \otimes \mathfrak{J}^{\mathsf{T}}), \varepsilon) \\
&\quad = (\mathbb{I} \otimes \mathfrak{J}) \, \mathtt{syq}((\varepsilon \otimes \varepsilon), \varepsilon) && \text{Proposition 2.1.4.iv} \\
&\quad = (\mathbb{I} \otimes \mathfrak{J}) ; \mathfrak{M}_2 && \text{by definition} \qquad \square
\end{aligned}$$

4.4 Boolean Algebra Using Lifted Operations

As we have now all the tools available, we may also lift the entire set of operations of a Boolean algebra. A Boolean algebra is quite frequently studied, starting from its signature

$$\langle X, \cdot, +, -, \mathbf{0}, \mathbf{1} \rangle.$$

In our terminology the signature would therefore be

$$\langle X, \cap, \cup, \overline{}, \perp\!\!\!\perp, \top\!\!\!\top \rangle.$$

However, we often distinguish between a subset $U \subseteq X$ in traditional form and the corresponding element e, considered as a point in the powerset $\mathbf{2}^X$. Below, this difference is visualized in a tiny example. The two, U and e, are related via the membership relation $\varepsilon : X \longrightarrow \mathbf{2}^X$ as shown in Fig. 4.11 together with the powerset ordering $\Omega = \overline{\varepsilon^{\mathsf{T}} ; \overline{\varepsilon}}$. Following this idea, we have yet another signature, namely

$$\langle \mathbf{2}^X, \mathfrak{M}_2, \mathfrak{J}_2, \mathcal{N}, \mathtt{syq}(\varepsilon, \perp\!\!\!\perp), \mathtt{syq}(\varepsilon, \top\!\!\!\top) \rangle.$$

Since it may not be immediate, we observe the 0-ary operators or elements $\overline{\varepsilon^{\mathsf{T}} ; \top\!\!\!\top} \approx \mathbf{0}$, $\overline{\overline{\varepsilon}^{\mathsf{T}} ; \top\!\!\!\top} \approx \mathbf{1}$ for which obviously, looking at Fig. 4.11,

$$\perp\!\!\!\perp = \varepsilon ; \overline{\varepsilon^{\mathsf{T}} ; \top\!\!\!\top} = \varepsilon ; \mathtt{syq}(\varepsilon, \perp\!\!\!\perp), \quad \top\!\!\!\top = \varepsilon ; \overline{\overline{\varepsilon}^{\mathsf{T}} ; \top\!\!\!\top} = \varepsilon ; \mathtt{syq}(\varepsilon, \top\!\!\!\top).$$

Summing up, this chapter was used to reformulate several traditional parts of mathematics in terms of point- and quantifier-free relation algebra. Some parts proved to be simple translations, while others were quite involved; in particular, when sharp factorizations were needed. The resulting formulae, however, turned out to be concise and directly intuitive.

$$
\begin{array}{c}
\\
\varepsilon = \begin{array}{c} \\ \mathrm{a} \\ \mathrm{b} \\ \mathrm{c} \\ \mathrm{d} \end{array}
\end{array}
\begin{array}{cccccccccccccccc}
\{\} & \{a\} & \{b\} & \{a,b\} & \{c\} & \{a,c\} & \{b,c\} & \{a,b,c\} & \{d\} & \{a,d\} & \{b,d\} & \{a,b,d\} & \{c,d\} & \{a,c,d\} & \{b,c,d\} & \{a,b,c,d\} \\
\end{array}
$$

$$
\varepsilon = \begin{array}{c} \mathrm{a} \\ \mathrm{b} \\ \mathrm{c} \\ \mathrm{d} \end{array}
\begin{pmatrix}
0 & 1 & 0 & 1 & 0 & 1 & 0 & 1 & 0 & 1 & 0 & 1 & 0 & 1 & 0 & 1 \\
0 & 0 & 1 & 1 & 0 & 0 & 1 & 1 & 0 & 0 & 1 & 1 & 0 & 0 & 1 & 1 \\
0 & 0 & 0 & 0 & 1 & 1 & 1 & 1 & 0 & 0 & 0 & 0 & 1 & 1 & 1 & 1 \\
0 & 0 & 0 & 0 & 0 & 0 & 0 & 0 & 1 & 1 & 1 & 1 & 1 & 1 & 1 & 1
\end{pmatrix}
\begin{pmatrix} 0 \\ 1 \\ 0 \\ 1 \end{pmatrix} = U = \varepsilon{:}e
$$

$$
e^{\mathsf{T}} = \mathtt{syq}(U,\varepsilon) = \begin{pmatrix} 0 & 0 & 0 & 0 & 0 & 0 & 0 & 0 & 0 & 0 & 1 & 0 & 0 & 0 & 0 & 0 \end{pmatrix}
$$

$$
\Omega = \begin{array}{r} \{\} \\ \{a\} \\ \{b\} \\ \{a,b\} \\ \{c\} \\ \{a,c\} \\ \{b,c\} \\ \{a,b,c\} \\ \{d\} \\ \{a,d\} \\ \{b,d\} \\ \{a,b,d\} \\ \{c,d\} \\ \{a,c,d\} \\ \{b,c,d\} \\ \{a,b,c,d\} \end{array}
\begin{pmatrix}
1 & 1 & 1 & 1 & 1 & 1 & 1 & 1 & 1 & 1 & 1 & 1 & 1 & 1 & 1 & 1 \\
0 & 1 & 0 & 1 & 0 & 1 & 0 & 1 & 0 & 1 & 0 & 1 & 0 & 1 & 0 & 1 \\
0 & 0 & 1 & 1 & 0 & 0 & 1 & 1 & 0 & 0 & 1 & 1 & 0 & 0 & 1 & 1 \\
0 & 0 & 0 & 1 & 0 & 0 & 0 & 1 & 0 & 0 & 0 & 1 & 0 & 0 & 0 & 1 \\
0 & 0 & 0 & 0 & 1 & 1 & 1 & 1 & 0 & 0 & 0 & 0 & 1 & 1 & 1 & 1 \\
0 & 0 & 0 & 0 & 0 & 1 & 0 & 1 & 0 & 0 & 0 & 0 & 0 & 1 & 0 & 1 \\
0 & 0 & 0 & 0 & 0 & 0 & 1 & 1 & 0 & 0 & 0 & 0 & 0 & 0 & 1 & 1 \\
0 & 0 & 0 & 0 & 0 & 0 & 0 & 1 & 0 & 0 & 0 & 0 & 0 & 0 & 0 & 1 \\
0 & 0 & 0 & 0 & 0 & 0 & 0 & 0 & 1 & 1 & 1 & 1 & 1 & 1 & 1 & 1 \\
0 & 0 & 0 & 0 & 0 & 0 & 0 & 0 & 0 & 1 & 0 & 1 & 0 & 1 & 0 & 1 \\
0 & 0 & 0 & 0 & 0 & 0 & 0 & 0 & 0 & 0 & 1 & 1 & 0 & 0 & 1 & 1 \\
0 & 0 & 0 & 0 & 0 & 0 & 0 & 0 & 0 & 0 & 0 & 1 & 0 & 0 & 0 & 1 \\
0 & 0 & 0 & 0 & 0 & 0 & 0 & 0 & 0 & 0 & 0 & 0 & 1 & 1 & 1 & 1 \\
0 & 0 & 0 & 0 & 0 & 0 & 0 & 0 & 0 & 0 & 0 & 0 & 0 & 1 & 0 & 1 \\
0 & 0 & 0 & 0 & 0 & 0 & 0 & 0 & 0 & 0 & 0 & 0 & 0 & 0 & 1 & 1 \\
0 & 0 & 0 & 0 & 0 & 0 & 0 & 0 & 0 & 0 & 0 & 0 & 0 & 0 & 0 & 1
\end{pmatrix}
\begin{pmatrix} 0 \\ 0 \\ 0 \\ 0 \\ 0 \\ 0 \\ 0 \\ 0 \\ 0 \\ 0 \\ 1 \\ 0 \\ 0 \\ 0 \\ 0 \\ 0 \end{pmatrix} = e
$$

Fig. 4.11 Subset U and corresponding point e in the powerset, related via ε

Exercises

Exercise 4.1 Prove that $(\mathbb{I} \otimes \mathbb{I}) : \mathfrak{M}_2 = \mathbb{I}$, resembling that $\forall x : x \wedge x = x$, and that $(\mathbb{I} \otimes \mathbb{I}) : \mathfrak{J}_2 = \mathbb{I}$, resembling that $x \vee x = x$ for all x.

Exercise 4.2 Prove $(\mathbb{I} \otimes \mathtt{syq}(\mathbb{T}, \varepsilon)) : \mathfrak{M}_2 = \mathbb{I}$ and $(\mathbb{I} \otimes \mathtt{syq}(\mathbb{L}, \varepsilon)) : \mathfrak{J}_2 = \mathbb{I}$, which resemble that $\forall x : x \wedge \mathbb{T} = x$, resp. $\forall x : x \vee \mathbb{L} = x$.

Exercise 4.3 Prove $(\mathbb{I} \otimes \mathcal{N}) : \mathfrak{M}_2 = \mathtt{syq}(\mathbb{L}, \varepsilon)$, resembling $\forall x : x \wedge \overline{x} = \mathbb{L}$ and also $(\mathbb{I} \otimes \mathcal{N}) : \mathfrak{J}_2 = \mathtt{syq}(\mathbb{T}, \varepsilon)$, resembling $\forall x : x \vee \overline{x} = \mathbb{T}$.

Exercise 4.4 If f is a map with $(\mathbb{I} \otimes f) : \mathfrak{M}_2 = \mathtt{syq}(\mathbb{L}, \varepsilon)$ and $(\mathbb{I} \otimes f) : \mathfrak{J}_2 = \mathtt{syq}(\mathbb{T}, \varepsilon)$, then $f = \mathcal{N}$.

Chapter 5
Applying Relations in Topology

Since its first appearence[1] in the book *Vorstudien zur Topologie* by Johann Benedict Listing of 1847, topology (then and for a long period termed ANALYSIS SITUS) has been given many facets; among the main ones are considerations of neighborhoods, open sets, and closed sets. We start here, giving the corresponding definitions lifted to point-free as well as quantifier-free versions, showing how they are interrelated, thus exhibiting their cryptomorphism and offering the possibility to transform one version into the other, not least visualizing them via TITUREL programs.

It has also been reported by Georg Faber that Karl von Staudt in Erlangen with his *Geometrie der Lage* of 1848 has made one of the greatest achievements of Geometry over thousands of years; see [Fab59]. von Staudt does no longer talk on the length of a line, nor on the degree of an angle. Instead, he talks on points on lines, and incidence—and thus works with relations.

Early in the twentieth century, topology has split into 'general topology' or 'point set theory', mainly invented by Georg Cantor and later developed further by Felix Hausdorff, and what we today call 'algebraic topology'.

[1]Citation: Es mag erlaubt sein, für diese Art Untersuchungen räumlicher Gebilde den Namen "Topologie" zu gebrauchen statt der von Leibniz vorgeschlagenen Benennung "geometria situs", welche an den Begriff des Maßes, der hier ganz untergeordnet ist, erinnert, und mit dem bereits für eine andere Art geometrischer Betrachtungen gebräuchlich gewordenen Namen "géométrie de position" collidiert.

G. Schmidt, M. Winter, *Relational Topology*, Lecture Notes
in Mathematics 2208, https://doi.org/10.1007/978-3-319-74451-3_5

5.1 General Properties of Kernel Forming

We consider some set X and its powerset $\mathbf{2}^X$, so that one automatically has the membership relation $\varepsilon : X \longrightarrow \mathbf{2}^X$, the powerset order $\Omega : \mathbf{2}^X \longrightarrow \mathbf{2}^X$, the powerset negation $\mathcal{N} : \mathbf{2}^X \longrightarrow \mathbf{2}^X$, and the binary powerset join and meet $\mathfrak{J}_2, \mathfrak{M}_2 : \mathbf{2}^X \times \mathbf{2}^X \longrightarrow \mathbf{2}^X$.

We recall here for convenience the definitions of a *closure operation* ρ as well as a *kernel operation* $\mathcal{K}$ with regard to some ordering Ω. The very general concept is that both are mappings which satisfy, respectively,

$$\begin{array}{lll} \rho \subseteq \Omega & \Omega ; \rho \subseteq \rho ; \Omega & \rho ; \rho \subseteq \rho, \\ \mathcal{K} \subseteq \Omega^\mathsf{T} & \Omega ; \mathcal{K} \subseteq \mathcal{K} ; \Omega & \mathcal{K} ; \mathcal{K} \subseteq \mathcal{K}. \end{array}$$

The first postulates express that ρ shall be expanding, resp. $\mathcal{K}$ contracting. The second postulates are common for both, requiring them to be monotonic (also often called isotonic, i.e. homomorphisms wrt. Ω). Following the third postulates, both have to be idempotent; in fact an equality because they are mappings. Kernel-forming will soon be recognized as being crytomorphic with a neighborhood topology—up to a trivial additional totality requirement and the distributivity in Proposition 5.1.2.iv below.

We first investigate in which way a monotone mapping f and the forming of binary meets with $\mathfrak{M}_2$ are related. The interpretation of the following proposition is that when going from a pair of subsets to the intersection of their f-images, one may also first obtain the intersection of the two sets and take its f-image and find oneself below—and having to follow Ω to catch up with the former.

Proposition 5.1.1 *For every monotone mapping f on a powerset we have with regard to meet forming $\mathfrak{M}_2$*

$$(f \otimes f) ; \mathfrak{M}_2 \subseteq \mathfrak{M}_2 ; f ; \Omega.$$

Proof After having reformulated the initial and tentative proof, we start rather unexpectedly:

$$\begin{array}{lll} \mathbb{I} \subseteq \mathfrak{M}_2 ; \mathfrak{M}_2^\mathsf{T} \subseteq \mathfrak{M}_2 ; \Omega ; \mathfrak{M}_2^\mathsf{T} & = \mathfrak{M}_2 ; (\Omega \oslash \Omega) & \text{Proposition 4.3.3.i} \\ \subseteq \mathfrak{M}_2 ; (f ; \Omega ; f^\mathsf{T} \oslash f ; \Omega ; f^\mathsf{T}) & & \text{since } f \text{ is monotonic} \\ = \mathfrak{M}_2 ; f ; (\Omega ; f^\mathsf{T} \oslash \Omega ; f^\mathsf{T}) & & \text{since } f \text{ is univalent} \\ = \mathfrak{M}_2 ; f ; (\Omega \oslash \Omega) ; (f^\mathsf{T} \otimes f^\mathsf{T}) & & \text{since } f \text{ is a map} \\ = \mathfrak{M}_2 ; f ; \Omega ; \mathfrak{M}_2^\mathsf{T} ; (f^\mathsf{T} \otimes f^\mathsf{T}) & & \text{again Proposition 4.3.3.i} \end{array}$$

Shunting the latter two mappings $(f \otimes f)$ and $\mathfrak{M}_2$ gives the result as presented above. □

Fig. 5.1 Sub-distributive monotone mapping satisfying $(f \otimes f) ; \mathfrak{M}_2 \not\subseteq \mathfrak{M}_2 ; f$ and $(f \otimes f) ; \mathfrak{M}_2 \subseteq \mathfrak{M}_2 ; f ; \Omega$

$f =$

	{}	{a}	{b}	{a,b}	{c}	{a,c}	{b,c}	{a,b,c}	{d}	{a,d}	{b,d}	{a,b,d}	{c,d}	{a,c,d}	{b,c,d}	{a,b,c,d}
{}	1	0	0	0	0	0	0	0	0	0	0	0	0	0	0	0
{a}	1	0	0	0	0	0	0	0	0	0	0	0	0	0	0	0
{b}	1	0	0	0	0	0	0	0	0	0	0	0	0	0	0	0
{a,b}	1	0	0	0	0	0	0	0	0	0	0	0	0	0	0	0
{c}	1	0	0	0	0	0	0	0	0	0	0	0	0	0	0	0
{a,c}	1	0	0	0	0	0	0	0	0	0	0	0	0	0	0	0
{b,c}	1	0	0	0	0	0	0	0	0	0	0	0	0	0	0	0
{a,b,c}	1	0	0	0	0	0	0	0	0	0	0	0	0	0	0	0
{d}	1	0	0	0	0	0	0	0	0	0	0	0	0	0	0	0
{a,d}	0	0	0	0	0	0	0	0	0	1	0	0	0	0	0	0
{b,d}	0	0	0	0	0	0	0	0	0	0	1	0	0	0	0	0
{a,b,d}	0	0	0	0	0	0	0	0	0	0	0	1	0	0	0	0
{c,d}	1	0	0	0	0	0	0	0	0	0	0	0	0	0	0	0
{a,c,d}	0	0	0	0	0	0	0	0	0	1	0	0	0	0	0	0
{b,c,d}	0	0	0	0	0	0	0	0	0	0	1	0	0	0	0	0
{a,b,c,d}	0	0	0	0	0	0	0	0	0	0	0	0	0	0	0	1

We may qualify the result of Proposition 5.1.1 as being sub-distributive: When starting from a pair with f on either side and intersecting afterwards, one will end above what one reaches when intersecting first and applying then f.

We consider the f in Fig. 5.1 as an example. Looking at $(\{a,d\},\{b,d\})$, we see that the left side results in $\{d\}$. The right side, however, produces as intersection $\{d\}$ with f-image $\{\}$, so that

$$(f \otimes f) ; \mathfrak{M}_2 \not\subseteq \mathfrak{M}_2 ; f.$$

This shall now be specialized to kernel forming $\mathcal{K}$.

Proposition 5.1.2 *Meet forming $\mathfrak{M}_2$, projections π, ρ, and kernel forming $\mathcal{K}$ are related as*

i) $(\mathcal{K} \otimes \mathcal{K}) ; \mathfrak{M}_2 \subseteq \mathfrak{M}_2 ; \mathcal{K} ; \Omega^\mathsf{T} \quad \Longrightarrow \quad (\mathcal{K} \otimes \mathcal{K}) ; \mathfrak{M}_2 = \mathfrak{M}_2 ; \mathcal{K}$,
ii) $\Omega ; \mathcal{K} \cap \Omega^\mathsf{T} = \mathcal{K}$,
iii) $\pi ; \mathcal{K} \cap \rho ; \mathcal{K} \subseteq \mathfrak{M}_2 ; \mathcal{K}, \quad$ *or* $\quad (\mathcal{K} \oslash \mathcal{K}) \subseteq \mathfrak{M}_2 ; \mathcal{K}$,
iv) $\pi ; \mathcal{K}^\mathsf{T} \cap \rho ; \mathcal{K}^\mathsf{T} \subseteq \mathfrak{M}_2 ; \mathcal{K}^\mathsf{T} \quad$ *or* $\quad (\mathcal{K}^\mathsf{T} \oslash \mathcal{K}^\mathsf{T}) \subseteq \mathfrak{M}_2 ; \mathcal{K}^\mathsf{T}$.

Proof

i) When adding the assumption $(\mathcal{K} \otimes \mathcal{K}) ; \mathfrak{M}_2 \subseteq \mathfrak{M}_2 ; \mathcal{K} ; \Omega^\mathsf{T}$ to Proposition 5.1.1, we have in total

$$(\mathcal{K} \otimes \mathcal{K}) ; \mathfrak{M}_2 \subseteq \mathfrak{M}_2 ; \mathcal{K} ; (\Omega \cap \Omega^\mathsf{T}) = \mathfrak{M}_2 ; \mathcal{K},$$

which means even equality, since both sides are mappings.

ii) Direction "$\supseteq$" is trivial because Ω is reflexive and $\mathcal{K} \subseteq \Omega^\mathsf{T}$. The other direction "$\subseteq$" is slightly more challenging. At position $(*)$ the effect of multiplying a

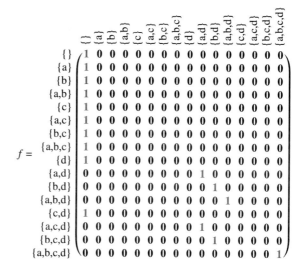

partial identity $J \subseteq \mathbb{I}$ from the right side is used, namely $Q{:}J \cap R = Q \cap R{:}J = (Q \cap R){:}J$. It is irrelevant whether one multiplies it to the left term or the right or the total intersection. The effect is always the same: Columns not selected by J will be annihilated.

$$\begin{aligned}
\Omega{:}\mathcal{K} \cap \Omega^{\mathsf{T}} &= \Omega{:}\mathcal{K}{:}(\mathcal{K} \cap \mathbb{I}) \cap \Omega^{\mathsf{T}} && \text{since for } \mathcal{K} \text{ idempotent } \mathcal{K}{:}(\mathcal{K} \cap \mathbb{I}) \\
& && = \mathcal{K}{:}\mathcal{K} \cap \mathcal{K} = \mathcal{K} \\
&= \Omega{:}\mathcal{K} \cap \Omega^{\mathsf{T}}{:}(\mathcal{K} \cap \mathbb{I}) && (*) \\
&\subseteq \Omega{:}\mathcal{K} \cap \Omega^{\mathsf{T}}{:}\mathcal{K} \\
&\subseteq \mathcal{K}{:}\Omega \cap \mathcal{K}{:}\Omega^{\mathsf{T}} && \text{using monotony of } \mathcal{K} \text{ twice} \\
&\subseteq \mathcal{K}{:}(\Omega \cap \Omega^{\mathsf{T}}) && \mathcal{K} \text{ is univalent} \\
&= \mathcal{K} && \text{antisymmetry}
\end{aligned}$$

iii) We have

$$\begin{aligned}
\pi{:}\mathcal{K} \cap \rho{:}\mathcal{K} &\subseteq \mathfrak{M}_2{:}\,\mathfrak{M}_2^{\mathsf{T}}{:}(\pi{:}\mathcal{K} \cap \rho{:}\mathcal{K}) \quad \text{since } \mathfrak{M}_2 \text{ is a mapping} \\
&\subseteq \mathfrak{M}_2{:}(\mathfrak{M}_2^{\mathsf{T}}{:}\pi{:}\mathcal{K} \cap \mathfrak{M}_2^{\mathsf{T}}{:}\rho{:}\mathcal{K}) = \mathfrak{M}_2{:}(\Omega{:}\mathcal{K} \cap \Omega{:}\mathcal{K}) \quad \text{Proposition 4.2.2.vi} \\
&= \mathfrak{M}_2{:}\Omega{:}\mathcal{K}
\end{aligned}$$

and also

$$\pi{:}\mathcal{K} \cap \rho{:}\mathcal{K} \subseteq \pi{:}\Omega^{\mathsf{T}} \cap \rho{:}\Omega^{\mathsf{T}} = \mathfrak{M}_2{:}\Omega^{\mathsf{T}} \quad \text{following Proposition 4.3.3.i,}$$

so that in total

$$\pi{:}\mathcal{K} \cap \rho{:}\mathcal{K} \subseteq \mathfrak{M}_2{:}\Omega{:}\mathcal{K} \cap \mathfrak{M}_2{:}\Omega^{\mathsf{T}} = \mathfrak{M}_2{:}(\Omega{:}\mathcal{K} \cap \Omega^{\mathsf{T}}) = \mathfrak{M}_2{:}\mathcal{K} \quad \text{using (ii).}$$

iv) follows from $\pi \cap \rho \subseteq \mathfrak{M}_2$ and univalency of $\mathcal{K}$ via shunting:

$$(\pi \cap \rho){:}\,\mathcal{K}^{\mathsf{T}}{:}\mathcal{K} \subseteq \pi \cap \rho \subseteq \mathfrak{M}_2 \quad \Longleftrightarrow \quad (\pi \cap \rho){:}\,\mathcal{K}^{\mathsf{T}} \subseteq \mathfrak{M}_2{:}\mathcal{K}^{\mathsf{T}} \qquad \square$$

Part (i) expresses under what condition $\mathcal{K}$, $\mathfrak{M}_2$ commute; appropriately modified, however, to cope with a binary and a unary mapping.

Lemma 5.1.3 *Any kernel forming operation $\mathcal{K}$ satisfies*

i) $\Omega{:}\,\mathcal{K} \cap \mathcal{K}^{\mathsf{T}} = \mathcal{K}^{\mathsf{T}}{:}\mathcal{K} \subseteq \mathbb{I}$,
ii) $\Omega{:}\,\mathcal{K}^{\mathsf{T}} \cap \Omega^{\mathsf{T}} = \mathcal{K}^{\mathsf{T}}{:}\mathcal{K}$.

Proof

i)

$$\begin{aligned}
\Omega{:}\mathcal{K} \cap \mathcal{K}^{\mathsf{T}} &\subseteq (\Omega \cap \mathcal{K}^{\mathsf{T}}{:}\mathcal{K}^{\mathsf{T}}){:}(\mathcal{K} \cap \Omega^{\mathsf{T}}{:}\mathcal{K}^{\mathsf{T}}) && \text{Dedekind rule} \\
&= (\Omega \cap \mathcal{K}^{\mathsf{T}}){:}(\mathcal{K} \cap \Omega^{\mathsf{T}}{:}\mathcal{K}^{\mathsf{T}}) \\
&\subseteq \mathcal{K}^{\mathsf{T}}{:}\mathcal{K} \subseteq \mathbb{I} && \text{idempotency and univalency.}
\end{aligned}$$

It remains to prove the other direction: The term $\mathcal{K}^{\mathsf{T}}{:}\mathcal{K}$ is contained in $\Omega{:}\mathcal{K}$ because $\mathcal{K} \subseteq \Omega^{\mathsf{T}}$. The term $\mathcal{K}^{\mathsf{T}}{:}\mathcal{K}$ is also contained in $\mathcal{K}^{\mathsf{T}}$ since shunting makes this statement equivalent with $\mathcal{K}^{\mathsf{T}}{:}\mathcal{K}{:}\mathcal{K} \subseteq \mathbb{I}$, where the latter holds for the idempotent and univalent $\mathcal{K}$.

ii) "$\supseteq$" is trivial: $\mathcal{K}^{\mathsf{T}}{:}\mathcal{K} \subseteq \mathbb{I} = \Omega \cap \Omega^{\mathsf{T}}$ and $\mathcal{K}^{\mathsf{T}}{:}\mathcal{K} \subseteq \Omega{:}\mathcal{K}^{\mathsf{T}}$ is via shunting $\mathcal{K}^{\mathsf{T}}{:}\mathcal{K}{:}\mathcal{K} = \mathcal{K}^{\mathsf{T}}{:}\mathcal{K} \subseteq \Omega$. On the other hand

$$\Omega{:}\mathcal{K}^{\mathsf{T}} \cap \Omega^{\mathsf{T}} \subseteq \Omega{:}\Omega \cap \Omega^{\mathsf{T}} = \Omega \cap \Omega^{\mathsf{T}} = \mathbb{I}$$

and similarly

$$\Omega{:}\mathcal{K}^{\mathsf{T}} \cap \Omega^{\mathsf{T}} \subseteq (\Omega \cap \Omega^{\mathsf{T}}{:}\mathcal{K}){:}(\mathcal{K}^{\mathsf{T}} \cap \Omega^{\mathsf{T}}{:}\Omega^{\mathsf{T}}) \subseteq \mathbb{I}{:}\mathcal{K}^{\mathsf{T}} = \mathcal{K}^{\mathsf{T}} \quad \text{observing } \mathcal{K} \subseteq \Omega^{\mathsf{T}}.$$

Together

$$\ldots \subseteq \mathcal{K}^{\mathsf{T}} \cap \mathbb{I} = \mathcal{K}^{\mathsf{T}}{:}\mathcal{K}^{\mathsf{T}} \cap \mathbb{I} \subseteq (\mathcal{K}^{\mathsf{T}} \cap \mathbb{I}{:}\mathcal{K}){:}(\mathcal{K}^{\mathsf{T}} \cap \mathcal{K}{:}\mathbb{I}) \subseteq \mathcal{K}^{\mathsf{T}}{:}\mathcal{K}.$$

□

The following counterplay between two relations $\mathcal{U}$ and $\mathcal{K}$ will prove helpful when we later study the concept of a neighborhood topology as related with the kernel operation.

Proposition 5.1.4 *Based on an arbitrary membership $\varepsilon : X \longrightarrow \mathbf{2}^X$, we consider any pair of transitions of the type*

$$\mathcal{U} \mapsto \mathcal{K} := \mathtt{syq}(\mathcal{U}, \varepsilon) : \mathbf{2}^X \longrightarrow \mathbf{2}^X \quad \textit{and} \quad \mathcal{K} \mapsto \mathcal{U} := \varepsilon{:}\mathcal{K}^{\mathsf{T}} : X \longrightarrow \mathbf{2}^X.$$

i) Such transitions are inverses of one another and $\mathcal{K}$ is always a mapping.
ii) The following two equivalences hold:

$$\begin{array}{lcl} \varepsilon{:}\mathcal{K}^{\mathsf{T}} \textit{ total} & \Longleftrightarrow & \mathcal{U} \textit{ total} \\ \mathcal{K}{:}\mathcal{K} = \mathcal{K} & \Longleftrightarrow & \mathcal{U} = \mathcal{U}{:}\mathtt{syq}(\varepsilon, \mathcal{U}) \end{array}$$

Proof

i) $\mathcal{K}$ as defined on the left is certainly a mapping, since formed as a symmetric quotient with a membership ε on the right side. The two are indeed inverses:

$$\begin{array}{ll} \varepsilon{:}[\mathtt{syq}(\mathcal{U}, \varepsilon)]^{\mathsf{T}} = \varepsilon{:}\mathtt{syq}(\varepsilon, \mathcal{U}) = \mathcal{U}, & \text{since } \mathtt{syq}(\varepsilon, X) \text{ is always surjective} \\ \mathtt{syq}(\varepsilon{:}\mathcal{K}^{\mathsf{T}}, \varepsilon) = \mathcal{K}{:}\mathtt{syq}(\varepsilon, \varepsilon) = \mathcal{K}{:}\mathbb{I} = \mathcal{K} & \text{since } \mathcal{K} \text{ is a mapping} \end{array}$$

ii) The first statement is trivial in view of the definitions. For "$\Longrightarrow$" in the second statement, we show using the definition of $\mathcal{U}$ and idempotency

$$\mathcal{U}{:}\mathtt{syq}(\varepsilon, \mathcal{U}) = \varepsilon{:}\mathcal{K}^{\mathsf{T}}{:}\mathcal{K}^{\mathsf{T}} = \varepsilon{:}\mathcal{K}^{\mathsf{T}} = \mathcal{U}.$$

$$\text{“}\Longleftarrow\text{”}:\quad \mathcal{K}{:}\mathcal{K} = \mathcal{K}{:}\mathtt{syq}(\mathcal{U},\varepsilon) = \mathtt{syq}(\mathcal{U}{:}\mathcal{K}^{\mathsf{T}},\varepsilon)$$
$$= \mathtt{syq}(\mathcal{U}{:}\mathtt{syq}(\varepsilon,\mathcal{U}),\varepsilon)$$
$$= \mathtt{syq}(\mathcal{U},\varepsilon) = \mathcal{K} \quad \text{by assumption}$$

□

5.2 Topology Via Neighborhoods and Kernel Forming

We recall the definition of a topology via a neighborhood system, as proposed by, e.g. [Fra60], Husain [Hus77], von Querenburg[2] [vQ79], or [Dob15], mentioning that in the classical definition a set X endowed with a system $\mathcal{U}(p)$ of subsets for every $p \in X$—called neighborhoods—is a **topological structure**, provided

- $p \in U$ for every neighborhood $U \in \mathcal{U}(p)$,
- if $U \in \mathcal{U}(p)$ and $V \supseteq U$, then $V \in \mathcal{U}(p)$,
- if $U_1, U_2 \in \mathcal{U}(p)$, then $U_1 \cap U_2 \in \mathcal{U}(p)$ and $X \in \mathcal{U}(p)$,
- for every $U \in \mathcal{U}(p)$ there is a $V \in \mathcal{U}(p)$ so that $U \in \mathcal{U}(y)$ for all $y \in V$.

Thus prepared, we present a relational definition of a topology in point-free form without quantifiers.

Definition 5.2.1 A relation $\mathcal{U} : X \longrightarrow \mathbf{2}^X$ will be called a **neighborhood topology** if it satisfies the following properties:

i) $\mathcal{U}{:}\mathbb{T} = \mathbb{T}$ and $\mathcal{U} \subseteq \varepsilon$,
ii) $\mathcal{U}{:}\Omega \subseteq \mathcal{U}$,
iii) $(\mathcal{U}\circledless\mathcal{U}){:}\,\mathfrak{M}_2 \subseteq \mathcal{U}$,
iv) $\mathcal{U} \subseteq \mathcal{U}{:}\overline{\varepsilon^{\mathsf{T}}{:}\overline{\mathcal{U}}}$. □

An example is given in Fig. 5.2 where (since it is finite and discrete) only the tightest neighborhood of every element is shown; others may be obtained as arbitrary supersets.

Definition 5.2.1 obviously resembles being total and assigning only subsets as neighborhoods to an element it is indeed contained in (i), being up-closed (ii), admitting binary meets (iii), and providing open subsets. Property (iv) is not so easily recognized as providing an open kernel for every neighborhood.

[2]One should not attempt to find a person named *Boto von Querenburg*! This is just the name of a community of authors at *Bo*chum University working on *To*pology, situated in the suburb of *Querenburg*. They provided an influential text, but—sadly—starting with metric spaces, as opposed to our relational approach.

$\mathcal{U} =$

	{}	{a}	{b}	{a,b}	{c}	{a,c}	{b,c}	{a,b,c}	{d}	{a,d}	{b,d}	{a,b,d}	{c,d}	{a,c,d}	{b,c,d}	{a,b,c,d}
a	0	1	0	1	0	1	0	1	0	1	0	1	0	1	0	1
b	0	0	0	0	0	0	1	1	0	0	0	0	0	0	1	1
c	0	0	0	0	1	1	1	1	0	0	0	0	1	1	1	1
d	0	0	0	0	0	0	0	0	0	0	0	0	1	1	1	1

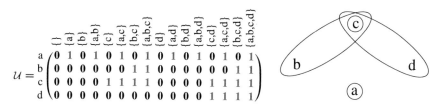

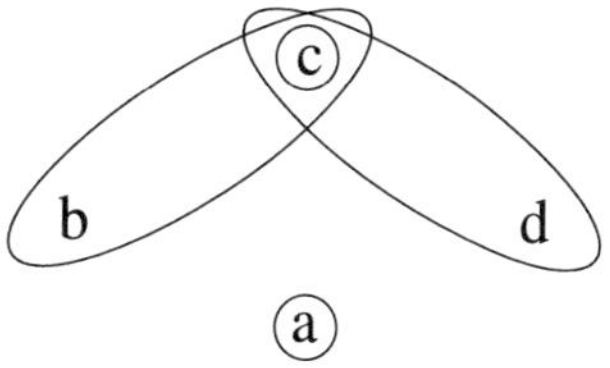

Fig. 5.2 A finite neighborhood topology and the basis of its open sets

Proposition 5.2.2 *In every neighborhood topology according to the minimalistic properties of Definition 5.2.1, some stronger ones are satisfied:*

i) $\mathcal{U};\Omega = \mathcal{U}$
ii) $(\mathcal{U}\otimes\mathcal{U});\mathfrak{M}_2 = \mathcal{U}$
iii) $(\mathcal{U}\otimes\mathcal{U}) = \mathcal{U};\mathfrak{M}_2^{\mathsf{T}}$
iv) $\mathcal{U} = \mathcal{U};\overline{\varepsilon^{\mathsf{T}};\overline{\mathcal{U}}}$
v) $\mathcal{U} = \mathcal{U};\overline{\mathcal{U}^{\mathsf{T}};\overline{\varepsilon}}$
vi) $\mathcal{U} = \mathcal{U};\mathtt{syq}(\varepsilon,\mathcal{U}) = \mathcal{U};\mathcal{K}^{\mathsf{T}}$

Proof

i) follows from Definition 5.2.1.ii since Ω is reflexive.
ii) In addition to Definition 5.2.1.iii:

$$\begin{aligned}
(\mathcal{U}\otimes\mathcal{U});\mathfrak{M}_2 &= [\mathcal{U};\pi^{\mathsf{T}} \cap \mathcal{U};\rho^{\mathsf{T}}];\mathfrak{M}_2 && \text{expanded}\\
&= [\mathcal{U};\Omega;\pi^{\mathsf{T}} \cap \mathcal{U};\Omega;\rho^{\mathsf{T}}];\mathfrak{M}_2 && \text{(i)}\\
&= [\mathcal{U};\mathfrak{M}_2^{\mathsf{T}};\pi;\pi^{\mathsf{T}} \cap \mathcal{U};\mathfrak{M}_2^{\mathsf{T}};\rho;\rho^{\mathsf{T}}];\mathfrak{M}_2 && \text{Proposition 4.2.2.vi}\\
&\supseteq [\mathcal{U};\mathfrak{M}_2^{\mathsf{T}} \cap \mathcal{U};\mathfrak{M}_2^{\mathsf{T}}];\mathfrak{M}_2 && \text{projections are total}\\
&= \mathcal{U};\mathfrak{M}_2^{\mathsf{T}};\mathfrak{M}_2 = \mathcal{U} && \text{meet-forming } \mathfrak{M}_2 \text{ is surjective}
\end{aligned}$$

iii) Direction "$\subseteq$" is a trivial variant of Definition 5.2.1.iii obtained via shunting. The other direction $\mathcal{U};\mathfrak{M}_2^{\mathsf{T}} \subseteq \mathcal{U};\pi^{\mathsf{T}} \cap \mathcal{U};\rho^{\mathsf{T}}$ splits into two similar parts that are shown with Proposition 4.2.2.vi after having shunted:

$$\mathcal{U};\mathfrak{M}_2^{\mathsf{T}} \subseteq \mathcal{U};\pi^{\mathsf{T}} \quad \Longleftarrow \quad \mathcal{U};\mathfrak{M}_2^{\mathsf{T}};\pi = \mathcal{U};\Omega = \mathcal{U}$$

iv) In addition to Definition 5.2.1.iv, $\mathcal{U};\overline{\varepsilon^{\mathsf{T}};\overline{\mathcal{U}}} \subseteq \mathcal{U} \iff \mathcal{U}^{\mathsf{T}};\overline{\mathcal{U}} \subseteq \varepsilon^{\mathsf{T}};\overline{\mathcal{U}}$ and $\mathcal{U} \subseteq \varepsilon$.
v) $\mathcal{U} = \mathcal{U};\Omega = \mathcal{U};\overline{\varepsilon^{\mathsf{T}};\overline{\varepsilon}} \subseteq \mathcal{U};\overline{\mathcal{U}^{\mathsf{T}};\overline{\varepsilon}}$, using (i), definition of Ω, and $\mathcal{U} \subseteq \varepsilon$. It remains to show the reverse direction:

$$\begin{aligned}
\mathcal{U};\overline{\mathcal{U}^{\mathsf{T}};\overline{\varepsilon}} &\subseteq \mathcal{U};\overline{\varepsilon^{\mathsf{T}};\overline{\mathcal{U}}};\overline{\mathcal{U}^{\mathsf{T}};\overline{\varepsilon}} && \text{Definition 5.2.1.iv}\\
&\subseteq \mathcal{U};\overline{\varepsilon^{\mathsf{T}};\overline{\varepsilon}} && \text{see below}\\
&= \mathcal{U};\Omega && \text{definition of } \Omega\\
&= \mathcal{U} && \text{(i)}
\end{aligned}$$

The postponed part:

$$
\begin{array}{lll}
& \overline{\varepsilon}^{\mathsf{T}}; \mathcal{U} \subseteq \overline{\varepsilon}^{\mathsf{T}}; \mathcal{U} & \text{is certainly satisfied} \\
\iff & \overline{\varepsilon}; \overline{\overline{\varepsilon}^{\mathsf{T}}; \mathcal{U}} \subseteq \overline{\mathcal{U}} & \\
\implies & \varepsilon^{\mathsf{T}}; \overline{\varepsilon}; \overline{\overline{\varepsilon}^{\mathsf{T}}; \mathcal{U}} \subseteq \varepsilon^{\mathsf{T}}; \overline{\mathcal{U}} & \\
\iff & \overline{\varepsilon^{\mathsf{T}}; \overline{\mathcal{U}}}; \overline{\mathcal{U}^{\mathsf{T}}; \overline{\varepsilon}} \subseteq \overline{\varepsilon^{\mathsf{T}}; \overline{\varepsilon}} & \text{Schröder equivalence}
\end{array}
$$

vi) We start from (iv) and get immediately

$$\mathcal{U} = \mathcal{U}; \overline{\varepsilon^{\mathsf{T}}; \overline{\mathcal{U}}} = \mathcal{U}; \overline{\varepsilon^{\mathsf{T}}; \overline{\varepsilon; \mathcal{K}^{\mathsf{T}}}} = \mathcal{U}; \overline{\varepsilon^{\mathsf{T}}; \overline{\varepsilon}}; \mathcal{K}^{\mathsf{T}} = \mathcal{U}; \Omega; \mathcal{K}^{\mathsf{T}} = \mathcal{U}; \mathcal{K}^{\mathsf{T}} = \mathcal{U}; \mathtt{syq}(\varepsilon, \mathcal{U}).$$

□

We will now study in which way the idea of Definition 5.2.1 may also be expressed in terms of conditions to be imposed on kernel forming $\mathcal{K}$ alone instead on the neighborhood $\mathcal{U}$.

Definition 5.2.3 A relation $\mathcal{K} : \mathbf{2}^X \longrightarrow \mathbf{2}^X$ is called a **kernel-mapping topology**, if

i) $\mathcal{K}$ is a kernel forming, i.e., $\mathcal{K} \subseteq \Omega^{\mathsf{T}}$, $\Omega; \mathcal{K} \subseteq \mathcal{K}; \Omega$, $\mathcal{K}; \mathcal{K} \subseteq \mathcal{K}$,
ii) $\varepsilon; \mathcal{K}^{\mathsf{T}}$ is total,
iii) $(\mathcal{K} \otimes \mathcal{K}); \mathfrak{M}_2 = \mathfrak{M}_2; \mathcal{K}$. □

Considering (ii) together with Proposition 5.1.4.ii, one will observe that it shall later guarantee totality of the corresponding neighborhood topology. From Proposition 5.1.1, we know that isotone mappings and forming the meet do not commute in general. The discussion of Proposition 5.1.2.i, has already shown in which way forming kernels and meets commute in (iii) (Fig. 5.3).

The following lemma may be helpful. It is intuitively clear when interpreted in the topology context.

It is mainly the counterplay of Proposition 5.1.4 with which we study how a neighborhood topology and a kernel-mapping topology are bijectively interrelated.

Proposition 5.2.4 *The properties imposed on a neighborhood topology $\mathcal{U}$ may also be expressed for $\mathcal{K}$, and vice versa:*

i) Given any neighborhood topology $\mathcal{U}$, the construct $\mathcal{K} := \mathtt{syq}(\mathcal{U}, \varepsilon)$ is a kernel-mapping topology.
ii) Given any kernel-mapping topology $\mathcal{K}$, the construct $\mathcal{U} := \varepsilon; \mathcal{K}^{\mathsf{T}}$ results in a neighborhood topology.

Proof

i) Given the proofs of Proposition 5.1.4.i,ii, it remains to prove that $\mathcal{K}$ is contracting, monotonic and idempotent. For this, we are going to use

$$\mathcal{K} = \mathtt{syq}(\mathcal{U}, \varepsilon) = \overline{\overline{\mathcal{U}}^{\mathsf{T}}; \varepsilon} \cap \overline{\mathcal{U}^{\mathsf{T}}; \overline{\varepsilon}}$$

	{}	{a}	{b}	{a,b}	{c}	{a,c}	{b,c}	{a,b,c}	{d}	{a,d}	{b,d}	{a,b,d}	{c,d}	{a,c,d}	{b,c,d}	{a,b,c,d}
{}	1	0	0	0	0	0	0	0	0	0	0	0	0	0	0	0
{a}	0	1	0	0	0	0	0	0	0	0	0	0	0	0	0	0
{b}	1	0	0	0	0	0	0	0	0	0	0	0	0	0	0	0
{a,b}	0	1	0	0	0	0	0	0	0	0	0	0	0	0	0	0
{c}	0	0	0	0	1	0	0	0	0	0	0	0	0	0	0	0
{a,c}	0	0	0	0	0	1	0	0	0	0	0	0	0	0	0	0
{b,c}	0	0	0	0	0	0	1	0	0	0	0	0	0	0	0	0
{a,b,c}	0	0	0	0	0	0	0	1	0	0	0	0	0	0	0	0
{d}	1	0	0	0	0	0	0	0	0	0	0	0	0	0	0	0
{a,d}	0	1	0	0	0	0	0	0	0	0	0	0	0	0	0	0
{b,d}	1	0	0	0	0	0	0	0	0	0	0	0	0	0	0	0
{a,b,d}	0	1	0	0	0	0	0	0	0	0	0	0	0	0	0	0
{c,d}	0	0	0	0	0	0	0	0	0	0	0	0	1	0	0	0
{a,c,d}	0	0	0	0	0	0	0	0	0	0	0	0	0	1	0	0
{b,c,d}	0	0	0	0	0	0	0	0	0	0	0	0	0	0	1	0
{a,b,c,d}	0	0	0	0	0	0	0	0	0	0	0	0	0	0	0	1
a	0	1	0	0	0	1	0	1	0	0	0	0	0	1	0	1
b	0	0	0	0	0	0	1	1	0	0	0	0	0	0	1	1
c	0	0	0	0	1	1	1	1	0	0	0	0	1	1	1	1
d	0	0	0	0	0	0	0	0	0	0	0	0	1	1	1	1

The lower part (rows a, b, c, d) $= \varepsilon_{\mathcal{O}} = \varepsilon \cap \mathbb{T} ; \mathcal{O}^{\mathsf{T}} \subseteq \mathcal{U}$

{}	1
{a}	1
{b}	0
{a,b}	0
{c}	1
{a,c}	1
{b,c}	1
{a,b,c}	1
{d}	0
{a,d}	0
{b,d}	0
{a,b,d}	0
{c,d}	1
{a,c,d}	1
{b,c,d}	1
{a,b,c,d}	1

{}	0
{a}	1
{b}	0
{a,b}	0
{c}	1
{a,c}	0
{b,c}	1
{a,b,c}	0
{d}	0
{a,d}	0
{b,d}	0
{a,b,d}	0
{c,d}	1
{a,c,d}	0
{b,c,d}	0
{a,b,c,d}	0

Fig. 5.3 Kernel forming, open sets, a basis, membership in open sets for Fig. 5.2

twice. Firstly, $\mathcal{K}$ is contracting, $\mathcal{K} \subseteq \Omega^{\mathsf{T}}$, because

$$\mathcal{K} \subseteq \overline{\overline{\mathcal{U}}^{\mathsf{T}} ; \varepsilon} \subseteq \overline{\overline{\varepsilon}^{\mathsf{T}} ; \varepsilon} = \Omega^{\mathsf{T}}$$

follows from $\mathcal{U} \subseteq \varepsilon$. Secondly, $\mathcal{K}$ is monotonic, $\Omega ; \mathcal{K} \subseteq \mathcal{K} ; \Omega$, because

$$\Omega ; \mathcal{K} \subseteq \Omega ; \overline{\mathcal{U}^{\mathsf{T}} ; \overline{\varepsilon}} = \Omega ; \overline{\Omega^{\mathsf{T}} ; \mathcal{U}^{\mathsf{T}} ; \overline{\varepsilon}} = \overline{\Omega^{\mathsf{T}} ; \mathcal{U}^{\mathsf{T}} ; \overline{\varepsilon}} = \overline{\mathcal{U}^{\mathsf{T}} ; \overline{\varepsilon}}.$$

Now we use $\mathcal{U} = \varepsilon ; \mathcal{K}^{\mathsf{T}}$ according to Proposition 5.1.4.i, and that $\mathcal{K}$ is a mapping, ending in $\Omega ; \mathcal{K} \subseteq \mathcal{K} ; \Omega$.

That $\mathcal{K}$ is idempotent follows using Propositions 5.1.4.ii and 5.2.2.vi.

The second condition that $\varepsilon ; \mathcal{K}^{\mathsf{T}}$ is total follows from Proposition 5.1.4.ii.

The third condition for a kernel-mapping topology:

$$\begin{aligned}
(\mathcal{K} \otimes \mathcal{K}) ; \mathfrak{M}_2 &= (\mathcal{K} \otimes \mathcal{K}) \, \mathtt{syq}((\varepsilon \otimes\!\!< \varepsilon), \varepsilon) && \\
&= \mathtt{syq}((\varepsilon \otimes\!\!< \varepsilon) ; (\mathcal{K}^{\mathsf{T}} \otimes \mathcal{K}^{\mathsf{T}}), \varepsilon) && \text{Proposition 2.1.4.i} \\
&= \mathtt{syq}((\varepsilon ; \mathcal{K}^{\mathsf{T}} \otimes\!\!< \varepsilon ; \mathcal{K}^{\mathsf{T}}), \varepsilon) && \text{since } \mathcal{K} \text{ is a mapping} \\
&= \mathtt{syq}((\mathcal{U} \otimes\!\!< \mathcal{U}), \varepsilon) && \\
&= \mathtt{syq}(\mathcal{U} ; \mathfrak{M}_2^{\mathsf{T}}, \varepsilon) && \text{Proposition 5.2.2.iii} \\
&= \mathfrak{M}_2 ; \mathtt{syq}(\mathcal{U}, \varepsilon) = \mathfrak{M}_2 ; \mathcal{K} &&
\end{aligned}$$

ii) The topology $\mathcal{U}$ is total in view of Definition 5.2.3.ii and Proposition 5.1.4.ii. Contraction $\mathcal{K} \subseteq \Omega^{\mathsf{T}}$ is equivalent with $\overline{\varepsilon}^{\mathsf{T}} ; \varepsilon \subseteq \overline{\mathcal{K}}$, further with $\overline{\varepsilon} ; \mathcal{K} \subseteq \overline{\varepsilon}$, and finally with $\mathcal{U} = \varepsilon ; \mathcal{K}^{\mathsf{T}} \subseteq \varepsilon$ as demanded. In order to prove $\mathcal{U} ; \Omega \subseteq \mathcal{U}$, we start with monotony, univalency, and shunting applied in

$$\varepsilon ; \mathcal{K}^{\mathsf{T}} ; \Omega ; \mathcal{K} \subseteq \varepsilon ; \mathcal{K}^{\mathsf{T}} ; \mathcal{K} ; \Omega \subseteq \varepsilon ; \Omega = \varepsilon \quad \Longleftrightarrow \quad \varepsilon ; \mathcal{K}^{\mathsf{T}} ; \Omega = \mathcal{U} ; \Omega \subseteq \mathcal{U} = \varepsilon ; \mathcal{K}^{\mathsf{T}}.$$

$$\begin{aligned}
&\mathcal{U} ; \mathfrak{M}_2^{\mathsf{T}} = \varepsilon ; \mathcal{K}^{\mathsf{T}} ; \mathfrak{M}_2^{\mathsf{T}} = \varepsilon ; \mathfrak{M}_2^{\mathsf{T}} ; (\mathcal{K}^{\mathsf{T}} \otimes \mathcal{K}^{\mathsf{T}}) && \\
&= (\varepsilon \mathbin{\otimes\!\!\!<} \varepsilon) ; (\mathcal{K}^{\mathsf{T}} \otimes \mathcal{K}^{\mathsf{T}}) && \text{Proposition 4.3.2.i} \\
&= (\varepsilon ; \mathcal{K}^{\mathsf{T}} \mathbin{\otimes\!\!\!<} \varepsilon ; \mathcal{K}^{\mathsf{T}}) = (\mathcal{U} \mathbin{\otimes\!\!\!<} \mathcal{U}) && \text{Proposition 4.3.2.ii}
\end{aligned}$$

For the last property, we use Proposition 5.1.4.ii. □

The f of Fig. 5.1 does not satisfy $(f \otimes f) ; \mathfrak{M}_2 = \mathfrak{M}_2 ; f$ and, thus, fails to satisfy the requirements for a kernel-mapping topology. Also the mapping $\mathcal{K}_0 := \overline{\mathbb{T} ; \varepsilon}$ which sends everything to the empty set would be contracting, isotonic, and idempotent without $\varepsilon ; \mathcal{K}_0^{\mathsf{T}}$ being total; however it would lead to $\mathcal{U}_0 = \mathbb{⊥}$ which cannot be a neighborhood system.

Qualifying a topology via a neighborhood system $\mathcal{U}$ or kernel mapping $\mathcal{K}$ has, thus, been shown to mean basically the same; $\mathcal{U}$ and $\mathcal{K}$ may be converted into one another. In what follows, we will use them interchangeably as required.

The open sets are often defined identifying a subset of all open sets as a so-called *basis* with the idea that finite intersections and arbitrary unions will then produce them all. It is often convenient to restrict such a basis to just the smallest ones, i.e., those that are not non-trivial unions. Observe that the empty set is also an open one and would be the minimal one when not explicitly excluded. For finite cases at least, it is possible to characterize a basis of open sets as follows. The topology of the real axis, for example, does not allow such atomic open sets since the basis mapping β below turns out to be the singleton injection—and, hence, fails to satisfy the assumption $\beta \subseteq U$ of (iii) of the following proposition.

Proposition 5.2.5 *When we consider* $\beta := \mathtt{syq}(\overline{\varepsilon ; \mathcal{U}^{\mathsf{T}}}, \varepsilon) : X \longrightarrow \mathbf{2}^X$,

i) β *is a mapping,*
ii) $\beta = \left[\mathtt{glb}_{\Omega}(\mathcal{U}^{\mathsf{T}})\right]^{\mathsf{T}}$,
iii) $\beta \subseteq \mathcal{U} \quad \Longrightarrow \quad \beta ; \Omega = \mathcal{U}$.

Proof

i) β is a mapping by construction through a symmetric quotient with a membership relation ε on the right side.
ii) See Definition 4.2.1, relating greatest lower bounds with symmetric quotients between the lower bound $\mathtt{lbd}_E(\varepsilon_1)$ and E when E happens to be a powerset ordering. This result has been proved as Prop. 9.10 of [Sch11].

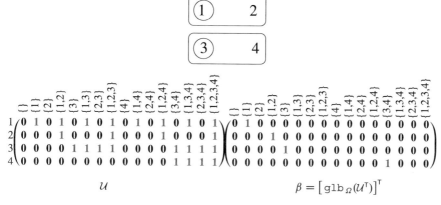

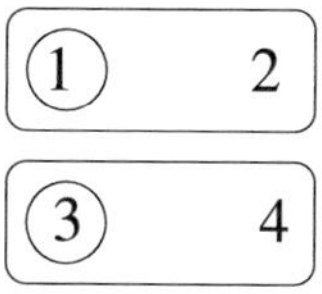

	{}	{1}	{2}	{1,2}	{3}	{1,3}	{2,3}	{1,2,3}	{4}	{1,4}	{2,4}	{1,2,4}	{3,4}	{1,3,4}	{2,3,4}	{1,2,3,4}
1	0	1	0	1	0	1	0	1	0	1	0	1	0	1	0	1
2	0	0	0	1	0	0	0	1	0	0	0	1	0	0	0	1
3	0	0	0	0	1	1	1	1	0	0	0	0	1	1	1	1
4	0	0	0	0	0	0	0	0	0	0	0	0	1	1	1	1

$\mathcal{U}$

	{}	{1}	{2}	{1,2}	{3}	{1,3}	{2,3}	{1,2,3}	{4}	{1,4}	{2,4}	{1,2,4}	{3,4}	{1,3,4}	{2,3,4}	{1,2,3,4}
1	0	1	0	0	0	0	0	0	0	0	0	0	0	0	0	0
2	0	0	0	1	0	0	0	0	0	0	0	0	0	0	0	0
3	0	0	0	0	1	0	0	0	0	0	0	0	0	0	0	0
4	0	0	0	0	0	0	0	0	0	0	0	0	1	0	0	0

$\beta = \left[\mathtt{glb}_{\Omega}(\mathcal{U}^{\mathsf{T}})\right]^{\mathsf{T}}$

Fig. 5.4 A topology $\mathcal{U}$ with its basis mapping β

iii) It is trivial that $\beta \subseteq \mathcal{U}$ implies $\beta;\Omega \subseteq \mathcal{U};\Omega = \mathcal{U}$. In addition, $\mathcal{U} \subseteq \beta;\Omega$ holds since

$$\beta;\Omega = \beta;\overline{\varepsilon^{\mathsf{T}};\overline{\varepsilon}} = \overline{\beta;\varepsilon^{\mathsf{T}};\overline{\varepsilon}}$$
$$= \overline{\overline{\overline{\mathcal{U};\overline{\varepsilon}^{\mathsf{T}}};\overline{\varepsilon}}} \quad \text{because always } \varepsilon;\mathtt{syq}(\varepsilon, X) = X$$

and

$$\mathcal{U};\overline{\varepsilon}^{\mathsf{T}} \subseteq \mathcal{U};\overline{\varepsilon}^{\mathsf{T}} \quad \Longleftrightarrow \quad \overline{\mathcal{U};\overline{\varepsilon}^{\mathsf{T}}};\overline{\varepsilon} \subseteq \overline{\mathcal{U}} \quad \Longleftrightarrow \quad \mathcal{U} \subseteq \overline{\overline{\mathcal{U};\overline{\varepsilon}^{\mathsf{T}}};\overline{\varepsilon}} = \beta;\Omega.$$

□

We have illustrated this result for a simple topology in Fig. 5.4.

5.3 Qualifying a Topology Via Its Open Sets

The next idea for topologies was to define them via their open sets. We recall that in our exposition, any subset offers two ways to be described, namely as a vector $v = v;\mathbb{T}$, or alternately as a subidentity or partial diagonal $s \subseteq \mathbb{I}$. In a similar way as we could $\mathcal{U}, \mathcal{K}$ let more or less represent each other mutually, we here have the versions $\mathcal{O}_V, \mathcal{O}_D$ positioned against the former two:

$$\begin{matrix} \mathcal{U} & & \mathcal{O}_D \\ & \Longleftrightarrow & \\ \mathcal{K} & & \mathcal{O}_V \end{matrix}$$

The transitions up and down between $\mathcal{U}$ and $\mathcal{K}$ on the left of the diagram above have already been mentioned. Toggling between the vector $\mathcal{O}_V : \mathbf{2}^X \longrightarrow \mathbb{1}$ and the corresponding partial identity $\mathcal{O}_D : \mathbf{2}^X \longrightarrow \mathbf{2}^X$ is completely trivial and doesn't need any topological consideration:

$$\mathcal{O}_V = \mathcal{O}_D ; \mathbb{T} \qquad \mathcal{O}_D = \mathbb{I} \cap \mathcal{O}_V ; \mathbb{T}$$

Given any pair of a vector and a partial identity, we obtain the following results; we have, however, maintained the notations $\mathcal{O}_V, \mathcal{O}_D$ reminding us of open sets since we will apply it only in this context.

Proposition 5.3.1 *Given $\mathcal{O}_V$ resp. $\mathcal{O}_D$, two other relations*

$$\varepsilon_{\mathcal{O}} := \varepsilon \cap \mathbb{T} ; \mathcal{O}_V^{\mathsf{T}} = \varepsilon ; \mathcal{O}_D \quad \textit{and} \quad \omega := \mathtt{syq}(\varepsilon_{\mathcal{O}}, \varepsilon)$$

are introduced for technical reasons. They satisfy the following properties:

i) ω *is a mapping that satisfies* $\omega^{\mathsf{T}} \subseteq \Omega$.
ii) $\varepsilon_{\mathcal{O}} ; \omega = \varepsilon \cap \mathbb{T} ; \omega$
iii) $\varepsilon ; \omega^{\mathsf{T}} = \varepsilon_{\mathcal{O}}$
iv) $\varepsilon_{\mathcal{O}} = \varepsilon \cap \mathbb{T} ; \varepsilon_{\mathcal{O}}$

Proof

i) The mapping property follows from the definition as a symmetric quotient with a membership on the right side; furthermore

$$\omega^{\mathsf{T}} = \mathtt{syq}(\varepsilon, \varepsilon_{\mathcal{O}}) \subseteq \overline{\varepsilon^{\mathsf{T}} ; \overline{\varepsilon_{\mathcal{O}}}} \subseteq \overline{\varepsilon^{\mathsf{T}} ; \overline{\varepsilon}} = \Omega.$$

ii) $\varepsilon_{\mathcal{O}} ; \omega = \varepsilon_{\mathcal{O}} ; \mathtt{syq}(\varepsilon_{\mathcal{O}}, \varepsilon) = \varepsilon \cap \mathbb{T} ; \mathtt{syq}(\varepsilon_{\mathcal{O}}, \varepsilon)$ Proposition 2.1.1.i
$= \varepsilon \cap \mathbb{T} ; \omega$ by definition

iii) $\varepsilon ; \omega^{\mathsf{T}} = \varepsilon ; \mathtt{syq}(\varepsilon, \varepsilon_{\mathcal{O}}) = \varepsilon_{\mathcal{O}}$

iv) $\varepsilon_{\mathcal{O}} = \varepsilon \cap \mathbb{T} ; \mathcal{O}_V^{\mathsf{T}} = \varepsilon \cap (\mathbb{T} ; \varepsilon \cap \mathbb{T} ; \mathcal{O}_V^{\mathsf{T}})$ □
$= \varepsilon \cap \mathbb{T} ; (\varepsilon \cap \mathbb{T} ; \mathcal{O}_V^{\mathsf{T}})$ masking
$= \varepsilon \cap \mathbb{T} ; \varepsilon_{\mathcal{O}}$

The global situation with several methods of characterizing a topology is best visualized with Fig. 5.5.

To relate the two, $\mathcal{O}_D, \mathcal{O}_V$, with $\varepsilon_{\mathcal{O}}$ (i.e. the membership ε, but restricted to membership in sets qualified as being open) is a little complicated:

$$\varepsilon_{\mathcal{O}} = \varepsilon \cap \mathbb{T} ; \mathcal{O}_V^{\mathsf{T}} = \varepsilon ; \mathcal{O}_D, \qquad \mathcal{O}_V = \varepsilon_{\mathcal{O}}^{\mathsf{T}} ; \mathbb{T} \cup \mathtt{syq}(\varepsilon, \mathbb{L})$$

The disturbing term $\mathtt{syq}(\varepsilon, \mathbb{L}) = \overline{\varepsilon^{\mathsf{T}} ; \mathbb{T}}$ in the definition of $\mathcal{O}_V$ above owes its existence to the fact that also the empty set is by definition an open set, but does not contain any element. The marking of the empty set would not be shown in $\varepsilon_{\mathcal{O}}^{\mathsf{T}} ; \mathbb{T}$. Therefore, one may encounter some technicalities when always adding

ε

	{}	{a}	{b}	{a,b}	{c}	{a,c}	{b,c}	{a,b,c}	{d}	{a,d}	{b,d}	{a,b,d}	{c,d}	{a,c,d}	{b,c,d}	{a,b,c,d}
a	0	1	0	1	0	1	0	1	0	1	0	1	0	1	0	1
b	0	0	1	1	0	0	1	1	0	0	1	1	0	0	1	1
c	0	0	0	0	1	1	1	1	0	0	0	0	1	1	1	1
d	0	0	0	0	0	0	0	0	1	1	1	1	1	1	1	1

$\mathcal{U}$

	{}	{a}	{b}	{a,b}	{c}	{a,c}	{b,c}	{a,b,c}	{d}	{a,d}	{b,d}	{a,b,d}	{c,d}	{a,c,d}	{b,c,d}	{a,b,c,d}
a	0	1	0	1	0	1	0	1	0	1	0	1	0	1	0	1
b	0	0	0	0	0	0	1	1	0	0	0	0	0	0	1	1
c	0	0	0	0	1	1	1	1	0	0	0	0	1	1	1	1
d	0	0	0	0	0	0	0	0	0	0	0	0	1	1	1	1

$\varepsilon_{\mathcal{O}} := \varepsilon \cap \mathbb{T}{:}\mathcal{O}_V^{\mathsf{T}} = \varepsilon{:}\mathcal{K} \cap \varepsilon$

	{}	{a}	{b}	{a,b}	{c}	{a,c}	{b,c}	{a,b,c}	{d}	{a,d}	{b,d}	{a,b,d}	{c,d}	{a,c,d}	{b,c,d}	{a,b,c,d}
a	0	1	0	0	0	1	0	1	0	0	0	0	0	1	0	1
b	0	0	0	0	0	0	1	1	0	0	0	0	0	0	1	1
c	0	0	0	0	1	1	1	1	0	0	0	0	1	1	1	1
d	0	0	0	0	0	0	0	0	0	0	0	0	1	1	1	1

$\mathcal{K} := \mathtt{syq}(\mathcal{U}, \varepsilon)$ and $\mathcal{O}_V$

	{}	{a}	{b}	{a,b}	{c}	{a,c}	{b,c}	{a,b,c}	{d}	{a,d}	{b,d}	{a,b,d}	{c,d}	{a,c,d}	{b,c,d}	{a,b,c,d}	$\mathcal{O}_V$
{}	1	0	0	0	0	0	0	0	0	0	0	0	0	0	0	0	1
{a}	0	1	0	0	0	0	0	0	0	0	0	0	0	0	0	0	1
{b}	1	0	0	0	0	0	0	0	0	0	0	0	0	0	0	0	0
{a,b}	0	1	0	0	0	0	0	0	0	0	0	0	0	0	0	0	0
{c}	0	0	0	0	1	0	0	0	0	0	0	0	0	0	0	0	1
{a,c}	0	0	0	0	0	1	0	0	0	0	0	0	0	0	0	0	1
{b,c}	0	0	0	0	0	0	1	0	0	0	0	0	0	0	0	0	1
{a,b,c}	0	0	0	0	0	0	0	1	0	0	0	0	0	0	0	0	1
{d}	1	0	0	0	0	0	0	0	0	0	0	0	0	0	0	0	0
{a,d}	0	1	0	0	0	0	0	0	0	0	0	0	0	0	0	0	0
{b,d}	1	0	0	0	0	0	0	0	0	0	0	0	0	0	0	0	0
{a,b,d}	0	1	0	0	0	0	0	0	0	0	0	0	0	0	0	0	0
{c,d}	0	0	0	0	0	0	0	0	0	0	0	0	1	0	0	0	1
{a,c,d}	0	0	0	0	0	0	0	0	0	0	0	0	0	1	0	0	1
{b,c,d}	0	0	0	0	0	0	0	0	0	0	0	0	0	0	1	0	1
{a,b,c,d}	0	0	0	0	0	0	0	0	0	0	0	0	0	0	0	1	1

Fig. 5.5 A topology in different representations; $\mathcal{O}_D$ indicated as diagonal of $\mathcal{K}$

resp. deleting it. The following table summarizes the mutual transformations of the topology versions considered into one another.

	$\mathcal{O}_V$	$\mathcal{O}_D$	$\varepsilon_{\mathcal{O}}$	ω
$\mathcal{O}_V =$	$\times$	$\mathcal{O}_D{:}\mathbb{T}$	$\varepsilon_{\mathcal{O}}^{\mathsf{T}}{:}\mathbb{T} \cup \overline{\varepsilon^{\mathsf{T}}{:}\mathbb{T}}$	$\omega^{\mathsf{T}}{:}\mathbb{T}$
$\mathcal{O}_D =$	$\mathbb{I} \cap \mathcal{O}_V{:}\mathbb{T}$	$\times$	$\mathbb{I} \cap (\varepsilon_{\mathcal{O}}^{\mathsf{T}}{:}\mathbb{T} \cup \overline{\varepsilon^{\mathsf{T}}{:}\mathbb{T}})$	$\mathbb{I} \cap \omega$
$\varepsilon_{\mathcal{O}} =$	$\varepsilon \cap \mathbb{T}{:}\mathcal{O}_V^{\mathsf{T}}$	$\varepsilon{:}\mathcal{O}_D$	$\times$	$\varepsilon{:}\omega^{\mathsf{T}}$
$\omega =$	$\mathtt{syq}(\varepsilon \cap \mathbb{T}{:}\mathcal{O}_V^{\mathsf{T}}, \varepsilon)$	$\mathtt{syq}(\varepsilon{:}\mathcal{O}_D, \varepsilon)$	$\mathtt{syq}(\varepsilon_{\mathcal{O}}, \varepsilon)$	$\times$

And this is the reason why we have decided for an alternative approach when relating the 'membership-in-open-sets' relation $\varepsilon_{\mathcal{O}}$ forth and back with the $\mathcal{U}, \mathcal{K}$ side. The mapping ω reproduces open environments. It does not, however, map environments to their open kernel, as $\mathcal{K}$, but to the empty set (Fig. 5.6).

A topology may also be characterized when the open sets are given as a vector or its equivalent partial identity. One will recognize that in the following two definitions the conditions (i) define the empty as well as the full set to be open. The second conditions (ii) demand that arbitrary unions and the third (iii) that finite intersections of open sets are open again.

Definition 5.3.2 A vector $\mathcal{O}_V$ along $\mathbf{2}^X$ will be called an **open set topology** provided

i) $\mathtt{syq}(\varepsilon, \mathbb{L}) \subseteq \mathcal{O}_V \qquad \mathtt{syq}(\varepsilon, \mathbb{T}) \subseteq \mathcal{O}_V$,
ii) $v \subseteq \mathcal{O}_V \implies \mathtt{syq}(\varepsilon, \varepsilon{:}v) \subseteq \mathcal{O}_V$ for all vectors $v \subseteq \mathbf{2}^X$,
iii) $\mathfrak{M}_2^{\mathsf{T}}{:}(\mathcal{O}_V \otimes \mathcal{O}_V) \subseteq \mathcal{O}_V$. □

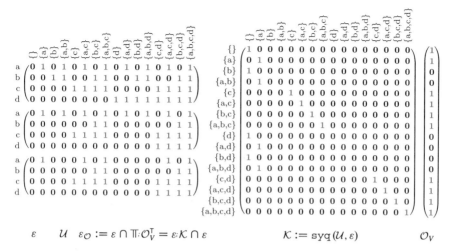

	{}	{a}	{b}	{a,b}	{c}	{a,c}	{b,c}	{a,b,c}	{d}	{a,d}	{b,d}	{a,b,d}	{c,d}	{a,c,d}	{b,c,d}	{a,b,c,d}
{}	1	0	0	0	0	0	0	0	0	0	0	0	0	0	0	0
{a}	0	1	0	0	0	0	0	0	0	0	0	0	0	0	0	0
{b}	1	0	0	0	0	0	0	0	0	0	0	0	0	0	0	0
{a,b}	0	1	0	0	0	0	0	0	0	0	0	0	0	0	0	0
{c}	0	0	0	0	1	0	0	0	0	0	0	0	0	0	0	0
{a,c}	0	0	0	0	0	1	0	0	0	0	0	0	0	0	0	0
{b,c}	0	0	0	0	0	0	1	0	0	0	0	0	0	0	0	0
{a,b,c}	0	0	0	0	0	0	0	1	0	0	0	0	0	0	0	0
{d}	1	0	0	0	0	0	0	0	0	0	0	0	0	0	0	0
{a,d}	0	1	0	0	0	0	0	0	0	0	0	0	0	0	0	0
{b,d}	1	0	0	0	0	0	0	0	0	0	0	0	0	0	0	0
{a,b,d}	0	1	0	0	0	0	0	0	0	0	0	0	0	0	0	0
{c,d}	0	0	0	0	0	0	0	0	0	0	0	0	1	0	0	0
{a,c,d}	0	0	0	0	0	0	0	0	0	0	0	0	0	1	0	0
{b,c,d}	0	0	0	0	0	0	0	0	0	0	0	0	0	0	1	0
{a,b,c,d}	0	0	0	0	0	0	0	0	0	0	0	0	0	0	0	1

	{}	{a}	{b}	{a,b}	{c}	{a,c}	{b,c}	{a,b,c}	{d}	{a,d}	{b,d}	{a,b,d}	{c,d}	{a,c,d}	{b,c,d}	{a,b,c,d}
{}	1	0	0	0	0	0	0	0	0	0	0	0	0	0	0	0
{a}	0	1	0	0	0	0	0	0	0	0	0	0	0	0	0	0
{b}	1	0	0	0	0	0	0	0	0	0	0	0	0	0	0	0
{a,b}	1	0	0	0	0	0	0	0	0	0	0	0	0	0	0	0
{c}	0	0	0	0	1	0	0	0	0	0	0	0	0	0	0	0
{a,c}	0	0	0	0	0	1	0	0	0	0	0	0	0	0	0	0
{b,c}	0	0	0	0	0	0	1	0	0	0	0	0	0	0	0	0
{a,b,c}	0	0	0	0	0	0	0	1	0	0	0	0	0	0	0	0
{d}	1	0	0	0	0	0	0	0	0	0	0	0	0	0	0	0
{a,d}	1	0	0	0	0	0	0	0	0	0	0	0	0	0	0	0
{b,d}	1	0	0	0	0	0	0	0	0	0	0	0	0	0	0	0
{a,b,d}	1	0	0	0	0	0	0	0	0	0	0	0	0	0	0	0
{c,d}	0	0	0	0	0	0	0	0	0	0	0	0	1	0	0	0
{a,c,d}	0	0	0	0	0	0	0	0	0	0	0	0	0	1	0	0
{b,c,d}	0	0	0	0	0	0	0	0	0	0	0	0	0	0	1	0
{a,b,c,d}	0	0	0	0	0	0	0	0	0	0	0	0	0	0	0	1

Fig. 5.6 $\mathcal{K}$ as opposed to ω

The following definition means largely the same as the one just given. The only difference rests in the representation as a partial identity as opposed to a column vector which leads to minor technical changes.

Definition 5.3.3 A partial identity $\mathcal{O}_D$ on $\mathbf{2}^X$ is an **open diagonal topology** provided

i) $\mathtt{syq}(\varepsilon, \bot\!\!\!\bot) \subseteq \mathcal{O}_D ; \top\!\!\!\top \qquad \mathtt{syq}(\varepsilon, \top\!\!\!\top) \subseteq \mathcal{O}_D ; \top\!\!\!\top$,
ii) $v \subseteq \mathcal{O}_D ; \top\!\!\!\top \implies \mathtt{syq}(\varepsilon, \varepsilon ; v) \subseteq \mathcal{O}_D ; \top\!\!\!\top$ for all vectors $v \subseteq \mathbf{2}^X$,
iii) $(\mathcal{O}_D \otimes \mathcal{O}_D) ; \mathfrak{M}_2 \subseteq \mathfrak{M}_2 ; \mathcal{O}_D$. □

One will observe that these two versions mean the same, however, formulated with a vector $\mathcal{O}_V$ or a partial identity $\mathcal{O}_D$, respectively. Only the equivalence of 5.3.3.iii and 5.3.2.iii may need a bit of an explanation. First, we show "5.3.3.iii $\implies$ 5.3.2.iii":

$$\begin{aligned}
&\mathfrak{M}_2^{\mathsf{T}} ; (\mathcal{O}_V \circledcirc \mathcal{O}_V) = \mathfrak{M}_2^{\mathsf{T}} ; (\mathcal{O}_D ; \top\!\!\!\top \circledcirc \mathcal{O}_D ; \top\!\!\!\top) && \text{by definition} \\
&= \mathfrak{M}_2^{\mathsf{T}} ; (\mathcal{O}_D \otimes \mathcal{O}_D) ; \top\!\!\!\top && \text{Proposition 3.1.6.i} \\
&\subseteq \mathcal{O}_D ; \mathfrak{M}_2^{\mathsf{T}} \subseteq \mathcal{O}_D ; \top\!\!\!\top = \mathcal{O}_V && \text{shunted version of Definition 5.3.3.iii}
\end{aligned}$$

Also in reverse direction "5.3.2.iii $\implies$ 5.3.3.iii" —as a finger exercise:

$$\begin{aligned}
&(\mathcal{O}_D \otimes \mathcal{O}_D) ; \mathfrak{M}_2 \\
&\quad = (\mathbb{I} \cap \mathcal{O}_V ; \top\!\!\!\top \otimes \mathbb{I} \cap \mathcal{O}_V ; \top\!\!\!\top) ; \mathfrak{M}_2 && \text{by definition} \\
&\quad = \big[(\mathbb{I} \otimes \mathbb{I}) \cap (\mathcal{O}_V ; \top\!\!\!\top \circledcirc \mathcal{O}_V ; \top\!\!\!\top) \big] ; \mathfrak{M}_2 && \text{Proposition 3.1.6.iii} \\
&\quad = \big[\mathbb{I} \cap (\mathcal{O}_V \circledcirc \mathcal{O}_V) ; \top\!\!\!\top \big] ; \mathfrak{M}_2
\end{aligned}$$

$$
\begin{array}{ll}
= \left[\mathbb{I} \cap \mathbb{T}{:}(\mathcal{O}_V^{\mathsf{T}} \textcircled{<} \mathcal{O}_V^{\mathsf{T}})\right}{:}\, \mathfrak{M}_2 & \text{transposing a partial identity} \\
\subseteq \mathfrak{M}_2 \cap \mathbb{T}{:}(\mathcal{O}_V^{\mathsf{T}} \textcircled{<} \mathcal{O}_V^{\mathsf{T}}){:}\, \mathfrak{M}_2 & \\
\subseteq \mathfrak{M}_2{:}\mathbb{I} \cap \mathbb{T}{:}\mathcal{O}_V^{\mathsf{T}} & \text{Definition 5.3.2.iii} \\
= \mathfrak{M}_2{:}\mathbb{I} \cap \mathfrak{M}_2{:}\mathbb{T}{:}\mathcal{O}_V^{\mathsf{T}} & \text{since } \mathfrak{M}_2 \text{ is a map} \\
= \mathfrak{M}_2{:}(\mathbb{I} \cap \mathbb{T}{:}\mathcal{O}_V^{\mathsf{T}}) & \\
= \mathfrak{M}_2{:}\mathcal{O}_D &
\end{array}
$$

The following is a slight variant of the last two definitions.

Definition 5.3.4 A relation $\varepsilon_{\mathcal{O}} : X \longrightarrow \mathbf{2}^X$ will be called a **membership-in-open-sets topology** provided

i) $\varepsilon_{\mathcal{O}}{:}\mathbb{T} = \mathbb{T} \qquad \varepsilon_{\mathcal{O}} = \varepsilon \cap \mathbb{T}{:}\varepsilon_{\mathcal{O}}$,
ii) $v \subseteq \varepsilon_{\mathcal{O}}^{\mathsf{T}}{:}\mathbb{T} \implies \mathrm{syq}(\varepsilon, \varepsilon{:}v) \subseteq \varepsilon_{\mathcal{O}}^{\mathsf{T}}{:}\mathbb{T} \cup \overline{\varepsilon^{\mathsf{T}}{:}\mathbb{T}}$ (i.e. $= \omega^{\mathsf{T}}{:}\mathbb{T}$) for all $v \subseteq \mathbf{2}^X$,
iii) $(\varepsilon_{\mathcal{O}} \textcircled{<} \varepsilon_{\mathcal{O}}) \subseteq \varepsilon_{\mathcal{O}}{:}\, \mathfrak{M}_2^{\mathsf{T}}$. □

Remark In view of $\mathcal{O}_V{:}\mathbb{T} = \varepsilon_{\mathcal{O}}^{\mathsf{T}}{:}\mathbb{T} \cup \overline{\varepsilon^{\mathsf{T}}{:}\mathbb{T}}$, one may ask why in (ii) the condition does not start with $v \subseteq \varepsilon_{\mathcal{O}}^{\mathsf{T}}{:}\mathbb{T} \cup \overline{\varepsilon^{\mathsf{T}}{:}\mathbb{T}}$, while this extended version is used after the symmetric quotient. When taking $v := \mathbb{L}$, the latter will be needed. However, $\varepsilon{:}\, v$ will not be changed when v is replaced by $v \cup \overline{\varepsilon^{\mathsf{T}}{:}\mathbb{T}}$. □

We will show that Definition 5.3.4 is cryptomorphic with Definition 5.3.2. First we prove rather tricky that the postulates of Definition 5.3.4.iii imply those of Definition 5.3.2.iii:

$$
\begin{array}{ll}
\mathfrak{M}_2^{\mathsf{T}}{:}(\mathcal{O}_V{:}\mathbb{T} \textcircled{>} \mathcal{O}_V{:}\mathbb{T}) & \text{willfully enlarged to} \\
\quad \subseteq (\varepsilon^{\mathsf{T}}{:}\mathbb{T} \cup \overline{\varepsilon^{\mathsf{T}}{:}\mathbb{T}}) \cap \left[\mathfrak{M}_2^{\mathsf{T}}{:}(\mathcal{O}_V{:}\mathbb{T} \textcircled{>} \mathcal{O}_V{:}\mathbb{T}) \cup \overline{\varepsilon^{\mathsf{T}}{:}\mathbb{T}}\right] & \text{with } \varepsilon^{\mathsf{T}}{:}\mathbb{T} \cup \overline{\varepsilon^{\mathsf{T}}{:}\mathbb{T}} = \mathbb{T} \\
\quad = \left[\varepsilon^{\mathsf{T}}{:}\mathbb{T} \cap \mathfrak{M}_2^{\mathsf{T}}{:}(\mathcal{O}_V{:}\mathbb{T} \textcircled{>} \mathcal{O}_V{:}\mathbb{T})\right] \cup \overline{\varepsilon^{\mathsf{T}}{:}\mathbb{T}} & \text{distributivity} \\
\quad = \left[\varepsilon^{\mathsf{T}} \cap \mathfrak{M}_2^{\mathsf{T}}{:}(\mathcal{O}_V{:}\mathbb{T} \textcircled{>} \mathcal{O}_V{:}\mathbb{T})\right]{:}\mathbb{T} \cup \overline{\varepsilon^{\mathsf{T}}{:}\mathbb{T}} & \text{masking} \\
\quad = \mathfrak{M}_2^{\mathsf{T}}{:}\left[\mathfrak{M}_2{:}\varepsilon^{\mathsf{T}} \cap (\mathcal{O}_V{:}\mathbb{T} \textcircled{>} \mathcal{O}_V{:}\mathbb{T})\right]{:}\mathbb{T} \cup \overline{\varepsilon^{\mathsf{T}}{:}\mathbb{T}} & \text{destroy and append} \\
\quad = \mathfrak{M}_2^{\mathsf{T}}{:}\left[(\varepsilon^{\mathsf{T}} \textcircled{>} \varepsilon^{\mathsf{T}}) \cap (\mathcal{O}_V{:}\mathbb{T} \textcircled{>} \mathcal{O}_V{:}\mathbb{T})\right]{:}\mathbb{T} \cup \overline{\varepsilon^{\mathsf{T}}{:}\mathbb{T}} & \text{Proposition 4.3.2.i} \\
\quad = \mathfrak{M}_2^{\mathsf{T}}{:}\left[\pi{:}\varepsilon^{\mathsf{T}} \cap \rho{:}\varepsilon^{\mathsf{T}} \cap \pi{:}\mathcal{O}_V{:}\mathbb{T} \cap \rho{:}\mathcal{O}_V{:}\mathbb{T}\right]{:}\mathbb{T} \cup \overline{\varepsilon^{\mathsf{T}}{:}\mathbb{T}} & \text{expanded} \\
\quad = \mathfrak{M}_2^{\mathsf{T}}{:}(\varepsilon^{\mathsf{T}} \cap \mathcal{O}_V{:}\mathbb{T} \textcircled{>} \varepsilon^{\mathsf{T}} \cap \mathcal{O}_V{:}\mathbb{T}){:}\, \mathbb{T} \cup \overline{\varepsilon^{\mathsf{T}}{:}\mathbb{T}} & \text{shuffled} \\
\quad = \mathfrak{M}_2^{\mathsf{T}}{:}(\varepsilon_{\mathcal{O}}^{\mathsf{T}} \textcircled{>} \varepsilon_{\mathcal{O}}^{\mathsf{T}}){:}\, \mathbb{T} \cup \overline{\varepsilon^{\mathsf{T}}{:}\mathbb{T}} & \text{Proposition 5.3.1.i} \\
\quad \subseteq \varepsilon_{\mathcal{O}}^{\mathsf{T}}{:}\mathbb{T} \cup \overline{\varepsilon^{\mathsf{T}}{:}\mathbb{T}} = \mathcal{O}_V{:}\mathbb{T} & \text{Definition 5.3.4.iii}
\end{array}
$$

and then "Definition 5.3.2.iii $\implies$ Definition 5.3.4.iii":

$$
\begin{array}{ll}
(\varepsilon_{\mathcal{O}} \textcircled{<} \varepsilon_{\mathcal{O}}) = (\varepsilon \cap \mathbb{T}{:}\mathcal{O}_V^{\mathsf{T}} \textcircled{<} \varepsilon \cap \mathbb{T}{:}\mathcal{O}_V^{\mathsf{T}}) & \text{by definition} \\
\quad = (\varepsilon \textcircled{<} \varepsilon) \cap (\mathbb{T}{:}\mathcal{O}_V^{\mathsf{T}} \textcircled{<} \mathbb{T}{:}\mathcal{O}_V^{\mathsf{T}}) & \\
\quad \subseteq \varepsilon{:}\, \mathfrak{M}_2^{\mathsf{T}} \cap \mathbb{T}{:}\mathcal{O}_V^{\mathsf{T}}{:}\, \mathfrak{M}_2^{\mathsf{T}} & \text{Proposition 4.3.2.i and Definition 5.3.2.iii} \\
\quad = (\varepsilon \cap \mathbb{T}{:}\, \mathcal{O}_V^{\mathsf{T}}){:}\, \mathfrak{M}_2^{\mathsf{T}} = \varepsilon_{\mathcal{O}}{:}\, \mathfrak{M}_2^{\mathsf{T}}. &
\end{array}
$$

One may also find it difficult to see how to obtain $\mathtt{syq}(\varepsilon, \mathbb{T}) \subseteq \mathcal{O}_V$ of Definition 5.3.2.i, but

$$\mathbb{T} = \varepsilon_{\mathcal{O}} ; \mathbb{T} = (\varepsilon \cap \mathbb{T} ; \varepsilon_{\mathcal{O}}) ; \mathbb{T} = \varepsilon ; (\mathbb{T} \cap \varepsilon_{\mathcal{O}}^{\mathsf{T}} ; \mathbb{T}) = \varepsilon ; \varepsilon_{\mathcal{O}}^{\mathsf{T}} ; \mathbb{T}$$

allows us to choose $v := \varepsilon_{\mathcal{O}}^{\mathsf{T}} ; \mathbb{T}$, so that

$$\mathcal{O}_V = \varepsilon_{\mathcal{O}}^{\mathsf{T}} ; \mathbb{T} \cup \overline{\varepsilon^{\mathsf{T}} ; \mathbb{T}} \supseteq \mathtt{syq}(\varepsilon, \varepsilon ; \varepsilon_{\mathcal{O}}^{\mathsf{T}} ; \mathbb{T}) = \mathtt{syq}(\varepsilon, \mathbb{T}).$$

The cryptomorphy of the topology Definitions 5.2.1, 5.2.3 and 5.3.2 has, thus, slightly informally been established.

One may wonder why in Definitions 5.3.2.ii, 5.3.3.ii, and 5.3.4.ii explicit quantifications over subsets have occurred in contrast to our policy of avoiding quantifiers—at the cost of lifting to a higher algebraic level. Also here will we lift to a quantifier-free version, which is shown in the next proposition.

Proposition 5.3.5 *The requirements (ii) of Definition 5.3.2 (and correspondingly also in Definitions 5.3.3, and 5.3.4) may be replaced by considering the higher membership relation and join forming*

$$\varepsilon_1 : \mathbf{2}^X \longrightarrow \mathbf{2}^{(\mathbf{2}^X)} \qquad \mathfrak{J} := \mathtt{syq}(\varepsilon ; \varepsilon_1, \varepsilon) : \mathbf{2}^{(\mathbf{2}^X)} \longrightarrow \mathbf{2}^X,$$

according to Definition 4.2.1, by demanding

$$\overline{\mathfrak{J}^{\mathsf{T}} ; \overline{\varepsilon_1^{\mathsf{T}} ; \overline{\mathcal{O}_V}}} \subseteq \mathcal{O}_V.$$

Proof We consider points $p := \mathtt{syq}(\varepsilon_1, v) \subseteq \mathbf{2}^{(\mathbf{2}^X)}$ related to $v = \varepsilon_1 ; p \subseteq \mathcal{O}_V$ as corresponding subset. Obviously, $p \subseteq \overline{\varepsilon_1^{\mathsf{T}} ; \overline{\mathcal{O}_V}}$ is equivalent with $v = \varepsilon_1 p \subseteq \mathcal{O}_V$. The condition $\mathtt{syq}(\varepsilon, \varepsilon ; v) \subseteq \mathcal{O}_V$ of Definition 5.3.2.ii is now transformed accordingly:

$$\mathtt{syq}(\varepsilon, \varepsilon ; v) = \mathtt{syq}(\varepsilon, \varepsilon ; \varepsilon_1 ; p) = \mathtt{syq}(\varepsilon, \varepsilon ; \varepsilon_1) ; p = \mathfrak{J}^{\mathsf{T}} ; p \subseteq \mathcal{O}_V$$

□

Now, the relationship between $\mathcal{U}, \mathcal{K}$ and the vector $\mathcal{O}_V$ describing the open sets shall be investigated.

Proposition 5.3.6 *Given an open set topology $\mathcal{O}_V$, the construct*

$$\mathcal{U} := \varepsilon ; (\Omega \cap \mathcal{O}_V ; \mathbb{T}) = (\varepsilon \cap \mathbb{T} ; \mathcal{O}_V^{\mathsf{T}}) ; \Omega = \varepsilon ; \mathcal{O}_D ; \Omega = \varepsilon_{\mathcal{O}} ; \Omega$$

will be a neighborhood topology.

Proof The variants mentioned in the statement of the proposition are obvious; in what follows we focus mainly on the first. The numbering in the proof follows that of Definition 5.2.1.

i) We have $\mathbb{T} = \varepsilon{:}\mathtt{syq}(\varepsilon, \mathbb{T}) \subseteq \varepsilon{:}\mathcal{O}_V$ following Definition 5.3.2.i and may proceed with

$$\begin{aligned} \mathcal{U}{:}\mathbb{T} &= \varepsilon{:}(\Omega \cap \mathcal{O}_V{:}\mathbb{T}){:}\mathbb{T} && \text{by definition of } \mathcal{U} \\ &= \varepsilon{:}(\Omega{:}\mathbb{T} \cap \mathcal{O}_V{:}\mathbb{T}) && \text{masking} \\ &= \varepsilon{:}\mathcal{O}_V{:}\mathbb{T} && \text{since } \Omega{:}\mathbb{T} = \mathbb{T} \\ &= \mathbb{T}{:}\mathbb{T} = \mathbb{T} && \text{see above} \\ \mathcal{U} &= \varepsilon{:}(\Omega \cap \mathcal{O}_V{:}\mathbb{T}) \subseteq \varepsilon{:}\Omega = \varepsilon \end{aligned}$$

ii)

$$\mathcal{U}{:}\Omega = (\varepsilon \cap \mathbb{T}{:}\mathcal{O}_V^{\mathsf{T}}){:}\Omega{:}\Omega = \mathcal{U}$$

iii) Here it appears more convenient to use the condition on $\mathcal{O}_D$.

$$\begin{aligned} &(\mathcal{U} \otimes \mathcal{U}){:}\,\mathfrak{M}_2 \\ &= (\varepsilon{:}\mathcal{O}_D{:}\Omega \otimes \varepsilon{:}\mathcal{O}_D{:}\Omega){:}\,\mathfrak{M}_2 && \text{by definition} \\ &= (\varepsilon \otimes \varepsilon){:}\,(\mathcal{O}_D{:}\Omega \otimes \mathcal{O}_D{:}\Omega){:}\,\mathfrak{M}_2 && \text{sharply factorized, Corollary 3.2.3} \\ &= (\varepsilon \otimes \varepsilon){:}\,(\mathcal{O}_D \otimes \mathcal{O}_D){:}\,(\Omega \otimes \Omega){:}\,\mathfrak{M}_2 && \text{since } (\mathcal{O}_D \otimes \mathcal{O}_D) \text{ is univalent} \\ &= (\varepsilon \otimes \varepsilon){:}\,(\mathcal{O}_D \otimes \mathcal{O}_D){:}\,\mathfrak{M}_2{:}\Omega && \text{Proposition 4.3.2.ii} \\ &\subseteq (\varepsilon \otimes \varepsilon){:}\,\mathfrak{M}_2{:}\mathcal{O}_D{:}\Omega && \text{Definition 5.3.3.iii} \\ &= \varepsilon{:}\mathcal{O}_D{:}\Omega && \text{Proposition 4.3.2.i} \\ &= \mathcal{U} && \text{by definition} \end{aligned}$$

iv)

$$\begin{aligned} &\mathcal{U}{:}\mathtt{syq}(\varepsilon, \mathcal{U}) \\ &= \varepsilon{:}(\Omega \cap \mathcal{O}_V{:}\mathbb{T}){:}\mathtt{syq}(\varepsilon, \mathcal{U}) && \text{by definition} \\ &= \varepsilon{:}(\Omega{:}\mathtt{syq}(\varepsilon, \mathcal{U}) \cap \mathcal{O}_V{:}\mathbb{T}) && \text{masking} \\ &= \varepsilon{:}(\overline{\varepsilon^{\mathsf{T}}{:}\overline{\varepsilon}}{:}\mathtt{syq}(\varepsilon, \mathcal{U}) \cap \mathcal{O}_V{:}\mathbb{T}) && \text{definition of } \Omega \\ &= \varepsilon{:}(\overline{\varepsilon^{\mathsf{T}}{:}\overline{\varepsilon{:}\mathtt{syq}(\varepsilon, \mathcal{U})}} \cap \mathcal{O}_V{:}\mathbb{T}) && \text{the } \mathtt{syq} \text{ is a transposed mapping} \\ &= \varepsilon{:}(\overline{\varepsilon^{\mathsf{T}}{:}\overline{\mathcal{U}}} \cap \mathcal{O}_V{:}\mathbb{T}) && \text{property of the symmetric quotient} \\ &= \varepsilon{:}(\overline{\varepsilon^{\mathsf{T}}{:}\overline{\varepsilon{:}(\Omega \cap \mathcal{O}_V{:}\mathbb{T})}} \cap \mathcal{O}_V{:}\mathbb{T}) && \text{expanded} \\ &\supseteq \varepsilon{:}(\Omega \cap \mathcal{O}_V{:}\mathbb{T} \cap \mathcal{O}_V{:}\mathbb{T}) && \text{Schröder rule} \\ &= \mathcal{U} && \text{by definition} \end{aligned}$$

□

One may also go from $\mathcal{U}$ (always connected with its $\mathcal{K}$) to $\mathcal{O}_V$:

Proposition 5.3.7 *Given any neighborhood topology $\mathcal{U}$ together with its kernel mapping $\mathcal{K}$, the construct $\mathcal{O}_V := \mathcal{K}^{\mathsf{T}}{:}\mathbb{T}$ is an open set topology.*

Proof The numbering follows Definition 5.3.2.

i)

$$
\begin{aligned}
&\mathcal{O}_V = \mathcal{K}^\top{;}\mathbb{T} = \mathtt{syq}(\varepsilon,\mathcal{U}){;}\mathbb{T} = (\overline{\overline{\varepsilon}^\top{;}\mathcal{U}} \cap \overline{\varepsilon^\top{;}\overline{\mathcal{U}}}){;}\mathbb{T}\\
&\supseteq (\mathbb{I} \cap \overline{\varepsilon^\top{;}\overline{\mathcal{U}}}){;}\mathbb{T} \quad \text{since } \mathbb{I} \subseteq \Omega^\top = \overline{\overline{\varepsilon}^\top{;}\varepsilon} \subseteq \overline{\overline{\varepsilon}^\top{;}\mathcal{U}} \text{ follows from } \mathcal{U} \subseteq \varepsilon\\
&= \overline{(\mathbb{I} \cap \varepsilon^\top{;}\overline{\mathcal{U}}){;}\mathbb{T}} \quad \text{because } (\mathbb{I} \cap \overline{\Delta}){;}\mathbb{T} = \overline{(\mathbb{I} \cap \Delta){;}\mathbb{T}} \text{ for any homogeneous } \Delta\\
&\supseteq \overline{\varepsilon^\top{;}\overline{\mathcal{U}}{;}\mathbb{T}} \supseteq \overline{\varepsilon^\top{;}\mathbb{T}} = \mathtt{syq}(\varepsilon, \perp\!\!\!\perp)
\end{aligned}
$$

In order to prove the second inclusion, we introduce g as notation for the point $g := \mathtt{syq}(\varepsilon, \mathbb{T})$ and start showing

$$\Omega{;}g = \overline{\varepsilon^\top{;}\overline{\varepsilon{;}g}} = \overline{\varepsilon^\top{;}\overline{\varepsilon{;}\mathtt{syq}(\varepsilon, \mathbb{T})}} = \overline{\varepsilon^\top{;}\overline{\mathbb{T}}} = \overline{\varepsilon^\top{;}\perp\!\!\!\perp} = \mathbb{T}.$$

Now we get $g \subseteq \mathcal{O}_V$:

$$
\begin{aligned}
g &= \mathtt{syq}(\varepsilon, \mathbb{T}) = \mathtt{syq}(\varepsilon,\mathcal{U}{;}\mathbb{T}) = \mathtt{syq}(\varepsilon,\mathcal{U}{;}\Omega{;}g)\\
&= \mathtt{syq}(\varepsilon,\mathcal{U}{;}g) = \mathtt{syq}(\varepsilon,\mathcal{U}){;}g = \mathcal{K}^\top{;}g \subseteq \mathcal{K}^\top{;}\mathbb{T}
\end{aligned}
$$

ii) We prove in advance that $v \subseteq \mathcal{O}_V = \mathcal{K}^\top{;}\mathbb{T}$ implies $\mathcal{K}^\top{;}v = v$:

$$v = \mathcal{K}^\top{;}\mathbb{T} \cap v \subseteq (\mathcal{K}^\top \cap v{;}\mathbb{T}){;}(\mathbb{T} \cap \mathcal{K}{;}v) \subseteq \mathcal{K}^\top{;}\mathcal{K}{;}v \subseteq v,$$

since $\mathcal{K}$ is univalent, i.e. an equality. Therefore with idempotency

$$\mathcal{K}^\top{;}v = \mathcal{K}^\top{;}\mathcal{K}^\top{;}\mathcal{K}{;}v = \mathcal{K}^\top{;}\mathcal{K}{;}v = v.$$

Now follows $\mathcal{U}{;}v = \varepsilon{;}\mathcal{K}^\top{;}v = \varepsilon{;}v$, so that $\mathtt{syq}(\varepsilon,\mathcal{U}{;}v) = \mathtt{syq}(\varepsilon, \varepsilon{;}v) =: p$, which is necessarily a point; it represents the union in the powerset. For p, we prove

$$\varepsilon{;}p = \varepsilon{;}\mathtt{syq}(\varepsilon, \varepsilon{;}v) = \varepsilon{;}v = \mathcal{U}{;}v \subseteq \mathcal{U}{;}\Omega{;}p = \mathcal{U}{;}p \subseteq \varepsilon{;}p,$$

using

$$\varepsilon{;}v \subseteq \varepsilon{;}p \quad\iff\quad \varepsilon^\top{;}\overline{\varepsilon{;}p} \subseteq \overline{v} \quad\iff\quad v \subseteq \Omega{;}p = \overline{\varepsilon^\top{;}\overline{\varepsilon}}{;}p = \overline{\varepsilon^\top{;}\overline{\varepsilon{;}p}}.$$

This allows us to reason

$$
\begin{aligned}
p &= \mathtt{syq}(\varepsilon, \varepsilon{;}v) = \mathtt{syq}(\varepsilon, \varepsilon{;}\mathcal{K}^\top{;}v) = \mathtt{syq}(\varepsilon,\mathcal{U}{;}v)\\
&= \mathtt{syq}(\varepsilon,\mathcal{U}{;}p) = \mathtt{syq}(\varepsilon,\mathcal{U}){;}p = \mathcal{K}^\top{;}p.
\end{aligned}
$$

In total, we have shown that $v \subseteq \mathcal{O}_V$ implies $\mathtt{syq}(\varepsilon, \varepsilon{;}v) \subseteq \mathcal{O}_V$.

iii)

$$\begin{aligned}\mathfrak{M}_2^{\mathsf{T}};(\mathcal{O}_V \otimes \mathcal{O}_V) &= \mathfrak{M}_2^{\mathsf{T}};(\mathcal{K}^{\mathsf{T}};\mathbb{T} \otimes \mathcal{K}^{\mathsf{T}};\mathbb{T}) \\ &= \mathfrak{M}_2^{\mathsf{T}};(\mathcal{K}^{\mathsf{T}} \otimes \mathcal{K}^{\mathsf{T}});\mathbb{T} \quad \text{Proposition 3.1.6.i} \\ &= \mathcal{K}^{\mathsf{T}};\mathfrak{M}_2^{\mathsf{T}};\mathbb{T} \quad \text{Definition 5.2.3.iii} \\ &\subseteq \mathcal{K}^{\mathsf{T}};\mathbb{T} = \mathcal{O}_V\end{aligned}$$

□

Having established the interrelationship, we proceed proving some additional formulae that quite intuitively characterize the different aspects of a topology.

Proposition 5.3.8

i) $\mathcal{O}_D = \mathcal{K}^{\mathsf{T}};\mathcal{K}$ $\quad \varepsilon_{\mathcal{O}} = \varepsilon;\mathcal{K}^{\mathsf{T}};\mathcal{K} = \mathcal{U};\mathcal{K}$

ii) $\mathcal{K};\omega = \mathcal{K}$ $\quad \omega;\mathcal{K} = \omega$ $\quad \mathbb{T};\omega = \mathbb{T};\mathcal{K}$

iii) $\varepsilon_{\mathcal{O}} = \mathcal{U};\omega^{\mathsf{T}}$

iv) $\varepsilon_{\mathcal{O}};\omega = \varepsilon_{\mathcal{O}} = \varepsilon_{\mathcal{O}};\omega^{\mathsf{T}}$

v) $\omega;\omega = \omega$

vi) $\omega^{\mathsf{T}};\omega = \mathcal{K}^{\mathsf{T}};\mathcal{K} = \omega \cap \omega^{\mathsf{T}};\mathbb{T}$

vii) $\varepsilon_{\mathcal{O}};\mathcal{K}^{\mathsf{T}} = \mathcal{U}$

viii) $(\omega \otimes \omega);\mathfrak{M}_2 = \mathtt{syq}((\varepsilon_{\mathcal{O}} \otimes \varepsilon_{\mathcal{O}}),\varepsilon)$

Proof

i) The first follows from the definition $\mathcal{O}_V := \mathcal{K}^{\mathsf{T}};\mathbb{T}$. The second:

$$\varepsilon_{\mathcal{O}} = \varepsilon;\mathcal{O}_D = \varepsilon;\mathcal{K}^{\mathsf{T}};\mathcal{K}$$

ii) We easily observe $\mathcal{K}^{\mathsf{T}};\mathcal{K};\mathcal{K}^{\mathsf{T}} = \mathcal{K}^{\mathsf{T}}$, so that

$$\begin{aligned}\mathcal{K};\omega &= \mathcal{K};\mathtt{syq}(\varepsilon_{\mathcal{O}},\varepsilon) = \mathtt{syq}(\varepsilon_{\mathcal{O}};\mathcal{K}^{\mathsf{T}},\varepsilon) \\ &= \mathtt{syq}(\varepsilon;\mathcal{K}^{\mathsf{T}};\mathcal{K};\mathcal{K}^{\mathsf{T}},\varepsilon) = \mathtt{syq}(\varepsilon;\mathcal{K}^{\mathsf{T}},\varepsilon) = \mathtt{syq}(\mathcal{U},\varepsilon) = \mathcal{K}.\end{aligned}$$

For the second statement, we prove just $\omega;\mathcal{K} \subseteq \omega$ from which equality follows since $\omega;\mathcal{K}$ as well as ω are mappings. Via shunting, $\omega;\mathcal{K} \subseteq \omega$ is equivalent with

$$\mathtt{syq}(\varepsilon_{\mathcal{O}},\varepsilon) = \omega \subseteq \omega;\mathcal{K}^{\mathsf{T}} = \mathtt{syq}(\varepsilon_{\mathcal{O}},\varepsilon);\mathcal{K}^{\mathsf{T}} = \mathtt{syq}(\varepsilon_{\mathcal{O}},\varepsilon;\mathcal{K}^{\mathsf{T}}) = \mathtt{syq}(\varepsilon_{\mathcal{O}},\mathcal{U}).$$

Expanding the outer symmetric quotients, we use $\mathcal{U} \subseteq \varepsilon$ to find out that it suffices to prove

$$\varepsilon_{\mathcal{O}}^{\mathsf{T}};\overline{\mathcal{U}} \subseteq \varepsilon_{\mathcal{O}}^{\mathsf{T}};\overline{\varepsilon}$$

which follows from

$$\begin{aligned}\varepsilon_{\mathcal{O}}; \overline{\varepsilon_{\mathcal{O}}^{\mathsf{T}}; \overline{\varepsilon}} &= \varepsilon; \mathcal{K}^{\mathsf{T}}; \mathcal{K}; \overline{\mathcal{K}^{\mathsf{T}}; \mathcal{K}; \varepsilon^{\mathsf{T}}; \overline{\varepsilon}} = \varepsilon; \mathcal{K}^{\mathsf{T}}; \overline{\mathcal{K}; \mathcal{K}^{\mathsf{T}}; \mathcal{K}; \varepsilon^{\mathsf{T}}; \overline{\varepsilon}} = \varepsilon; \mathcal{K}^{\mathsf{T}}; \overline{\mathcal{K}; \varepsilon^{\mathsf{T}}; \overline{\varepsilon}} \\ &= \varepsilon; \mathcal{K}^{\mathsf{T}}; \mathcal{K}; \overline{\varepsilon^{\mathsf{T}}; \overline{\varepsilon}} = \varepsilon; \mathcal{K}^{\mathsf{T}}; \mathcal{K}; \Omega = \varepsilon_{\mathcal{O}}; \Omega \subseteq \mathcal{U}; \Omega = \mathcal{U}.\end{aligned}$$

$$\mathbb{T}; \omega = \mathbb{T}; \omega; \mathcal{K} \subseteq \mathbb{T}; \mathcal{K} \qquad\qquad \mathbb{T}; \mathcal{K} = \mathbb{T}; \mathcal{K}; \omega \subseteq \mathbb{T}; \omega$$

iii) $\mathcal{U}; \omega^{\mathsf{T}} = \varepsilon; \mathcal{K}^{\mathsf{T}}; \omega^{\mathsf{T}} = \varepsilon; \omega^{\mathsf{T}} = \varepsilon; \mathtt{syq}(\varepsilon, \varepsilon_{\mathcal{O}}) = \varepsilon_{\mathcal{O}}$ using the second of (ii)
iv) $\varepsilon_{\mathcal{O}}; \omega = \mathcal{U}; \mathcal{K}; \omega = \mathcal{U}; \mathcal{K} = \varepsilon_{\mathcal{O}}$ employing (i,ii)

$$\begin{aligned}\varepsilon_{\mathcal{O}} &\subseteq \varepsilon_{\mathcal{O}}; \omega; \omega^{\mathsf{T}} && \omega \text{ is a mapping} \\ &= \varepsilon_{\mathcal{O}}; \omega^{\mathsf{T}} && \text{preceding result} \\ &= \varepsilon; \mathcal{K}^{\mathsf{T}}; \mathcal{K}; \omega^{\mathsf{T}} \\ &\subseteq \varepsilon; \omega^{\mathsf{T}} = \varepsilon_{\mathcal{O}} && \text{Proposition 5.3.1.iii}\end{aligned}$$

v)

$$\begin{aligned}\omega; \omega &= \omega; \mathtt{syq}(\varepsilon_{\mathcal{O}}, \varepsilon) = \mathtt{syq}(\varepsilon_{\mathcal{O}}; \omega^{\mathsf{T}}, \varepsilon) \\ &= \mathtt{syq}(\varepsilon_{\mathcal{O}}, \varepsilon) \quad \text{(iv)} \\ &= \omega\end{aligned}$$

vi) $\mathcal{K}^{\mathsf{T}}; \mathcal{K} = \omega^{\mathsf{T}}; \mathcal{K}^{\mathsf{T}}; \mathcal{K}; \omega \subseteq \omega^{\mathsf{T}}; \omega = \mathcal{K}^{\mathsf{T}}; \omega^{\mathsf{T}}; \omega; \mathcal{K} \subseteq \mathcal{K}^{\mathsf{T}}; \mathcal{K}$ using (ii) twice
The second equality follows with Proposition 2.1.4.i since $\omega^{\mathsf{T}}; \omega \subseteq \mathbb{I}$:

$$\omega^{\mathsf{T}}; \omega; \mathtt{syq}(\varepsilon, \varepsilon) = \omega^{\mathsf{T}}; \omega; \mathbb{T} \cap \mathtt{syq}(\varepsilon; \omega^{\mathsf{T}}; \omega, \varepsilon).$$

Now, ω is total and $\mathtt{syq}(\varepsilon, \varepsilon) = \mathbb{I}$, so that this means—observing $\varepsilon_{\mathcal{O}} = \varepsilon; \mathcal{K}^{\mathsf{T}}; \mathcal{K} = \varepsilon; \omega^{\mathsf{T}}; \omega$—in fact

$$\omega^{\mathsf{T}}; \omega = \omega^{\mathsf{T}}; \mathbb{T} \cap \omega.$$

vii)

$$\varepsilon_{\mathcal{O}}; \mathcal{K}^{\mathsf{T}} = \varepsilon; \mathcal{K}^{\mathsf{T}}; \mathcal{K}; \mathcal{K}^{\mathsf{T}} = \varepsilon; \mathcal{K}^{\mathsf{T}} = \mathcal{U}$$

viii)

$$\begin{aligned}(\omega \otimes \omega); \mathfrak{M}_2 &= (\omega \otimes \omega); \mathtt{syq}((\varepsilon \otimes \varepsilon), \varepsilon) = \mathtt{syq}((\varepsilon; \omega^{\mathsf{T}} \otimes \varepsilon; \omega^{\mathsf{T}}), \varepsilon) \\ &= \mathtt{syq}((\varepsilon_{\mathcal{O}} \otimes \varepsilon_{\mathcal{O}}), \varepsilon)\end{aligned}$$

□

Now follow some further transitions that might also be composed from preceding ones, but require slightly different techniques when executed directly.

Proposition 5.3.9 *Given the membership-in-open-sets topology $\varepsilon_{\mathcal{O}}$ according to Definition 5.3.4, one will obtain via $\mathcal{U} := \varepsilon_{\mathcal{O}} ; \varOmega$ a neighborhood topology.*

Proof

i) $\mathcal{U} ; \mathbb{T} = \varepsilon_{\mathcal{O}} ; \varOmega ; \mathbb{T} = \varepsilon_{\mathcal{O}} ; \mathbb{T} = \mathbb{T} \qquad \mathcal{U} = \varepsilon_{\mathcal{O}} ; \varOmega \subseteq \varepsilon ; \varOmega = \varepsilon$

ii) $\mathcal{U} ; \varOmega = \varepsilon_{\mathcal{O}} ; \varOmega ; \varOmega = \varepsilon_{\mathcal{O}} ; \varOmega = \mathcal{U}$ is completely trivial.

iii)
$$\begin{aligned}
(\mathcal{U} \varolessthan \mathcal{U}) &= (\varepsilon_{\mathcal{O}} ; \varOmega \varolessthan \varepsilon_{\mathcal{O}} ; \varOmega) \\
&= (\varepsilon_{\mathcal{O}} \varolessthan \varepsilon_{\mathcal{O}}) ; (\varOmega \otimes \varOmega) && \text{sharp factorization according to Corollary 3.2.3} \\
&\subseteq \varepsilon_{\mathcal{O}} ; \mathfrak{M}_2^{\mathsf{T}} ; (\varOmega \otimes \varOmega) && \text{Definition 5.3.4} \\
&\subseteq \varepsilon_{\mathcal{O}} ; \mathfrak{M}_2^{\mathsf{T}} ; \mathfrak{M}_2 ; \varOmega ; \mathfrak{M}_2^{\mathsf{T}} && \text{Proposition 4.3.8.iv} \\
&\subseteq \varepsilon_{\mathcal{O}} ; \varOmega ; \mathfrak{M}_2^{\mathsf{T}} = \mathcal{U} ; \mathfrak{M}_2^{\mathsf{T}}
\end{aligned}$$

iv)
$$\begin{aligned}
\mathcal{U} &= \varepsilon_{\mathcal{O}} ; \varOmega = (\varepsilon \cap \mathbb{T} ; \varepsilon_{\mathcal{O}}) ; \varOmega \\
&= \varepsilon ; (\varOmega \cap \varepsilon_{\mathcal{O}}^{\mathsf{T}} ; \mathbb{T}) && \text{using Definition 5.3.4.i, masking} \\
&= \varepsilon ; (\overline{\varOmega \cap \varepsilon_{\mathcal{O}}^{\mathsf{T}} ; \mathbb{T}} \cap \varepsilon_{\mathcal{O}}^{\mathsf{T}} ; \mathbb{T}) \\
&\subseteq \varepsilon ; (\overline{\varepsilon^{\mathsf{T}} ; \overline{\varepsilon ; (\varOmega \cap \varepsilon_{\mathcal{O}}^{\mathsf{T}} ; \mathbb{T})}} \cap \varepsilon_{\mathcal{O}}^{\mathsf{T}} ; \mathbb{T}) && \text{since } \varepsilon ; (\varOmega \cap \varepsilon_{\mathcal{O}}^{\mathsf{T}} ; \mathbb{T}) \subseteq \varepsilon ; (\varOmega \cap \varepsilon_{\mathcal{O}}^{\mathsf{T}} ; \mathbb{T}) \\
&= \varepsilon ; (\overline{\varepsilon^{\mathsf{T}} ; \overline{\mathcal{U}}} \cap \varepsilon_{\mathcal{O}}^{\mathsf{T}} ; \mathbb{T}) && \text{see first lines of this proof} \\
&= (\varepsilon \cap \mathbb{T} ; \varepsilon_{\mathcal{O}}) ; \overline{\varepsilon^{\mathsf{T}} ; \overline{\mathcal{U}}} && \text{masking} \\
&= \varepsilon_{\mathcal{O}} ; \overline{\varepsilon^{\mathsf{T}} ; \overline{\mathcal{U}}} && \text{Definition 5.3.4.i again} \\
&\subseteq \varepsilon_{\mathcal{O}} ; \varOmega ; \overline{\varepsilon^{\mathsf{T}} ; \overline{\mathcal{U}}} \\
&= \mathcal{U} ; \overline{\varepsilon^{\mathsf{T}} ; \overline{\mathcal{U}}} && \text{by definition}
\end{aligned}$$

□

As a further transition, we consider that from $\mathcal{O}_D$ to $\mathcal{U}$.

Proposition 5.3.10 *Given an open diagonal topology $\mathcal{O}_D$, the construct*

$$\mathcal{U} := \varepsilon ; \mathcal{O}_D ; \varOmega$$

constitutes a neighborhood topology.

Proof We follow the numbering of Definition 5.2.1.

i)

$$\mathcal{U} ; \mathbb{T} = \varepsilon ; \mathcal{O}_D ; \varOmega ; \mathbb{T} = \varepsilon ; \mathcal{O}_D ; \mathbb{T} \supseteq \varepsilon ; \mathtt{syq}(\varepsilon, \mathbb{T}) = \mathbb{T} \quad \text{using Definition 5.3.3.i}$$
$$\mathcal{U} = \varepsilon ; \mathcal{O}_D ; \varOmega \subseteq \varepsilon ; \varOmega = \varepsilon \quad \text{since } \mathcal{O}_D \text{ is a partial identity}$$

ii)

$$\mathcal{U} ; \varOmega = \varepsilon ; \mathcal{O}_D ; \varOmega ; \varOmega = \varepsilon ; \mathcal{O}_D ; \varOmega = \mathcal{U} \quad \text{is trivial}$$

iii)

$$
\begin{aligned}
&(\mathcal{U} \olessthan \mathcal{U}) ; \mathfrak{M}_2 \\
&= (\varepsilon ; \mathcal{O}_D ; \Omega \olessthan \varepsilon ; \mathcal{O}_D ; \Omega) ; \mathfrak{M}_2 \\
&= (\varepsilon \olessthan \varepsilon) ; (\mathcal{O}_D \otimes \mathcal{O}_D) ; (\Omega \otimes \Omega) ; \mathfrak{M}_2 && \text{sharply factorized} \\
&= (\varepsilon \olessthan \varepsilon) ; (\mathcal{O}_D \otimes \mathcal{O}_D) ; \mathfrak{M}_2 ; \Omega && \text{Proposition 4.3.8.iv} \\
&\subseteq (\varepsilon \olessthan \varepsilon) ; \mathfrak{M}_2 ; \mathcal{O}_D ; \Omega && \text{Proposition 5.3.3.iii} \\
&= \varepsilon ; \mathcal{O}_D ; \Omega && \text{Proposition 4.3.2.i} \\
&= \mathcal{U}
\end{aligned}
$$

iv) We start with the trivial fact

$$\varepsilon ; \mathcal{O}_D ; \Omega \subseteq \varepsilon ; \mathcal{O}_D ; \Omega \iff \varepsilon^{\mathsf{T}} ; \overline{\varepsilon ; \mathcal{O}_D ; \Omega} \subseteq \overline{\mathcal{O}_D ; \Omega} \iff \mathcal{O}_D ; \Omega \subseteq \overline{\varepsilon^{\mathsf{T}} ; \overline{\varepsilon ; \mathcal{O}_D ; \Omega}}.$$

This allows to estimate as follows:

$$
\begin{aligned}
\mathcal{U} &= \varepsilon ; \mathcal{O}_D ; \Omega = \varepsilon ; \mathcal{O}_D ; \mathcal{O}_D ; \Omega \subseteq \varepsilon ; \mathcal{O}_D ; \overline{\varepsilon^{\mathsf{T}} ; \overline{\varepsilon ; \mathcal{O}_D ; \Omega}} \\
&\subseteq \varepsilon ; \mathcal{O}_D ; \Omega ; \overline{\varepsilon^{\mathsf{T}} ; \overline{\varepsilon ; \mathcal{O}_D ; \Omega}} = \mathcal{U} ; \overline{\varepsilon^{\mathsf{T}} ; \overline{\mathcal{U}}}
\end{aligned}
$$

□

Now we investigate the reverse direction.

Proposition 5.3.11 *Given any topology via $\mathcal{U}$ or $\mathcal{K}$, we obtain an open diagonal topology with the construct $\mathcal{O}_D := \mathcal{K}^{\mathsf{T}} ; \mathcal{K}$.*

Proof

i) For the least element $\mathrm{syq}(\varepsilon, \bot\!\!\bot) =: n$ in the powerset, we have

$$n = \mathrm{syq}(\varepsilon, \bot\!\!\bot) = \overline{\varepsilon^{\mathsf{T}} ; \mathbb{T}} \subseteq \overline{\mathcal{U}^{\mathsf{T}} ; \mathbb{T}} = \overline{\mathcal{K} ; \varepsilon^{\mathsf{T}} ; \mathbb{T}} = \mathcal{K} ; \overline{\varepsilon^{\mathsf{T}} ; \mathbb{T}} = \mathcal{K} ; n.$$

This implies $n ; n^{\mathsf{T}} \subseteq \mathcal{K}$ when shunting the point n. Transposing gives $n ; n^{\mathsf{T}} \subseteq \mathcal{K}^{\mathsf{T}}$, shunting again $n \subseteq \mathcal{K}^{\mathsf{T}} ; n$, so that $n \subseteq \mathcal{K}^{\mathsf{T}} ; \mathbb{T} = \mathcal{K}^{\mathsf{T}} ; \mathcal{K} ; \mathbb{T} = \mathcal{O}_D ; \mathbb{T}$.

For the greatest element $g := \mathrm{syq}(\varepsilon, \mathbb{T}) = \overline{\overline{\varepsilon}^{\mathsf{T}} ; \mathbb{T}} : \mathbf{2}^X \longrightarrow \mathbb{1}$, we reason

$$
\begin{aligned}
\mathcal{U} ; g &= \mathcal{U} ; \Omega ; g = \mathcal{U} ; \overline{\varepsilon^{\mathsf{T}} ; \overline{\varepsilon}} ; g = \mathcal{U} ; \overline{\varepsilon^{\mathsf{T}} ; \overline{\varepsilon ; g}} = \mathcal{U} ; \overline{\varepsilon^{\mathsf{T}} ; \overline{\varepsilon ; \mathrm{syq}(\varepsilon, \mathbb{T})}} \\
&= \mathcal{U} ; \overline{\varepsilon^{\mathsf{T}} ; \overline{\mathbb{T}}} = \mathcal{U} ; \overline{\varepsilon^{\mathsf{T}} ; \bot\!\!\bot} = \mathcal{U} ; \overline{\bot\!\!\bot} = \mathcal{U} ; \mathbb{T} = \mathbb{T}
\end{aligned}
$$

followed by

$$
\begin{aligned}
g &= \mathrm{syq}(\varepsilon, \mathbb{T}) = \mathrm{syq}(\varepsilon, \mathcal{U} ; g) = \mathrm{syq}(\varepsilon, \mathcal{U}) ; g \\
&= \mathcal{K}^{\mathsf{T}} ; g \subseteq \mathcal{K}^{\mathsf{T}} ; \mathbb{T} = \mathcal{K}^{\mathsf{T}} ; \mathcal{K} ; \mathbb{T} = \mathcal{O}_D ; \mathbb{T}
\end{aligned}
$$

ii) Assuming $v \subseteq \mathcal{O}_D ; \mathbb{T} = \mathcal{K}^{\mathsf{T}} ; \mathcal{K} ; \mathbb{T}$, we get the equality $\mathcal{K}^{\mathsf{T}} ; \mathcal{K} ; v = v$, since

$$v \subseteq \mathcal{K}^{\mathsf{T}} ; \mathcal{K} ; \mathbb{T} \cap v = \mathcal{K}^{\mathsf{T}} ; \mathbb{T} \cap v \subseteq (\mathcal{K}^{\mathsf{T}} \cap v ; \mathbb{T}) ; (\mathbb{T} \cap \mathcal{K} ; v) \subseteq \mathcal{K}^{\mathsf{T}} ; \mathcal{K} ; v \subseteq v.$$

According to its definition, $e := \mathtt{syq}(\varepsilon, \varepsilon{:}v)$ is a point. Therefore, $\varepsilon{:}e = \varepsilon{:}v$, $v{:}e^\mathsf{T} \subseteq \Omega$ and finally

$$\begin{aligned}\mathcal{U}{:}v{:}e^\mathsf{T} &\subseteq \mathcal{U}{:}\Omega = \mathcal{U},\\ \mathcal{U}{:}v &\subseteq \mathcal{U}{:}e \subseteq \varepsilon{:}e = \varepsilon{:}\mathtt{syq}(\varepsilon, \varepsilon{:}v) = \varepsilon{:}v = \varepsilon{:}\mathcal{K}^\mathsf{T}{:}\mathcal{K}{:}v = \varepsilon{:}\mathcal{K}^\mathsf{T}{:}\mathcal{K}^\mathsf{T}{:}\mathcal{K}{:}v\\ &= \mathcal{U}{:}v,\end{aligned}$$

so that

$$\begin{aligned}e &= \mathtt{syq}(\varepsilon, \varepsilon{:}v) = \mathtt{syq}(\varepsilon, \mathcal{U}{:}e) = \mathtt{syq}(\varepsilon, \mathcal{U}){:}e\\ &= \mathcal{K}^\mathsf{T}{:}e \subseteq \mathcal{K}^\mathsf{T}{:}\mathbb{T} = \mathcal{K}^\mathsf{T}{:}\mathcal{K}{:}\mathbb{T} = \mathcal{O}_D{:}\mathbb{T}.\end{aligned}$$

iii) We obtain from Definition 5.2.3.iii shunting twice and transposing

$$(\mathcal{K}^\mathsf{T} \otimes \mathcal{K}^\mathsf{T}){:}\mathfrak{M}_2 \subseteq \mathfrak{M}_2{:}\mathcal{K}^\mathsf{T}$$

and may therefore reason as follows:

$(\mathcal{O}_D \otimes \mathcal{O}_D){:}\mathfrak{M}_2 = (\mathcal{K}^\mathsf{T}{:}\mathcal{K} \otimes \mathcal{K}^\mathsf{T}{:}\mathcal{K}){:}\mathfrak{M}_2$ by definition
$= (\mathcal{K}^\mathsf{T} \otimes \mathcal{K}^\mathsf{T}){:}(\mathcal{K} \otimes \mathcal{K}){:}\mathfrak{M}_2$ by sharp factorization according to Corollary 3.2.4.i setting $P := S := \mathcal{K}^\mathsf{T},\ Q := \mathbb{I},\ A := \mathcal{K}{:}\pi^\mathsf{T},\ B := \mathcal{K}{:}\rho^\mathsf{T}$
$= (\mathcal{K}^\mathsf{T} \otimes \mathcal{K}^\mathsf{T}){:}\mathfrak{M}_2{:}\mathcal{K}$ Definition 5.2.3.iii
$\subseteq \mathfrak{M}_2{:}\mathcal{K}^\mathsf{T}{:}\mathcal{K}$ see above
$= \mathfrak{M}_2{:}\mathcal{O}_D$

□

Of course, there are also all the widely symmetric concepts, namely

- the closure map $\mathcal{H} := \mathcal{N}{:}\mathcal{K}{:}\mathcal{N}$
- the closed sets diagonal $\mathcal{C}_D := \mathcal{H}^\mathsf{T}{:}\mathcal{H} = \mathcal{H} \cap \mathbb{I}$,
- the closed sets vector $\mathcal{C}_V := \mathcal{C}_D{:}\mathbb{T}$,
- the membership restricted to closed sets $\varepsilon_\mathcal{C} := \varepsilon \cap \mathbb{T}{:}\mathcal{C}_V^\mathsf{T} = \overline{\mathcal{U}}{:}\mathcal{K}{:}\mathcal{N}$.

As an example of how this might be used, we formulate what it means for a topology (given with membership-in-open-sets as well with membership-in-closed-sets relations $\varepsilon_\mathcal{O}, \varepsilon_\mathcal{C}$), to be *totally disconnected*. We may characterize this property by requiring

$$\overline{\mathbb{I}} \subseteq (\varepsilon_\mathcal{O} \cap \varepsilon_\mathcal{C}){:}\overline{\varepsilon}^\mathsf{T}.$$

This means that for any two distinct points there is a clopen set containing the first and its complement containing the second. Engelking [Eng78] lists further concepts of being disconnected.

5.4 Interior and Closure

Having learned how to proceed from a powerset element $a \in \mathbf{2}^X$ to its open kernel element via $\mathcal{K} : a \mapsto a^\circ$, we set this in correspondence with the well-known concept of the interior A° of some subset $A \subseteq X$ in a topological space. Or else: we study when a point x is an inner point of A. Interpreting it in plain words, we may say that x is contained in the interior A°, when it has a neighborhood u which is fully (i.e. with all its points p) contained in the given A, or lifting it gradually, if

$$A^\circ_x = \exists u : \mathcal{U}_{xu} \wedge [\forall p : \varepsilon_{pu} \rightarrow A_p] = \exists u : \mathcal{U}_{xu} \wedge \overline{\exists p : \varepsilon^\mathsf{T}_{up} \wedge \overline{A}_p} = [\mathcal{U} ; \overline{\varepsilon^\mathsf{T} ; \overline{A}}]_x.$$

Correspondingly for the closure A^- of some set A. An element belongs to the closure, when in every neighborhood some element p of A may be found:

$$A^-_x = \forall u : \mathcal{U}_{xu} \rightarrow [\exists p : \varepsilon_{pu} \wedge A_p] = \overline{\exists u : \mathcal{U}_{xu} \wedge \overline{\exists p : \varepsilon^\mathsf{T}_{up} \wedge A_p}} = [\overline{\mathcal{U} ; \overline{\varepsilon^\mathsf{T} ; A}}]_x$$

That is, we study when a point x is a tangent point or an accumulation point of a subset A.

Classically, one defines the *interior* A° of some subset $A \subseteq X$ as the largest open subset therein; respectively the closure as the smallest closed superset.

Proposition 5.4.1 *Given a neighborhood topology $\mathcal{U} : X \longrightarrow \mathbf{2}^X$ and any subset $A \subseteq X$, we obtain the largest open set contained in A, the interior, as*

$$A^\circ = \mathcal{U} ; \overline{\varepsilon^\mathsf{T} ; \overline{A}}; \qquad A^- = \overline{\mathcal{U} ; \overline{\varepsilon^\mathsf{T} ; A}}$$

correspondingly for the smallest closed set containing A, the closure of A.

Proof As always in our setting, the element or point a in the powerset and the subset A are related as follows

$$A = \varepsilon ; a \qquad a = \mathtt{syq}(\varepsilon, A).$$

First we apply the mapping $\mathcal{K}$ of a to the open kernel and obtain

$$\begin{aligned} a^\circ = \mathcal{K}^\mathsf{T} ; a &= \mathtt{syq}(\varepsilon, \mathcal{U}) ; a = \mathtt{syq}(\varepsilon, \mathcal{U} ; \Omega) ; a \\ &= \mathtt{syq}(\varepsilon, \mathcal{U} ; \overline{\varepsilon^\mathsf{T} ; \overline{\varepsilon}}) ; a = \mathtt{syq}(\varepsilon, \mathcal{U} ; \overline{\varepsilon^\mathsf{T} ; \overline{\varepsilon ; a}}) = \mathtt{syq}(\varepsilon, \mathcal{U} ; \overline{\varepsilon^\mathsf{T} ; \overline{A}}); \end{aligned}$$

afterwards we look for the corresponding vector, i.e. subset

$$A^\circ = \varepsilon ; a^\circ = \varepsilon ; \mathtt{syq}(\varepsilon, \mathcal{U} ; \overline{\varepsilon^\mathsf{T} ; \overline{A}}) = \mathcal{U} ; \overline{\varepsilon^\mathsf{T} ; \overline{A}};$$

similarly for A^-. Then with membership deletion

$$A^\circ = \mathcal{U};\overline{\varepsilon^\mathsf{T};\overline{A}} \subseteq \varepsilon;\overline{\varepsilon^\mathsf{T};\overline{A}} = A \qquad \text{and} \qquad A^- = \overline{\mathcal{U};\overline{\varepsilon^\mathsf{T};A}} \supseteq \overline{\varepsilon;\overline{\varepsilon^\mathsf{T};A}} = A.$$

Subsets X are considered as open when $X^\circ = X$. From $X \subseteq A$ and $X^\circ = X$ therefore follows with monotony $X = X^\circ \subseteq A^\circ$, making A° the greatest open subset of A. Correspondingly for the closure. □

We may also consider the production of the interior (or open kernel) and the closure for all subsets simultaneously by applying the operations $^\circ$ and $^-$ to the columns of the membership relation ε, i.e., obtaining

$$\begin{aligned}
\varepsilon^\circ &= \mathcal{U};\overline{\varepsilon^\mathsf{T};\overline{\varepsilon}} = \mathcal{U};\Omega = \mathcal{U} \\
\varepsilon^- &= \overline{\mathcal{U};\overline{\varepsilon^\mathsf{T};\varepsilon}} = \overline{\mathcal{U};\overline{\varepsilon^\mathsf{T};\varepsilon}};\mathcal{N};\mathcal{N} = \overline{\mathcal{U};\overline{\varepsilon^\mathsf{T};\varepsilon;\mathcal{N}}};\mathcal{N} = \overline{\mathcal{U};\overline{\varepsilon^\mathsf{T};\overline{\varepsilon}}};\mathcal{N} = \overline{\mathcal{U};\Omega};\mathcal{N} = \overline{\mathcal{U}};\mathcal{N}.
\end{aligned}$$

Indeed when looking at Fig. 5.2, e.g., every column of $\mathcal{U}$ shows just the open kernel or interior of the respective subset, i.e., $\varepsilon^\circ = U$. Furthermore, in anticipation of Chap. 7 the columns of a topological Aumann contact relation are the closure of the corresponding subset.

5.5 Separation

A major question is to which extent points or subsets may be *distinguished* or even *separated* by environments or open sets. This gave rise to several definitions which we recall here first in their traditional form: Let a topology on X be given via neighborhoods, open sets, kernel mapping as required. It is then called a

- T_0-space (sometimes a Kolmogorov space) if for any two points in X an open set exists that contains one of them but not the other, i.e., points are topologically distinguishable.
- T_1-space when any two points can be separated, i.e. if each lies in an open set which does not contain the other point:

$$\forall x, y : x \neq y \rightarrow \exists U, V \in \mathcal{O} : x \in U \wedge y \notin U \wedge y \in V \wedge x \notin V.$$

- T_2-space, i.e., a topology satisfying the Hausdorff property, when any two distinct points are contained in disjoint open sets, or when

$$\forall x, y : x \neq y \rightarrow \exists U, V \in \mathcal{O} : x \in U \wedge y \in V \wedge \emptyset = U \cap V.$$

Following our general guideline, we intend to lift these conditions to the relational level. We first discuss distinguishability of points. Any given topology

$\mathcal{U} : X \longrightarrow \mathbf{2}^X$ introduces the equivalence $\Xi := \mathtt{syq}(\mathcal{U}^\mathsf{T}, \mathcal{U}^\mathsf{T})$, the *topological non-distinguishability* of points. We convince ourselves that always

$$
\begin{aligned}
\Xi := \mathtt{syq}(\mathcal{U}^\mathsf{T}, \mathcal{U}^\mathsf{T}) &= \overline{\overline{\mathcal{U}} ; \mathcal{U}^\mathsf{T}} \cap \overline{\mathcal{U} ; \overline{\mathcal{U}}^\mathsf{T}} && \text{definition of the symmetric quotient}\\
&= \overline{\overline{\varepsilon_\mathcal{O} ; \mathcal{K}^\mathsf{T}} ; \mathcal{K} ; \varepsilon_\mathcal{O}^\mathsf{T}} \cap \overline{\varepsilon_\mathcal{O} ; \mathcal{K}^\mathsf{T} ; \overline{\mathcal{K} ; \varepsilon_\mathcal{O}^\mathsf{T}}} && \text{Proposition 5.3.8.vii}\\
&= \overline{\overline{\varepsilon_\mathcal{O}} ; \mathcal{K}^\mathsf{T} ; \mathcal{K} ; \varepsilon_\mathcal{O}^\mathsf{T}} \cap \overline{\varepsilon_\mathcal{O} ; \mathcal{K}^\mathsf{T} ; \mathcal{K} ; \overline{\varepsilon_\mathcal{O}^\mathsf{T}}}\\
&= \overline{\overline{\varepsilon_\mathcal{O}} ; \varepsilon_\mathcal{O}^\mathsf{T}} \cap \overline{\varepsilon_\mathcal{O} ; \overline{\varepsilon_\mathcal{O}^\mathsf{T}}} && \text{since } \varepsilon_\mathcal{O} ; \mathcal{K}^\mathsf{T} ; \mathcal{K} = \mathcal{U} ; \mathcal{K} = \varepsilon_\mathcal{O}\\
&= \mathtt{syq}(\varepsilon_\mathcal{O}^\mathsf{T}, \varepsilon_\mathcal{O}^\mathsf{T}).
\end{aligned}
$$

Let ξ be the natural projection according to the non-distinguishability Ξ, so that $\Xi = \xi ; \xi^\mathsf{T}$. When we divide Ξ out, thus anticipating the quotient topology discussed later in Sect. 6.1, it will satisfy the T_0-property. We have, namely,

$$
\begin{aligned}
&\mathtt{syq}(\varepsilon_{\mathcal{O}_\Xi}^\mathsf{T}, \varepsilon_{\mathcal{O}_\Xi}^\mathsf{T}) = \mathtt{syq}(\vartheta_{\xi^\mathsf{T}} ; \varepsilon_\mathcal{O}^\mathsf{T} ; \xi, \vartheta_{\xi^\mathsf{T}} ; \varepsilon_\mathcal{O}^\mathsf{T} ; \xi)\\
&= \mathtt{syq}(\varepsilon_\mathcal{O}^\mathsf{T} ; \xi, \varepsilon_\mathcal{O}^\mathsf{T} ; \xi) \qquad \vartheta_{\xi^\mathsf{T}} \text{ is a surjective map, Proposition 2.1.5.ii}\\
&= \xi^\mathsf{T} ; \mathtt{syq}(\varepsilon_\mathcal{O}^\mathsf{T}, \varepsilon_\mathcal{O}^\mathsf{T}) ; \xi \qquad \text{Proposition 2.1.4.iii directly and in transposed form}\\
&= \xi^\mathsf{T} ; \Xi ; \xi = \xi^\mathsf{T} ; \xi ; \xi^\mathsf{T} ; \xi = \mathbb{I}.
\end{aligned}
$$

For the following definition, we choose $\mathcal{U}$ as the most convenient one among the different topology definitions, but also $\mathcal{O}, \mathcal{K}, \varepsilon_\mathcal{O}$ might have been employed.

Definition 5.5.1 Let a topology $\mathcal{U}$ be given in relational form. It will be called a

i) T_0-**space** or a **Kolmogorov** space if $\quad \overline{\mathbb{I}} \subseteq \overline{\mathtt{syq}(\mathcal{U}^\mathsf{T}, \mathcal{U}^\mathsf{T})}$,
ii) T_1-**space** if $\quad \overline{\mathbb{I}} \subseteq \mathcal{U} ; \overline{\mathcal{U}}^\mathsf{T}$,
iii) T_2-**space** or a **Hausdorff** space if $\quad \overline{\mathbb{I}} \subseteq \mathcal{U} ; \mathcal{N} ; \mathcal{U}^\mathsf{T}$.

In all three cases inclusion means in fact equality. □

For (i), we might also have said $\mathtt{syq}(\mathcal{U}^\mathsf{T}, \mathcal{U}^\mathsf{T}) \subseteq \mathbb{I}$. Of course, we have the chain of implications

$$T_2\text{-space} \implies T_1\text{-space} \implies T_0\text{-space},$$

which can easily be proved observing

$$\mathcal{U} ; \mathcal{N} ; \mathcal{U}^\mathsf{T} \subseteq \mathcal{U} ; \mathcal{N} ; \varepsilon^\mathsf{T} = \mathcal{U} ; \overline{\varepsilon}^\mathsf{T} \subseteq \mathcal{U} ; \overline{\mathcal{U}}^\mathsf{T} \subseteq \overline{\mathtt{syq}(\mathcal{U}^\mathsf{T}, \mathcal{U}^\mathsf{T})}.$$

We establish equivalent versions using the membership-in-open-set topology definition.

Proposition 5.5.2 *A topology given as* $\mathcal{U}$*, resp.* $\varepsilon_\mathcal{O}$*, is a*

i) T_0*-space* $\iff$ $\overline{\mathbb{I}} \subseteq \overline{\mathrm{syq}(\varepsilon_\mathcal{O}^\mathsf{T}, \varepsilon_\mathcal{O}^\mathsf{T})}$,
ii) T_1*-space* $\iff$ $\overline{\mathbb{I}} \subseteq \varepsilon_\mathcal{O} ; \overline{\varepsilon}^\mathsf{T}$
iii) T_2*-space* $\iff$ $\overline{\mathbb{I}} \subseteq \varepsilon_\mathcal{O} ; \overline{\varepsilon^\mathsf{T} ; \varepsilon} ; \varepsilon_\mathcal{O}^\mathsf{T}$.

Proof

i) The T_0 case follows from the initial remark on distinguishability.
ii) For the T_1 case we have

$$\varepsilon_\mathcal{O} ; \overline{\varepsilon}^\mathsf{T} = \mathcal{U} ; \mathcal{K} ; \overline{\varepsilon}^\mathsf{T} = \mathcal{U} ; \mathrm{syq}(\mathcal{U}, \varepsilon) ; \overline{\varepsilon}^\mathsf{T} = \mathcal{U} ; \mathrm{syq}(\overline{\mathcal{U}}, \overline{\varepsilon}) ; \overline{\varepsilon}^\mathsf{T} = \mathcal{U} ; \overline{\mathcal{U}}^\mathsf{T}.$$

iii) The T_2-case is shown using that $\mathcal{U} = \varepsilon_\mathcal{O} ; \Omega$ and, obviously, $\overline{\varepsilon^\mathsf{T} ; \varepsilon} ; \Omega^\mathsf{T} = \overline{\varepsilon^\mathsf{T} ; \varepsilon}$:

$$\varepsilon_\mathcal{O} ; \overline{\varepsilon^\mathsf{T} ; \varepsilon} ; \varepsilon_\mathcal{O}^\mathsf{T} = \varepsilon_\mathcal{O} ; \Omega ; \overline{\varepsilon^\mathsf{T} ; \varepsilon} ; \Omega^\mathsf{T} ; \varepsilon_\mathcal{O}^\mathsf{T} = \mathcal{U} ; \overline{\varepsilon^\mathsf{T} ; \overline{\varepsilon} ; \mathcal{N}} ; \mathcal{U}^\mathsf{T} = \mathcal{U} ; \Omega ; \mathcal{N} ; \mathcal{U}^\mathsf{T} = \mathcal{U} ; \mathcal{N} ; \mathcal{U}^\mathsf{T}$$

□

By the way, any finite Hausdorff topology is necessarily discrete, i.e., satisfies $\mathcal{U} = \varepsilon$.

5.6 Continuity

For a mathematical structure, one routinely defines its structure-preserving mappings. Traditionally, this is often handled under the name of a homomorphism; it may be defined for relational structures as well as for algebraic ones in a more or less standard way; it is available for a homogeneous as well as for a heterogeneous structure. For topologies, however, the situation is different.

A neighborhood system requires relations between different sets (i.e. a heterogeneous setting as opposed to *homogeneous* relations *on a set*), with two neighborhood topologies $\mathcal{U}, \mathcal{U}'$ on sets X, X'. The continuity condition turns out to be a mixture of going forward and backwards as we will see (Fig. 5.7).

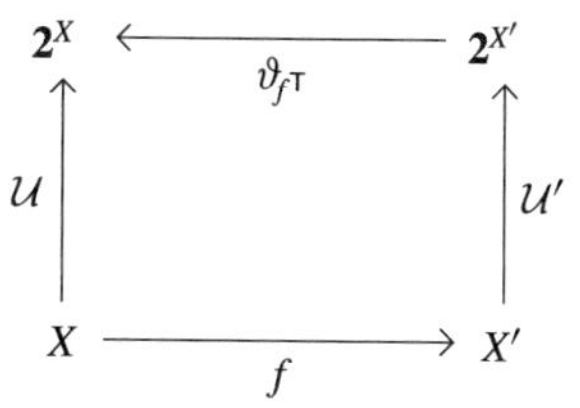

Fig. 5.7 Typing in case of the continuity condition

The standard—i.e. not yet lifted—definition of topological continuity for a neighborhood topology runs as follows: Let any two neighborhood topologies $\mathcal{U}, \mathcal{U}'$ be given on sets X, X', and consider a mapping $f : X \longrightarrow X'$. One says that f is continuous when

- for every point $p \in X$ and every neighborhood $V \in \mathcal{U}'(f(p))$, there exists a neighborhood $U \in \mathcal{U}(p)$ such that $f(U) \subseteq V$.

This definition has here only been recalled for convenience. Converting it gradually—but informally—to a point-free version, thus getting rid of quantifiers, is far from easy. Again, we *must not* execute quantification over subsets $U, V \subseteq X$ and move to quantifying over *points* $u, v \subseteq \mathbf{2}^X$ in the powerset.

For every $p \in X$ and every $V \in \mathcal{U}'(f(p))$,

there exists a $U \in \mathcal{U}(p)$ such that $f(U) \subseteq V$.

$$
\begin{aligned}
\iff\quad & \forall p \in X : \forall V \in \mathcal{U}'(f(p)) : \exists U \in \mathcal{U}(p) : f(U) \subseteq V \\
\iff\quad & \forall p \in X : \forall v \in \mathbf{2}^{X'} : \mathcal{U}'_{f(p),v} \rightarrow \big(\exists u : \mathcal{U}_{pu} \wedge \big[\forall y : \varepsilon_{yu} \rightarrow \varepsilon'_{f(y),v}\big]\big) \\
\iff\quad & \forall p : \forall v : (f;\mathcal{U}')_{pv} \rightarrow \big(\exists u : \mathcal{U}_{pu} \wedge \big[\forall y : \varepsilon_{yu} \rightarrow (f;\varepsilon')_{yv}\big]\big) \\
\iff\quad & \forall p : \forall v : (f;\mathcal{U}')_{pv} \rightarrow \big(\exists u : \mathcal{U}_{pu} \wedge \overline{\exists y : \varepsilon_{yu} \wedge \overline{(f;\varepsilon')_{yv}}}\big) \\
\iff\quad & \forall p : \forall v : (f;\mathcal{U}')_{pv} \rightarrow \big(\exists u : \mathcal{U}_{pu} \wedge \overline{\varepsilon^{\mathsf{T}};\overline{f;\varepsilon'}}_{uv}\big) \\
\iff\quad & \forall p : \forall v : (f;\mathcal{U}')_{pv} \rightarrow \big(\mathcal{U};\overline{\varepsilon^{\mathsf{T}};\overline{f;\varepsilon'}}\big)_{pv} \\
\iff\quad & f;\mathcal{U}' \subseteq \mathcal{U};\overline{\varepsilon^{\mathsf{T}};\overline{f;\varepsilon'}} \\
\iff\quad & f;\mathcal{U}' \subseteq \mathcal{U};\vartheta_{f^{\mathsf{T}}}^{\mathsf{T}}
\end{aligned}
$$

The last transition is correct since the right sides are equal:

$$
\begin{aligned}
\mathcal{U};\overline{\varepsilon^{\mathsf{T}};\overline{f;\varepsilon'}} &\subseteq \mathcal{U};\overline{\varepsilon^{\mathsf{T}};\overline{f;\varepsilon';\vartheta_{f^{\mathsf{T}}};\vartheta_{f^{\mathsf{T}}}^{\mathsf{T}}}} && \text{because } \vartheta_{f^{\mathsf{T}}} \text{ is total} \\
&= \mathcal{U};\overline{\varepsilon^{\mathsf{T}};\overline{f;\varepsilon';\mathtt{syq}(f;\varepsilon',\varepsilon);\vartheta_{f^{\mathsf{T}}}^{\mathsf{T}}}} && \text{by definition of } \vartheta_{f^{\mathsf{T}}} \\
&\subseteq \mathcal{U};\overline{\varepsilon^{\mathsf{T}};\overline{\varepsilon;\vartheta_{f^{\mathsf{T}}}^{\mathsf{T}}}} && \text{cancellation; always } A;\mathtt{syq}(A,B) \subseteq B \\
&= \mathcal{U};\overline{\varepsilon^{\mathsf{T}};\overline{\varepsilon}};\vartheta_{f^{\mathsf{T}}}^{\mathsf{T}} && \text{since } \vartheta_{f^{\mathsf{T}}} \text{ is a mapping} \\
&= \mathcal{U};\Omega;\vartheta_{f^{\mathsf{T}}}^{\mathsf{T}} = \mathcal{U};\vartheta_{f^{\mathsf{T}}}^{\mathsf{T}} && \text{Definition 5.2.1.ii} \\
&= \mathcal{U};\mathtt{syq}(\varepsilon, f;\varepsilon') \subseteq \mathcal{U};\overline{\varepsilon^{\mathsf{T}};\overline{f;\varepsilon'}}
\end{aligned}
$$

This idea is now turned into a definition.

Definition 5.6.1 Consider two neighborhood topologies $\mathcal{U} : X \longrightarrow \mathbf{2}^X$ and $\mathcal{U}' : X' \longrightarrow \mathbf{2}^{X'}$ as well as a mapping $f : X \longrightarrow X'$. We call

$$f \text{ (neighborhood-)}\textbf{continuous} \quad :\iff \quad f;\mathcal{U}' \subseteq \mathcal{U};\vartheta_{f^{\mathsf{T}}}^{\mathsf{T}}.$$

The equivalent $f;\mathcal{U}';\vartheta_{f^{\mathsf{T}}} \subseteq \mathcal{U}$ is obtained shunting the mapping $\vartheta_{f^{\mathsf{T}}}$. □

Observe that the mapping f cannot be shunted. The condition looks quite similar to a homomorphism condition, but it is definitely not a homomorphism. It allows, nevertheless, to be extended to iterated continuous mappings:

$$\mathcal{U} : X \longrightarrow \mathbf{2}^X, \qquad \mathcal{U}' : X' \longrightarrow \mathbf{2}^{X'}, \qquad \mathcal{U}'' : X'' \longrightarrow \mathbf{2}^{X''}$$
$$f : X \longrightarrow X', \qquad g : X' \longrightarrow X''$$
$$f; \mathcal{U}' \subseteq \mathcal{U}; \vartheta^{\mathsf{T}}_{f^{\mathsf{T}}}, \quad g; \mathcal{U}'' \subseteq \mathcal{U}'; \vartheta^{\mathsf{T}}_{g^{\mathsf{T}}}$$
$$\Longrightarrow f; g; \mathcal{U}'' \subseteq f; \mathcal{U}'; \vartheta^{\mathsf{T}}_{g^{\mathsf{T}}} \subseteq \mathcal{U}; \vartheta^{\mathsf{T}}_{f^{\mathsf{T}}}; \vartheta^{\mathsf{T}}_{g^{\mathsf{T}}} = \mathcal{U}; (\vartheta_{g^{\mathsf{T}}}; \vartheta_{f^{\mathsf{T}}})^{\mathsf{T}} = \mathcal{U}; (\vartheta_{g^{\mathsf{T}}f^{\mathsf{T}}})^{\mathsf{T}} = \mathcal{U}; (\vartheta_{(f;g)^{\mathsf{T}}})^{\mathsf{T}}$$

What is not possible is "rolling the condition" to the same extent as for homomorphisms—except what has been shown above wrt. to rolling based on the mapping $\vartheta_{f^{\mathsf{T}}}$ alone.[3]

We proceed defining continuity concepts for the other topology versions, and prove afterwards that they all mean the same.

Definition 5.6.2 Given sets X and X' with topologies, we consider a mapping $f : X \longrightarrow X'$ together with its inverse image mapping $\vartheta_{f^{\mathsf{T}}} : \mathbf{2}^{X'} \longrightarrow \mathbf{2}^X$. Then we say that f is

i)	(open-kernel-map-)**continuous**	$:\Longleftrightarrow$	$\mathcal{K}_2^{\mathsf{T}}; \vartheta_{f^{\mathsf{T}}} \subseteq \overline{\varepsilon_2{}^{\mathsf{T}}; f^{\mathsf{T}}; \overline{\varepsilon_1}}; \mathcal{K}_1^{\mathsf{T}}$
ii)	(open-diagonal-)**continuous**	$:\Longleftrightarrow$	$\mathcal{O}_{D_2}; \vartheta_{f^{\mathsf{T}}} \subseteq \vartheta_{f^{\mathsf{T}}}; \mathcal{O}_{D_1}$
iii)	(open-set-)**continuous**	$:\Longleftrightarrow$	$\vartheta^{\mathsf{T}}_{f^{\mathsf{T}}}; \mathcal{O}_{V_2} \subseteq \mathcal{O}_{V_1}$
iv)	(membership-in-open-sets-) **continuous**	$:\Longleftrightarrow$	$f; \varepsilon_{\mathcal{O}_2}; \vartheta_{f^{\mathsf{T}}} \subseteq \varepsilon_{\mathcal{O}_1}$

□

The second and third definition precisely meet the classical form which says that inverse images of open sets shall be open again. In the first definition, one can recognize some sort of a homomorphism with respect to the converse of kernel-forming; however not with $\vartheta_{f^{\mathsf{T}}}$ on the right side, but with a residual slightly above.

One will observe that in the following proposition first a direct equivalence is proved and afterwards four statements cyclically.

Proposition 5.6.3 *In view of the transitions between topology concepts as presented with Propositions 5.2.4, 5.3.6, 5.3.7, 5.3.9, 5.3.10, 5.3.11, the diverse continuity conditions mean all the same:*

i) (neighborhood-) ***continuous***	$\Longleftrightarrow$	*(open-kernel-map-)****continuous***
ii) (neighborhood-) ***continuous***	$\Longrightarrow$	*(open-diagonal-)****continuous***
iii) (open-diagonal-) ***continuous***	$\Longrightarrow$	*(open-set-)****continuous***
iv) (open-set-) ***continuous***	$\Longrightarrow$	*(membership-in-open-sets-)****cont.***
v) (membership-in-open-sets-) ***cont.***	$\Longrightarrow$	*(neighborhood-)****continuous***

[3] One may then wish to apply the language of simulation as explained in [dRE98] and [Sch11] Prop. 19.17, calling $\mathcal{U}'^{\mathsf{T}}$ an $f^{\mathsf{T}}, \vartheta_{f^{\mathsf{T}}}$-$L^{\mathsf{T}}$-simulation of $\mathcal{U}^{\mathsf{T}}$—or else an $\vartheta^{\mathsf{T}}_{f^{\mathsf{T}}}, f^{\mathsf{T}}$-$U$-simulation of $\mathcal{U}^{\mathsf{T}}$.

Proof

i)

$$
\begin{array}{ll}
f;\mathcal{U}_2;\vartheta_{f^\mathsf{T}} \subseteq \mathcal{U}_1 = \varepsilon_1;\mathcal{K}_1^\mathsf{T} & \text{assumption and expansion of } \mathcal{U}_1 \\
\iff \quad f;\varepsilon_2;\mathcal{K}_2^\mathsf{T};\vartheta_{f^\mathsf{T}};\mathcal{K}_1 \subseteq \varepsilon_1 & \text{expanding } \mathcal{U}_2 \text{ and shunting} \\
\iff \quad \varepsilon_2{}^\mathsf{T};f^\mathsf{T};\overline{\varepsilon_1} \subseteq \overline{\mathcal{K}_2^\mathsf{T};\vartheta_{f^\mathsf{T}};\mathcal{K}_1} & \text{Schröder rule} \\
\iff \quad \mathcal{K}_2^\mathsf{T};\vartheta_{f^\mathsf{T}};\mathcal{K}_1 \subseteq \overline{\varepsilon_2{}^\mathsf{T};f^\mathsf{T};\overline{\varepsilon_1}} & \text{negated} \\
\iff \quad \mathcal{K}_2^\mathsf{T};\vartheta_{f^\mathsf{T}} \subseteq \overline{\varepsilon_2{}^\mathsf{T};f^\mathsf{T};\overline{\varepsilon_1}};\mathcal{K}_1^\mathsf{T} & \text{shunting again}
\end{array}
$$

ii)

$$
\begin{array}{l}
\overline{\varepsilon_2^\mathsf{T};\overline{\mathcal{U}_2}} \subseteq \overline{\varepsilon_2^\mathsf{T};f^\mathsf{T};f;\overline{\mathcal{U}_2}} = \overline{\varepsilon_2^\mathsf{T};f^\mathsf{T};\overline{f;\mathcal{U}_2}} = \overline{\vartheta_{f^\mathsf{T}};\varepsilon_1^\mathsf{T};\overline{f;\mathcal{U}_2}} \\
\subseteq \overline{\vartheta_{f^\mathsf{T}};\varepsilon_1^\mathsf{T};\overline{\mathcal{U}_1;\vartheta_{f^\mathsf{T}}^\mathsf{T}}} = \vartheta_{f^\mathsf{T}};\overline{\varepsilon_1^\mathsf{T};\overline{\mathcal{U}_1}};\vartheta_{f^\mathsf{T}}^\mathsf{T} \\
\Longrightarrow \quad \mathcal{O}_{D_2} = \mathbb{I} \cap \overline{\varepsilon_2^\mathsf{T};\overline{\mathcal{U}_2}} \subseteq \vartheta_{f^\mathsf{T}};\vartheta_{f^\mathsf{T}}^\mathsf{T} \cap \vartheta_{f^\mathsf{T}};\overline{\varepsilon_1^\mathsf{T};\overline{\mathcal{U}_1}};\vartheta_{f^\mathsf{T}}^\mathsf{T} \\
\qquad = \vartheta_{f^\mathsf{T}};(\mathbb{I} \cap \overline{\varepsilon_1^\mathsf{T};\overline{\mathcal{U}_1}});\vartheta_{f^\mathsf{T}}^\mathsf{T} = \vartheta_{f^\mathsf{T}};\mathcal{O}_{D_1};\vartheta_{f^\mathsf{T}}^\mathsf{T}
\end{array}
$$

iii)

$$
\begin{array}{l}
\vartheta_{f^\mathsf{T}}^\mathsf{T};\mathcal{O}_{V_2} = \vartheta_{f^\mathsf{T}}^\mathsf{T};\mathcal{O}_{D_2};\mathbb{T} = \vartheta_{f^\mathsf{T}}^\mathsf{T};\mathcal{O}_{D_2}^\mathsf{T};\mathbb{T} \\
\qquad \subseteq \mathcal{O}_{D_1}^\mathsf{T};\vartheta_{f^\mathsf{T}}^\mathsf{T};\mathbb{T} = \mathcal{O}_{D_1};\vartheta_{f^\mathsf{T}}^\mathsf{T};\mathbb{T} \subseteq \mathcal{O}_{D_1};\mathbb{T} = \mathcal{O}_{V_1}
\end{array}
$$

iv)

$$
\begin{array}{ll}
f;\varepsilon_{\mathcal{O}_2};\vartheta_{f^\mathsf{T}} = f;(\varepsilon_2 \cap \mathbb{T};\mathcal{O}_{V_2}^\mathsf{T});\vartheta_{f^\mathsf{T}} = (f;\varepsilon_2 \cap f;\mathbb{T};\mathcal{O}_{V_2}^\mathsf{T});\vartheta_{f^\mathsf{T}} & \\
\qquad = (\varepsilon_1;\vartheta_{f^\mathsf{T}}^\mathsf{T} \cap \mathbb{T};\mathcal{O}_{V_2}^\mathsf{T});\vartheta_{f^\mathsf{T}} & \text{following Proposition 2.2.5.ii.} \\
\qquad = \varepsilon_1 \cap \mathbb{T};\mathcal{O}_{V_2}^\mathsf{T};\vartheta_{f^\mathsf{T}} & \text{destroy and append} \\
\qquad \subseteq \varepsilon_1 \cap \mathbb{T};\mathcal{O}_{V_1}^\mathsf{T} = \varepsilon_{\mathcal{O}_1} &
\end{array}
$$

v)

$$
\begin{array}{ll}
f;\mathcal{U}_2;\vartheta_{f^\mathsf{T}} = f;\varepsilon_{\mathcal{O}_2};\Omega_2;\vartheta_{f^\mathsf{T}} & \\
\qquad \subseteq f;\varepsilon_{\mathcal{O}_2};\vartheta_{f^\mathsf{T}};\Omega_1 & \text{Proposition 2.2.10.i} \\
\qquad \subseteq \varepsilon_{\mathcal{O}_1};\Omega_1 & \text{continuity condition} \\
\qquad = \mathcal{U}_1 &
\end{array}
$$

□

This is yet another situation where structure comparison mainly takes place in reverse direction, i.e. with $f^\mathsf{T}, \vartheta_{f^\mathsf{T}}$ and only the latter of the two is a mapping. "Rolling the homomorphism" may, thus, only be applied in a very restricted form.

In order to achieve completeness of information, we mention in addition that a mapping is said to be a **homeomorphism** in case it is a continuous mapping in both directions. If $f : X \longrightarrow Y$ is such a homeomorphic mapping, we say that X and Y are homeomorphic.

In this chapter we have investigated multiple equivalent definitions for a topology. The proofs of their equivalence seem longer and sometimes more sophisticated than the element-wise versions usually presented in text books. As mentioned, the algebraic proofs in this chapter add a higher level of preciseness to the argument. In addition, the theorems remain valid if we move from set-theoretic relations to non-standard models of the axioms for relations. For example, all results of this chapter also apply to certain fuzzy relations, i.e., to matrices that use a Boolean algebra as coefficients instead of $\mathbf{0}$, $\mathbf{1}$.

Exercises

Exercise 5.1 Prove that the transitions $v \mapsto \mathbb{I} \cap v;\mathbb{T}$ and $d \mapsto d;\mathbb{T}$, mapping vectors to partial identities and vice versa, are inverse to each other.

Exercise 5.2 Assume a topology $\mathcal{U} : X \longrightarrow \mathbf{2}^X$ and its corresponding open set vector $\mathcal{O}_V : \mathbf{2}^X \longrightarrow \mathbb{1}$. Prove that the transitions $\mathcal{O}_V \mapsto \varepsilon;(\Omega \cap \mathcal{O}_V;\mathbb{T})$ from a vector to a relation and $\mathcal{U} \mapsto \mathtt{syq}(\varepsilon, \mathcal{U});\mathbb{T}$ from a relation to a vector are inverse to each other.

Exercise 5.3 Prove that $\mathcal{U};\overline{\mathcal{U}^{\mathsf{T}};\overline{\mathcal{U}}} = \mathcal{U}$ for any neighborhood topology—and, even more, for an arbitrary relation R.

Chapter 6
Construction of Topologies

We investigate three frequently applied methods of constructing a topology from other given topologies, namely the relative topology, the quotient topology, and the product topology.

6.1 Quotient Topology

An arbitrary equivalence relation Ξ (and a natural projection ξ for it) on the space X with topology $\mathcal{U}$ shall be our starting configuration. Other necessary denotations are introduced with Fig. 6.1. In the traditional way, a quotient set X_Ξ is then obtained—uniquely determined up to isomorphism. By generic means according to Definition 2.2.1, also the membership relation ε_Ξ is defined up to isomorphism. The respective typing is presented in Fig. 6.1.

In a quotient topology, a subset shall be open precisely when its inverse image under the quotient map ξ is. In a first attempt, the authors were mislead to try the seemingly obvious

$$\mathcal{U}_\Xi := \xi^\mathsf{T} ; \mathcal{U} ; \vartheta^\mathsf{T}_{\xi^\mathsf{T}} : X_\Xi \longrightarrow \mathbf{2}^{X_\Xi}.$$

Figure 6.2 illustrates that this does not work: The set $\{[1], [3]\}$ would in this attempt be open but $\{1, 2, 3\}$ would not. One has to concentrate on open sets first, with $\varepsilon_\mathcal{O}$ and $\varepsilon_{\mathcal{O}_\Xi} := \xi^\mathsf{T} ; \varepsilon_\mathcal{O} ; \vartheta^\mathsf{T}_{\xi^\mathsf{T}} : X_\Xi \longrightarrow \mathbf{2}^{X_\Xi}$ including greater neighborhoods only later.

G. Schmidt, M. Winter, *Relational Topology*, Lecture Notes
in Mathematics 2208, https://doi.org/10.1007/978-3-319-74451-3_6

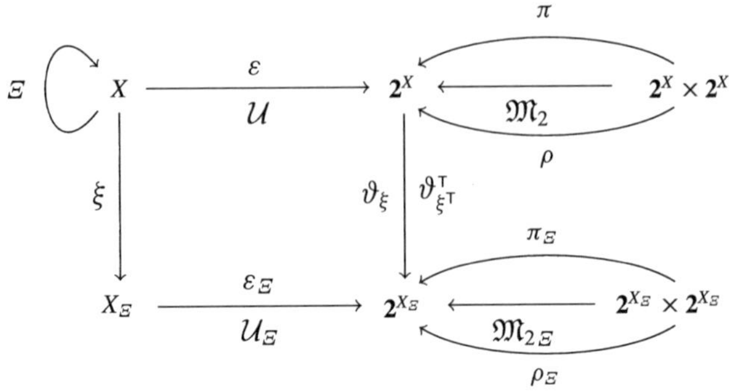

Fig. 6.1 Quotient of a topology

$\mathcal{U} =$

	{}	{1}	{2}	{1,2}	{3}	{1,3}	{2,3}	{1,2,3}	{4}	{1,4}	{2,4}	{1,2,4}	{3,4}	{1,3,4}	{2,3,4}	{1,2,3,4}	{5}	{1,5}	{2,5}	{1,2,5}	{3,5}	{1,3,5}	{2,3,5}	{1,2,3,5}	{4,5}	{1,4,5}	{2,4,5}	{1,2,4,5}	{3,4,5}	{1,3,4,5}	{2,3,4,5}	{1,2,3,4,5}
1	0	0	0	0	0	0	0	0	0	0	0	0	0	0	0	0	0	1	0	1	0	1	0	1	0	1	0	1	0	1	0	1
2	0	0	1	1	0	0	1	1	0	0	1	1	0	0	1	1	0	0	1	1	0	0	1	1	0	0	1	1	0	0	1	1
3	0	0	0	0	1	1	1	1	0	0	0	0	1	1	1	1	0	0	0	0	1	1	1	1	0	0	0	0	1	1	1	1
4	0	0	0	0	0	0	0	0	1	1	1	1	1	1	1	1	0	0	0	0	0	0	0	0	1	1	1	1	1	1	1	1
5	0	0	0	0	0	0	0	0	0	0	0	0	0	0	0	0	0	1	0	1	0	1	0	1	0	1	0	1	0	1	0	1

	1	2	3	4	5
1	1	1	0	0	0
2	1	1	0	0	0
3	0	0	1	0	0
4	0	0	0	1	0
5	0	0	0	0	1

Ξ

	{}	{[1]}	{[3]}	{[1],[3]}	{[4]}	{[1],[4]}	{[3],[4]}	{[1],[3],[4]}	{[5]}	{[1],[5]}	{[3],[5]}	{[1],[3],[5]}	{[4],[5]}	{[1],[4],[5]}	{[3],[4],[5]}	{[1],[3],[4],[5]}
[1]	0	1	0	1	0	1	0	1	0	1	0	1	0	1	0	1
[3]	0	0	1	1	0	0	1	1	0	0	1	1	0	0	1	1
[4]	0	0	0	0	1	1	1	1	0	0	0	0	1	1	1	1
[5]	0	0	0	0	0	0	0	0	0	1	0	1	0	1	0	1

$\xi^{\mathsf{T}}; \mathcal{U}; \vartheta_\xi = \xi^{\mathsf{T}}; \mathcal{U}; \vartheta^{\mathsf{T}}_{\xi^{\mathsf{T}}}$

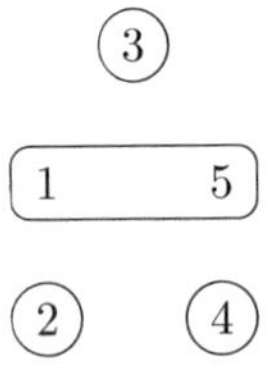

$\mathcal{U}$

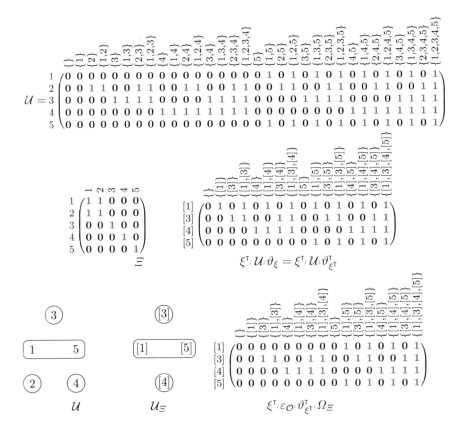

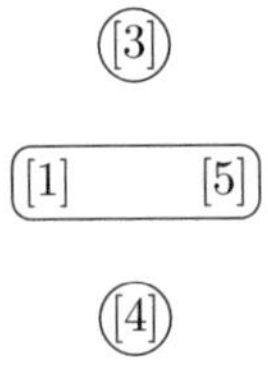

$\mathcal{U}_\Xi$

	{}	{[1]}	{[3]}	{[1],[3]}	{[4]}	{[1],[4]}	{[3],[4]}	{[1],[3],[4]}	{[5]}	{[1],[5]}	{[3],[5]}	{[1],[3],[5]}	{[4],[5]}	{[1],[4],[5]}	{[3],[4],[5]}	{[1],[3],[4],[5]}
[1]	0	0	0	0	0	0	0	0	0	1	0	1	0	1	0	1
[3]	0	0	1	1	0	0	1	1	0	0	1	1	0	0	1	1
[4]	0	0	0	0	1	1	1	1	0	0	0	0	1	1	1	1
[5]	0	0	0	0	0	0	0	0	0	1	0	1	0	1	0	1

$\xi^{\mathsf{T}}; \varepsilon_{\mathcal{O}}; \vartheta^{\mathsf{T}}_{\xi^{\mathsf{T}}}; \Omega_\Xi$

Fig. 6.2 Quotient of a topology indicated via the open set bases

Proposition 6.1.1 *Assume an open set topology on the set X given as a vector $\mathcal{O}_V$ along $\mathbf{2}^X$ and an equivalence $\Xi : X \longrightarrow X$ on that set. We consider its natural projection $\xi : X \longrightarrow X_\Xi$ as well as the membership $\varepsilon_\Xi : X_\Xi \longrightarrow \mathbf{2}^{X_\Xi}$ on the quotient. Then, the following is a topology*

$$\mathcal{O}_{V_\Xi} = \vartheta_{\xi^\mathsf{T}} ; \mathcal{O}_V \subseteq \mathbf{2}^{X_\Xi}$$

and quotient forming by the natural projection $\xi : X \longrightarrow X_\Xi$ is continuous.

Proof We convince ourselves in advance that the following hold

$$\begin{aligned}
\varepsilon_\Xi &= \xi^\mathsf{T} ; \xi ; \varepsilon_\Xi \quad \text{since the natural projection } \xi \text{ is a surjective mapping}\\
&= \xi^\mathsf{T} ; \varepsilon ; \mathtt{syq}(\varepsilon, \xi ; \varepsilon_\Xi) \quad \text{trivial property of the symmetric quotient}\\
&\subseteq \xi^\mathsf{T} ; \varepsilon ; \mathtt{syq}(\xi^\mathsf{T} ; \varepsilon, \varepsilon_\Xi) \quad \text{shifting a surjective mapping, Proposition 2.1.5.i}\\
&\subseteq \varepsilon_\Xi \quad \text{cancellation,}
\end{aligned}$$

resulting in equality and, interpreting the symmetric quotients, in

$$\varepsilon_\Xi = \xi^\mathsf{T} ; \varepsilon ; \vartheta^\mathsf{T}_{\xi^\mathsf{T}} = \xi^\mathsf{T} ; \varepsilon ; \vartheta_\xi .$$

Furthermore

$$\varepsilon ; \vartheta^\mathsf{T}_{\xi^\mathsf{T}} = \varepsilon ; \mathtt{syq}(\varepsilon, \xi ; \varepsilon_\Xi) = \xi ; \varepsilon_\Xi ,$$

besides the standard property $\varepsilon_\Xi ; \vartheta^\mathsf{T}_\xi = \xi^\mathsf{T} ; \varepsilon$ of an existential image.

Now follow the proofs of the topology properties numbered as in Definition 5.3.2:

i) We start from $\mathtt{syq}(\varepsilon, \mathbb{I\!L}) \subseteq \mathcal{O}_V$, obtain $\vartheta_{\xi^\mathsf{T}} ; \mathtt{syq}(\varepsilon, \mathbb{I\!L}) \subseteq \vartheta_{\xi^\mathsf{T}} ; \mathcal{O}_V = \mathcal{O}_{V_\Xi}$, where

$$\begin{aligned}
\vartheta_{\xi^\mathsf{T}} ; \mathtt{syq}(\varepsilon, \mathbb{I\!L}) &= \mathtt{syq}(\varepsilon ; \vartheta^\mathsf{T}_{\xi^\mathsf{T}}, \mathbb{I\!L}) = \mathtt{syq}(\xi ; \varepsilon_\Xi, \mathbb{I\!L}) = \mathtt{syq}(\xi ; \varepsilon_\Xi, \xi ; \mathbb{I\!L})\\
&= \mathtt{syq}(\varepsilon_\Xi, \mathbb{I\!L}) \quad \text{Proposition 2.1.5.ii}
\end{aligned}$$

Analogously $\mathtt{syq}(\varepsilon_\Xi, \mathbb{T}) \subseteq \mathcal{O}_{V_\Xi}$.

ii) Assume $v \subseteq \mathcal{O}_{V_\Xi} = \vartheta_{\xi^\mathsf{T}} ; \mathcal{O}_V$, which gives via shunting $\vartheta^\mathsf{T}_{\xi^\mathsf{T}} ; v \subseteq \mathcal{O}_V$. Since $\mathcal{O}_V$ is an open-set-topology then with Definition 5.3.2.ii

$$\mathtt{syq}(\varepsilon, \varepsilon ; \vartheta^\mathsf{T}_{\xi^\mathsf{T}} ; v) \subseteq \mathcal{O}_V .$$

Consequently

$$\begin{aligned}
\mathtt{syq}(\varepsilon_\Xi, \varepsilon_\Xi ; v) &= \mathtt{syq}(\xi ; \varepsilon_\Xi, \xi ; \varepsilon_\Xi ; v) \quad \text{Proposition 2.1.5.ii; } \xi \text{ is surjective}\\
&= \mathtt{syq}(\varepsilon ; \vartheta^\mathsf{T}_{\xi^\mathsf{T}}, \varepsilon ; \vartheta^\mathsf{T}_{\xi^\mathsf{T}} ; v) \quad \text{see above}\\
&= \vartheta_{\xi^\mathsf{T}} ; \mathtt{syq}(\varepsilon, \varepsilon ; \vartheta^\mathsf{T}_{\xi^\mathsf{T}} ; v) \subseteq \vartheta_{\xi^\mathsf{T}} ; \mathcal{O}_V = \mathcal{O}_{V_\Xi} .
\end{aligned}$$

iii) We start trying to express $\mathfrak{M}_{2\,\Xi}$ by already given constructs:

$$
\begin{aligned}
&(\vartheta_{\xi^\mathsf{T}} \otimes \vartheta_{\xi^\mathsf{T}}) ; \mathfrak{M}_2 ; \vartheta_{\xi^\mathsf{T}}^\mathsf{T} = (\vartheta_{\xi^\mathsf{T}} \otimes \vartheta_{\xi^\mathsf{T}}) ; \mathsf{syq}((\varepsilon \olessthan \varepsilon), \varepsilon) ; \vartheta_{\xi^\mathsf{T}}^\mathsf{T} \quad \text{by definition}\\
&= \mathsf{syq}((\varepsilon \olessthan \varepsilon) ; (\vartheta_{\xi^\mathsf{T}} \otimes \vartheta_{\xi^\mathsf{T}})^\mathsf{T}, \varepsilon ; \vartheta_{\xi^\mathsf{T}}^\mathsf{T})\\
&= \mathsf{syq}((\varepsilon ; \vartheta_{\xi^\mathsf{T}}^\mathsf{T} \olessthan \varepsilon ; \vartheta_{\xi^\mathsf{T}}^\mathsf{T}), \varepsilon ; \vartheta_{\xi^\mathsf{T}}^\mathsf{T}) \quad \text{since } \vartheta_{\xi^\mathsf{T}} \text{ is a map}\\
&= \mathsf{syq}((\xi ; \varepsilon_\Xi \olessthan \xi ; \varepsilon_\Xi), \xi ; \varepsilon_\Xi) \quad \text{see above}\\
&= \mathsf{syq}(\xi ; (\varepsilon_\Xi \olessthan \varepsilon_\Xi), \xi ; \varepsilon_\Xi) \quad \text{since } \xi \text{ is univalent}\\
&= \mathsf{syq}((\varepsilon_\Xi \olessthan \varepsilon_\Xi), \varepsilon_\Xi) \quad \text{Proposition 2.1.5.ii, since } \xi \text{ is a surjective map}\\
&= \mathfrak{M}_{2\,\Xi} \quad \text{by definition}
\end{aligned}
$$

$$
\begin{aligned}
&\mathfrak{M}_{2\,\Xi}^\mathsf{T} ; (\mathcal{O}_{V_\Xi} \ogreaterthan \mathcal{O}_{V_\Xi})\\
&= \vartheta_{\xi^\mathsf{T}} ; \mathfrak{M}_2^\mathsf{T} ; (\vartheta_{\xi^\mathsf{T}}^\mathsf{T} \otimes \vartheta_{\xi^\mathsf{T}}^\mathsf{T}) ; (\vartheta_{\xi^\mathsf{T}} ; \mathcal{O}_V \ogreaterthan \vartheta_{\xi^\mathsf{T}} ; \mathcal{O}_V) \quad \text{see above}\\
&= \vartheta_{\xi^\mathsf{T}} ; \mathfrak{M}_2^\mathsf{T} ; (\vartheta_{\xi^\mathsf{T}}^\mathsf{T} \otimes \vartheta_{\xi^\mathsf{T}}^\mathsf{T}) ; (\vartheta_{\xi^\mathsf{T}} \otimes \vartheta_{\xi^\mathsf{T}}) (\mathcal{O}_V \ogreaterthan \mathcal{O}_V) \quad \text{Proposition 3.1.8.ii}\\
&\subseteq \vartheta_{\xi^\mathsf{T}} ; \mathfrak{M}_2^\mathsf{T} ; (\vartheta_{\xi^\mathsf{T}}^\mathsf{T} ; \vartheta_{\xi^\mathsf{T}} \otimes \vartheta_{\xi^\mathsf{T}}^\mathsf{T} ; \vartheta_{\xi^\mathsf{T}}) ; (\mathcal{O}_V \ogreaterthan \mathcal{O}_V) \quad \text{Proposition 3.1.7.i}\\
&\subseteq \vartheta_{\xi^\mathsf{T}} ; \mathfrak{M}_2^\mathsf{T} ; (\mathcal{O}_V \ogreaterthan \mathcal{O}_V) \quad \text{because } \vartheta_{\xi^\mathsf{T}}^\mathsf{T} ; \vartheta_{\xi^\mathsf{T}} \subseteq \mathbb{I}\\
&\subseteq \vartheta_{\xi^\mathsf{T}} ; \mathcal{O}_V \quad \text{since } \mathcal{O}_V \text{ was supposed to be an open-set-vector topology}\\
&= \mathcal{O}_{V_\Xi}
\end{aligned}
$$

The quotient forming ξ thus introduced is continuous:

$$\vartheta_{\xi^\mathsf{T}}^\mathsf{T} ; \mathcal{O}_{V_\Xi} = \vartheta_{\xi^\mathsf{T}}^\mathsf{T} ; \vartheta_{\xi^\mathsf{T}} ; \mathcal{O}_V \subseteq \mathcal{O}_V \qquad \square$$

6.2 Relative Topology

We assume a neighborhood topology $\mathcal{U} : X \longrightarrow \mathbf{2}^X$ and some subset $Y \subseteq X$. A standard construction then allows to define a topology in a simple way also on an extruded version Y' of Y (Fig. 6.3).

In Fig. 6.4, the subset $\{a, c, e\} \subseteq X$ has got a copy $Y' := \{a{\rightarrow}, c{\rightarrow}, e{\rightarrow}\}$. It is, thus, extruded from X as explained on p. 26.

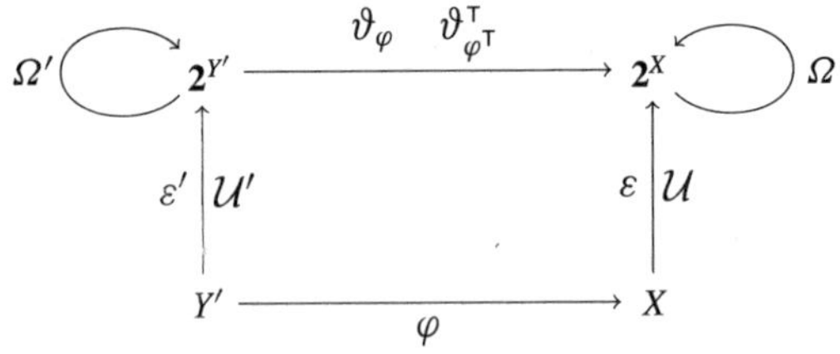

Fig. 6.3 Relative topology for injection φ

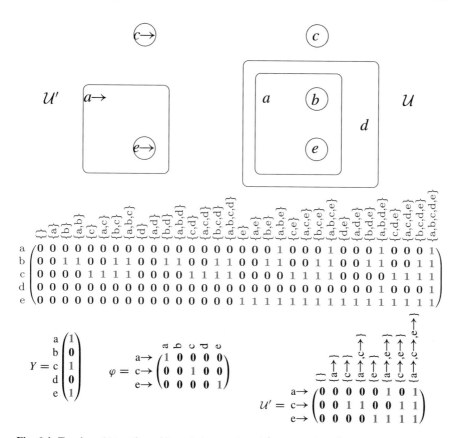

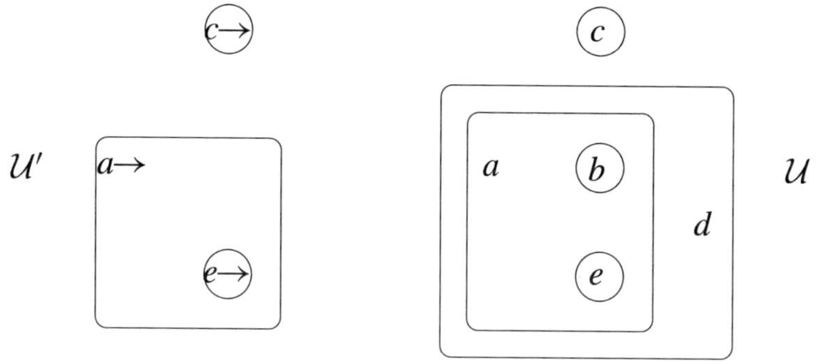

	{}	{a}	{b}	{a,b}	{c}	{a,c}	{b,c}	{a,b,c}	{d}	{a,d}	{b,d}	{a,b,d}	{c,d}	{a,c,d}	{b,c,d}	{a,b,c,d}	{e}	{a,e}	{b,e}	{a,b,e}	{c,e}	{a,c,e}	{b,c,e}	{a,b,c,e}	{d,e}	{a,d,e}	{b,d,e}	{a,b,d,e}	{c,d,e}	{a,c,d,e}	{b,c,d,e}	{a,b,c,d,e}
a	0	0	0	0	0	0	0	0	0	0	0	0	0	0	0	0	0	0	0	1	0	0	0	1	0	0	0	1	0	0	0	1
b	0	0	1	1	0	0	1	1	0	0	1	1	0	0	1	1	0	0	1	1	0	0	1	1	0	0	1	1	0	0	1	1
c	0	0	0	0	1	1	1	1	0	0	0	0	1	1	1	1	0	0	0	0	1	1	1	1	0	0	0	0	1	1	1	1
d	0	0	0	0	0	0	0	0	0	0	0	0	0	0	0	0	0	0	0	0	0	0	0	0	0	0	0	1	0	0	0	1
e	0	0	0	0	0	0	0	0	0	0	0	0	0	0	0	0	1	1	1	1	1	1	1	1	1	1	1	1	1	1	1	1

$$Y = \begin{array}{c} \text{a} \\ \text{b} \\ \text{c} \\ \text{d} \\ \text{e} \end{array} \begin{pmatrix} 1 \\ 0 \\ 1 \\ 0 \\ 1 \end{pmatrix} \qquad \varphi = \begin{array}{c} \\ \text{a}\rightarrow \\ \text{c}\rightarrow \\ \text{e}\rightarrow \end{array} \begin{array}{c} \begin{array}{ccccc} \text{a} & \text{b} & \text{c} & \text{d} & \text{e} \end{array} \\ \begin{pmatrix} 1 & 0 & 0 & 0 & 0 \\ 0 & 0 & 1 & 0 & 0 \\ 0 & 0 & 0 & 0 & 1 \end{pmatrix} \end{array}$$

$\mathcal{U}' =$	{}	{a→}	{c→}	{a→,c→}	{e→}	{a→,e→}	{c→,e→}	{a→,c→,e→}
a→	0	0	0	0	0	1	0	1
c→	0	0	1	1	0	0	1	1
e→	0	0	0	0	1	1	1	1

Fig. 6.4 Topology $\mathcal{U}$ together with a relative topology $\mathcal{U}'$ on extrusion Y' of Y

Proposition 6.2.1 *Given the neighborhood topology $\mathcal{U} : X \longrightarrow \mathbf{2}^X$ and some non-empty subset $Y \subseteq X$, one will obtain the so-called* ***relative (or subset-)topology*** *$\mathcal{U}' : Y' \longrightarrow \mathbf{2}^{Y'}$ with*

$$\mathcal{U}' := \varphi ; \mathcal{U} ; \vartheta_{\varphi^\mathsf{T}},$$

where $\varphi : Y' \longrightarrow X$ is the injection map for the extruded set and $\vartheta_{\varphi^\mathsf{T}} = \mathtt{syq}(\varphi;\varepsilon, \varepsilon')$ the inverse image of φ. The injection φ then turns out to be continuous.

Proof Continuity results immediately via shunting applied to direction "$\supseteq$" of this definition. We follow the numbering of Definition 5.2.1.

i)

$$\begin{aligned} \mathcal{U}' ; \mathbb{T} &= \varphi ; \mathcal{U} ; \vartheta_{\varphi^\mathsf{T}} ; \mathbb{T} = \varphi ; \mathcal{U} ; \mathbb{T} = \varphi ; \mathbb{T} = \mathbb{T} \\ \mathcal{U}' &= \varphi ; \mathcal{U} ; \vartheta_{\varphi^\mathsf{T}} \subseteq \varphi ; \varepsilon ; \vartheta_{\varphi^\mathsf{T}} = \varphi ; \varepsilon ; \mathtt{syq}(\varphi ; \varepsilon, \varepsilon') \subseteq \varepsilon' \end{aligned}$$

ii) We show $\mathcal{U}';\Omega' = \varphi;\mathcal{U};\vartheta_{\varphi^{\mathsf{T}}};\Omega' = \varphi;\mathcal{U};\Omega;\vartheta_{\varphi^{\mathsf{T}}} = \varphi;\mathcal{U};\vartheta_{\varphi^{\mathsf{T}}} = \mathcal{U}'$, for which we have used Proposition 4.3.6.

iii)

$$\begin{array}{ll}
(\mathcal{U}' \olessthan \mathcal{U}');\mathfrak{M}'_2 & \\
= (\varphi;\mathcal{U};\vartheta_{\varphi^{\mathsf{T}}} \olessthan \varphi;\mathcal{U};\vartheta_{\varphi^{\mathsf{T}}});\mathfrak{M}'_2 & \text{by definition} \\
= \varphi;(\mathcal{U};\vartheta_{\varphi^{\mathsf{T}}} \olessthan \mathcal{U};\vartheta_{\varphi^{\mathsf{T}}});\mathfrak{M}'_2 & \text{since } \varphi \text{ is univalent} \\
= \varphi;(\mathcal{U} \olessthan \mathcal{U});(\vartheta_{\varphi^{\mathsf{T}}} \otimes \vartheta_{\varphi^{\mathsf{T}}});\mathfrak{M}'_2 & \text{sharply factorized due to Corollary 3.2.3} \\
= \varphi;(\mathcal{U} \olessthan \mathcal{U});\mathfrak{M}_2;\vartheta_{\varphi^{\mathsf{T}}} & \text{Proposition 4.3.4.i} \\
\subseteq \varphi;\mathcal{U};\vartheta_{\varphi^{\mathsf{T}}} & \text{since } \mathcal{U} \text{ is a topology, Definition 5.2.1} \\
= \mathcal{U}' & \text{by definition}
\end{array}$$

iv)

$$\begin{array}{ll}
\mathcal{U}' = \varphi;\mathcal{U};\vartheta_{\varphi^{\mathsf{T}}} \subseteq \varphi;\mathcal{U};\overline{\varepsilon^{\mathsf{T}};\overline{\mathcal{U}}};\vartheta_{\varphi^{\mathsf{T}}} & \text{Definition 5.2.1.iv} \\
\subseteq \varphi;\mathcal{U};\vartheta_{\varphi^{\mathsf{T}}};\overline{\varepsilon'^{\mathsf{T}};\overline{\varphi;\mathcal{U};\vartheta_{\varphi^{\mathsf{T}}}}} & \text{see below} \\
= \mathcal{U}';\overline{\varepsilon'^{\mathsf{T}};\overline{\mathcal{U}'}} &
\end{array}$$

The remaining part uses that $\vartheta_{\varphi^{\mathsf{T}}}$ is a surjective map due to Proposition 2.2.8.ii:

$$\begin{array}{ll}
\overline{\varepsilon^{\mathsf{T}};\overline{\mathcal{U}}};\vartheta_{\varphi^{\mathsf{T}}} \subseteq \overline{\varepsilon^{\mathsf{T}};\varphi^{\mathsf{T}};\varphi;\overline{\mathcal{U};\vartheta_{\varphi^{\mathsf{T}}};\vartheta^{\mathsf{T}}_{\varphi^{\mathsf{T}}}}};\vartheta_{\varphi^{\mathsf{T}}} & \text{since } \varphi \text{ is univalent, } \vartheta_{\varphi^{\mathsf{T}}} \text{is total} \\
= \overline{\varepsilon^{\mathsf{T}};\varphi^{\mathsf{T}};\overline{\varphi;\mathcal{U};\vartheta_{\varphi^{\mathsf{T}}}}};\vartheta^{\mathsf{T}}_{\varphi^{\mathsf{T}}};\vartheta_{\varphi^{\mathsf{T}}} & \text{the two are mappings} \\
= \overline{\vartheta_{\varphi^{\mathsf{T}}};\varepsilon'^{\mathsf{T}};\overline{\varphi;\mathcal{U};\vartheta_{\varphi^{\mathsf{T}}}}} & \text{since } \vartheta_{\varphi^{\mathsf{T}}} \text{ is a surjective map, Proposition 2.2.5.ii} \\
= \vartheta_{\varphi^{\mathsf{T}}};\overline{\varepsilon'^{\mathsf{T}};\overline{\varphi;\mathcal{U};\vartheta_{\varphi^{\mathsf{T}}}}} & \square
\end{array}$$

In a first attempt one might have taken $\vartheta^{\mathsf{T}}_{\varphi}$ instead of $\vartheta_{\varphi^{\mathsf{T}}}$, but then φ would not be continuous as theory demands; cf. Definition 5.6.1.

6.3 Product Topology

To the standard constructions belongs also the product topology for two given topologies $\mathcal{U}', \mathcal{U}''$. For the introduction, we consider the typing diagram Fig. 6.5. Of course, there exist many topologies on $X' \times X''$; we are, however, only interested in those that behave sufficiently well with regard to the initially given ones.

We find out that $\mathbf{2}^{X'} \times \mathbf{2}^{X''}$ is rather tiny when compared with $\mathbf{2}^{X' \times X''}$. Assume cardinalities $|X'| = 5, |X''| = 7$; then $|\mathbf{2}^{X'} \times \mathbf{2}^{X''}| = 2^{12}$ as opposed to $|\mathbf{2}^{X' \times X''}| = 2^{35}$.

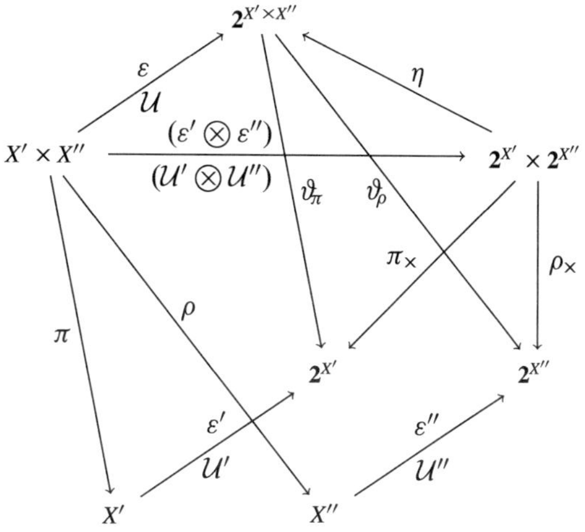

Fig. 6.5 Product of topologies

There is in particular the so-called **Tychonoff**-topology that has the fewest open sets and is usually identified with the name of being the product topology. The following proposition expresses the set-theoretical situation using relation-algebraic concepts. An illustration is given with Fig. 6.6.

Proposition 6.3.1 *Consider the three memberships and singleton injections*

$$\begin{aligned} \varepsilon', \sigma' &: X' \longrightarrow \mathbf{2}^{X'}, \\ \varepsilon'', \sigma'' &: X'' \longrightarrow \mathbf{2}^{X''}, \\ \varepsilon, \sigma &: X' \times X'' \longrightarrow \mathbf{2}^{X' \times X''}. \end{aligned}$$

Then the relation

$$\eta := \mathtt{syq}((\varepsilon' \bigotimes \varepsilon''), \varepsilon) : \mathbf{2}^{X'} \times \mathbf{2}^{X''} \longrightarrow \mathbf{2}^{X' \times X''},$$

is a mapping and the following hold

i) $(\sigma' \bigotimes \sigma'') ; \eta = \sigma$,
ii) $\varepsilon ; \eta^\mathsf{T} = (\varepsilon' \bigotimes \varepsilon'')$,
iii) $(\varepsilon' \bigotimes \varepsilon'') ; \eta = \varepsilon \cap \mathbb{T} ; \eta$,
iv) $(\varepsilon' \bigotimes \varepsilon'') ; \eta ; \eta^\mathsf{T} = (\varepsilon' \bigotimes \varepsilon'')$,
v) $(\varepsilon' \bigotimes \varepsilon'') ; \eta ; \Omega = \varepsilon$,
vi) $\eta ; \eta^\mathsf{T} = \mathtt{syq}((\varepsilon' \bigotimes \varepsilon''), (\varepsilon' \bigotimes \varepsilon'')) = \mathbb{I} \cup v ; v^\mathsf{T}$
where $v := \pi_\times ; \overline{\varepsilon'^\mathsf{T} ; \mathbb{T}} \cup \rho_\times ; \overline{\varepsilon''^\mathsf{T} ; \mathbb{T}}$.

$$\varepsilon' = \begin{array}{c} \\ 1 \\ 2 \end{array} \begin{array}{cccc} \{\} & \{1\} & \{2\} & \{1,2\} \\ 0 & 1 & 0 & 1 \\ 0 & 0 & 1 & 1 \end{array} \qquad \varepsilon'' = \begin{array}{c} \\ a \\ b \\ c \end{array} \begin{array}{cccccccc} \{\} & \{a\} & \{b\} & \{a,b\} & \{c\} & \{a,c\} & \{b,c\} & \{a,b,c\} \\ 0 & 1 & 0 & 1 & 0 & 1 & 0 & 1 \\ 0 & 0 & 1 & 1 & 0 & 0 & 1 & 1 \\ 0 & 0 & 0 & 0 & 1 & 1 & 1 & 1 \end{array}$$

$$(\varepsilon' \otimes \varepsilon'') = $$

Columns: (\{\},\{\}), (\{\},\{a\}), (\{\},\{b\}), (\{\},\{a,b\}), (\{\},\{c\}), (\{\},\{a,c\}), (\{\},\{b,c\}), (\{\},\{a,b,c\}), (\{1\},\{\}), (\{1\},\{a\}), (\{1\},\{b\}), (\{1\},\{a,b\}), (\{1\},\{c\}), (\{1\},\{a,c\}), (\{1\},\{b,c\}), (\{1\},\{a,b,c\}), (\{2\},\{\}), (\{2\},\{a\}), (\{2\},\{b\}), (\{2\},\{a,b\}), (\{2\},\{c\}), (\{2\},\{a,c\}), (\{2\},\{b,c\}), (\{2\},\{a,b,c\}), (\{1,2\},\{\}), (\{1,2\},\{a\}), (\{1,2\},\{b\}), (\{1,2\},\{a,b\}), (\{1,2\},\{c\}), (\{1,2\},\{a,c\}), (\{1,2\},\{b,c\}), (\{1,2\},\{a,b,c\})

$$\begin{array}{r} (1,a) \\ (1,b) \\ (1,c) \\ (2,a) \\ (2,b) \\ (2,c) \end{array} \begin{pmatrix} 0&0&0&0&0&0&0&0&0&1&0&1&0&1&0&1&0&0&0&0&0&0&0&0&0&1&0&1&0&1&0&1 \\ 0&0&0&0&0&0&0&0&0&0&1&1&0&0&1&1&0&0&0&0&0&0&0&0&0&0&1&1&0&0&1&1 \\ 0&0&0&0&0&0&0&0&0&0&0&0&1&1&1&1&0&0&0&0&0&0&0&0&0&0&0&0&1&1&1&1 \\ 0&0&0&0&0&0&0&0&0&0&0&0&0&0&0&0&0&1&0&1&0&1&0&1&0&1&0&1&0&1&0&1 \\ 0&0&0&0&0&0&0&0&0&0&0&0&0&0&0&0&0&0&1&1&0&0&1&1&0&0&1&1&0&0&1&1 \\ 0&0&0&0&0&0&0&0&0&0&0&0&0&0&0&0&0&0&0&0&1&1&1&1&0&0&0&0&1&1&1&1 \end{pmatrix}$$

$$\eta ; \eta^{\mathsf{T}} = \mathtt{syq}(\,(\varepsilon' \otimes \varepsilon''),\, (\varepsilon' \otimes \varepsilon'')\,) =$$

$$\begin{array}{r} (\{\},\{\}) \\ (\{\},\{a\}) \\ (\{\},\{b\}) \\ (\{\},\{a,b\}) \\ (\{\},\{c\}) \\ (\{\},\{a,c\}) \\ (\{\},\{b,c\}) \\ (\{\},\{a,b,c\}) \\ (\{1\},\{\}) \\ (\{1\},\{a\}) \\ (\{1\},\{b\}) \\ (\{1\},\{a,b\}) \\ (\{1\},\{c\}) \\ (\{1\},\{a,c\}) \\ (\{1\},\{b,c\}) \\ (\{1\},\{a,b,c\}) \\ (\{2\},\{\}) \\ (\{2\},\{a\}) \\ (\{2\},\{b\}) \\ (\{2\},\{a,b\}) \\ (\{2\},\{c\}) \\ (\{2\},\{a,c\}) \\ (\{2\},\{b,c\}) \\ (\{2\},\{a,b,c\}) \\ (\{1,2\},\{\}) \\ (\{1,2\},\{a\}) \\ (\{1,2\},\{b\}) \\ (\{1,2\},\{a,b\}) \\ (\{1,2\},\{c\}) \\ (\{1,2\},\{a,c\}) \\ (\{1,2\},\{b,c\}) \\ (\{1,2\},\{a,b,c\}) \end{array} \begin{pmatrix} 1&1&1&1&1&1&1&1&1&0&0&0&0&0&0&0&1&0&0&0&0&0&0&0&1&0&0&0&0&0&0&0 \\ 1&1&1&1&1&1&1&1&1&0&0&0&0&0&0&0&1&0&0&0&0&0&0&0&1&0&0&0&0&0&0&0 \\ 1&1&1&1&1&1&1&1&1&0&0&0&0&0&0&0&1&0&0&0&0&0&0&0&1&0&0&0&0&0&0&0 \\ 1&1&1&1&1&1&1&1&1&0&0&0&0&0&0&0&1&0&0&0&0&0&0&0&1&0&0&0&0&0&0&0 \\ 1&1&1&1&1&1&1&1&1&0&0&0&0&0&0&0&1&0&0&0&0&0&0&0&1&0&0&0&0&0&0&0 \\ 1&1&1&1&1&1&1&1&1&0&0&0&0&0&0&0&1&0&0&0&0&0&0&0&1&0&0&0&0&0&0&0 \\ 1&1&1&1&1&1&1&1&1&0&0&0&0&0&0&0&1&0&0&0&0&0&0&0&1&0&0&0&0&0&0&0 \\ 1&1&1&1&1&1&1&1&1&0&0&0&0&0&0&0&1&0&0&0&0&0&0&0&1&0&0&0&0&0&0&0 \\ 1&1&1&1&1&1&1&1&1&0&0&0&0&0&0&0&1&0&0&0&0&0&0&0&1&0&0&0&0&0&0&0 \\ 0&0&0&0&0&0&0&0&0&1&0&0&0&0&0&0&0&0&0&0&0&0&0&0&0&0&0&0&0&0&0&0 \\ 0&0&0&0&0&0&0&0&0&0&1&0&0&0&0&0&0&0&0&0&0&0&0&0&0&0&0&0&0&0&0&0 \\ 0&0&0&0&0&0&0&0&0&0&0&1&0&0&0&0&0&0&0&0&0&0&0&0&0&0&0&0&0&0&0&0 \\ 0&0&0&0&0&0&0&0&0&0&0&0&1&0&0&0&0&0&0&0&0&0&0&0&0&0&0&0&0&0&0&0 \\ 0&0&0&0&0&0&0&0&0&0&0&0&0&1&0&0&0&0&0&0&0&0&0&0&0&0&0&0&0&0&0&0 \\ 0&0&0&0&0&0&0&0&0&0&0&0&0&0&1&0&0&0&0&0&0&0&0&0&0&0&0&0&0&0&0&0 \\ 0&0&0&0&0&0&0&0&0&0&0&0&0&0&0&1&0&0&0&0&0&0&0&0&0&0&0&0&0&0&0&0 \\ 1&1&1&1&1&1&1&1&1&0&0&0&0&0&0&0&1&0&0&0&0&0&0&0&1&0&0&0&0&0&0&0 \\ 0&0&0&0&0&0&0&0&0&0&0&0&0&0&0&0&0&1&0&0&0&0&0&0&0&0&0&0&0&0&0&0 \\ 0&0&0&0&0&0&0&0&0&0&0&0&0&0&0&0&0&0&1&0&0&0&0&0&0&0&0&0&0&0&0&0 \\ 0&0&0&0&0&0&0&0&0&0&0&0&0&0&0&0&0&0&0&1&0&0&0&0&0&0&0&0&0&0&0&0 \\ 0&0&0&0&0&0&0&0&0&0&0&0&0&0&0&0&0&0&0&0&1&0&0&0&0&0&0&0&0&0&0&0 \\ 0&0&0&0&0&0&0&0&0&0&0&0&0&0&0&0&0&0&0&0&0&1&0&0&0&0&0&0&0&0&0&0 \\ 0&0&0&0&0&0&0&0&0&0&0&0&0&0&0&0&0&0&0&0&0&0&1&0&0&0&0&0&0&0&0&0 \\ 0&0&0&0&0&0&0&0&0&0&0&0&0&0&0&0&0&0&0&0&0&0&0&1&0&0&0&0&0&0&0&0 \\ 1&1&1&1&1&1&1&1&1&0&0&0&0&0&0&0&1&0&0&0&0&0&0&0&1&0&0&0&0&0&0&0 \\ 0&0&0&0&0&0&0&0&0&0&0&0&0&0&0&0&0&0&0&0&0&0&0&0&0&1&0&0&0&0&0&0 \\ 0&0&0&0&0&0&0&0&0&0&0&0&0&0&0&0&0&0&0&0&0&0&0&0&0&0&1&0&0&0&0&0 \\ 0&0&0&0&0&0&0&0&0&0&0&0&0&0&0&0&0&0&0&0&0&0&0&0&0&0&0&1&0&0&0&0 \\ 0&0&0&0&0&0&0&0&0&0&0&0&0&0&0&0&0&0&0&0&0&0&0&0&0&0&0&0&1&0&0&0 \\ 0&0&0&0&0&0&0&0&0&0&0&0&0&0&0&0&0&0&0&0&0&0&0&0&0&0&0&0&0&1&0&0 \\ 0&0&0&0&0&0&0&0&0&0&0&0&0&0&0&0&0&0&0&0&0&0&0&0&0&0&0&0&0&0&1&0 \\ 0&0&0&0&0&0&0&0&0&0&0&0&0&0&0&0&0&0&0&0&0&0&0&0&0&0&0&0&0&0&0&1 \end{pmatrix}$$

Fig. 6.6 Membership products illustrating Proposition 6.3.1

Proof The mapping property is trivial since a membership is positioned on the right of a symmetric quotient.

i)

$$\begin{aligned}(\sigma' \otimes \sigma'') ; \eta &= (\sigma' \otimes \sigma'')\, \mathtt{syq}((\varepsilon' \otimes \varepsilon''), \varepsilon) \\ &= \mathtt{syq}((\varepsilon' \otimes \varepsilon'') ; (\sigma' \otimes \sigma'')^{\mathsf{T}}, \varepsilon) = \mathtt{syq}((\varepsilon' ; \sigma'^{\mathsf{T}} \otimes \varepsilon'' ; \sigma''^{\mathsf{T}}), \varepsilon) \\ &= \mathtt{syq}((\mathbb{I} \otimes \mathbb{I}), \varepsilon) = \mathtt{syq}(\mathbb{I}, \varepsilon) = \sigma \quad \text{Proposition 2.2.2.i}\end{aligned}$$

ii)

$$\varepsilon ; \eta^{\mathsf{T}} = \varepsilon ; \mathtt{syq}(\varepsilon, (\varepsilon' \otimes \varepsilon'')) = (\varepsilon' \otimes \varepsilon'')$$

iii)

$$\begin{aligned}(\varepsilon' \otimes \varepsilon'') ; \eta &= (\varepsilon' \otimes \varepsilon'') ; \mathtt{syq}((\varepsilon' \otimes \varepsilon''), \varepsilon) \\ &= \varepsilon \cap \mathbb{T} ; \mathtt{syq}((\varepsilon' \otimes \varepsilon''), \varepsilon) = \varepsilon \cap \mathbb{T} ; \eta \quad \text{Proposition 2.1.1}\end{aligned}$$

iv)

$$(\varepsilon' \otimes \varepsilon'') ; \eta ; \eta^{\mathsf{T}} = (\varepsilon \cap \mathbb{T} ; \eta) ; \eta^{\mathsf{T}} = \varepsilon ; \eta^{\mathsf{T}} \cap \mathbb{T} = (\varepsilon' \otimes \varepsilon'')$$

v)

$$\varepsilon = \sigma ; \Omega = (\sigma' \otimes \sigma'') ; \eta ; \Omega \subseteq (\varepsilon' \otimes \varepsilon'') ; \eta ; \Omega = \varepsilon ; \eta^{\mathsf{T}} ; \eta ; \Omega \subseteq \varepsilon ; \Omega = \varepsilon.$$

vi) The last identity is proved using Proposition 3.1.9; we have namely

$$\begin{aligned}\overline{(\varepsilon' \otimes \varepsilon'')^{\mathsf{T}} ; \overline{(\varepsilon' \otimes \varepsilon'')}} &= (\varepsilon' \backslash \varepsilon' \otimes \varepsilon'' \backslash \varepsilon'') \cup \pi_{\times} ; \overline{\varepsilon'^{\mathsf{T}} ; \mathbb{T}} \cup \rho_{\times} ; \overline{\varepsilon''^{\mathsf{T}} ; \mathbb{T}} \\ &= (\Omega' \otimes \Omega'') \cup \pi_{\times} ; \overline{\varepsilon'^{\mathsf{T}} ; \mathbb{T}} \cup \rho_{\times} ; \overline{\varepsilon''^{\mathsf{T}} ; \mathbb{T}}.\end{aligned}$$

Analogously

$$\overline{\overline{(\varepsilon' \otimes \varepsilon'')}^{\mathsf{T}} ; (\varepsilon' \otimes \varepsilon'')} = (\Omega'^{\mathsf{T}} \otimes \Omega''^{\mathsf{T}}) \cup \overline{\mathbb{T} ; \varepsilon'} ; \pi_{\times}^{\mathsf{T}} \cup \overline{\mathbb{T} ; \varepsilon''} ; \rho_{\times}^{\mathsf{T}}. \qquad \square$$

Figure 6.6 shows that precisely the elements projected on empty sets prevent η from being injective.

The current investigation may be seen in correspondence with others known from different fields. Stochastics works with the product probability. Higher-dimensional integration has first been insufficiently conceived: the "rectangle-based" Riemann integral did not suffice. In both cases, sophisticated constructions had to be applied in order to catch up with the difficult situation.

Proposition 6.3.2 *Given one topology* $\mathcal{U}' : X' \longrightarrow \mathbf{2}^{X'}$ *as well as another* $\mathcal{U}'' : X'' \longrightarrow \mathbf{2}^{X''}$ *and assuming the context of Proposition 6.3.1 as well as Fig. 6.5,*

$$\mathcal{U} := (\mathcal{U}' \otimes \mathcal{U}'') ; \eta ; \Omega : X' \times X'' \longrightarrow \mathbf{2}^{X' \times X''}$$

is a topology, often called the ***product topology****.*

Proof We again follow the numbering of Definition 5.2.1.

i)

$$\begin{aligned} &\mathcal{U} ; \mathbb{T} = (\mathcal{U}' \otimes \mathcal{U}'') ; \eta ; \Omega ; \mathbb{T} = (\mathcal{U}' \otimes \mathcal{U}'') ; \mathbb{T} \\ &= (\mathcal{U}' ; \mathbb{T} \circledcirc \mathcal{U}'' ; \mathbb{T}) = (\mathbb{T} \circledcirc \mathbb{T}) = \mathbb{T} \quad \text{using Proposition 3.1.6.i} \\ &\mathcal{U} = (\mathcal{U}' \otimes \mathcal{U}'') ; \eta ; \Omega \subseteq (\varepsilon' \otimes \varepsilon'') ; \eta ; \Omega = \varepsilon \quad \text{see Proposition 6.3.1} \end{aligned}$$

ii)

$$\mathcal{U} ; \Omega = (\mathcal{U}' \otimes \mathcal{U}'') ; \eta ; \Omega ; \Omega = (\mathcal{U}' \otimes \mathcal{U}'') ; \eta ; \Omega = \mathcal{U}$$

iii)

$$\begin{aligned} &(\mathcal{U} \circledcirc \mathcal{U}) ; \mathfrak{M}_2 = ((\mathcal{U}' \otimes \mathcal{U}'') ; \eta ; \Omega \circledcirc (\mathcal{U}' \otimes \mathcal{U}'') ; \eta ; \Omega) ; \mathfrak{M}_2 \\ &= ((\mathcal{U}' \otimes \mathcal{U}'') ; \eta \circledcirc (\mathcal{U}' \otimes \mathcal{U}'') ; \eta) ; (\Omega \otimes \Omega) ; \mathfrak{M}_2 \\ &\qquad\qquad \text{sharply factorized, Corollary 3.2.3} \\ &= ((\mathcal{U}' \otimes \mathcal{U}'') ; \eta \circledcirc (\mathcal{U}' \otimes \mathcal{U}'') ; \eta) ; \mathfrak{M}_2 ; \Omega \quad \text{Proposition 4.3.8.iv} \\ &= ((\mathcal{U}' \otimes \mathcal{U}'') \circledcirc (\mathcal{U}' \otimes \mathcal{U}'')) ; (\eta \otimes \eta) ; \mathfrak{M}_2 ; \Omega \quad \text{see remark A below} \\ &= ((\mathcal{U}' \otimes \mathcal{U}'') \circledcirc (\mathcal{U}' \otimes \mathcal{U}'')) ; \mathfrak{K}^{\mathsf{T}} ; (\mathfrak{M}_2' \otimes \mathfrak{M}_2'') ; \eta ; \Omega \quad \text{remark B below} \\ &= ((\mathcal{U}' \circledcirc \mathcal{U}') \otimes (\mathcal{U}'' \circledcirc \mathcal{U}'')) ; (\mathfrak{M}_2' \otimes \mathfrak{M}_2'') ; \eta ; \Omega \quad \text{Proposition 3.3.7} \\ &= ((\mathcal{U}' \circledcirc \mathcal{U}') ; \mathfrak{M}_2' \otimes (\mathcal{U}'' \circledcirc \mathcal{U}'') ; \mathfrak{M}_2'') ; \eta ; \Omega \\ &= (\mathcal{U}' \otimes \mathcal{U}'') ; \eta ; \Omega \quad \text{Definition 5.2.1.iii} \\ &= \mathcal{U} \end{aligned}$$

A: This step is justified as a sharp factorization according to Corollary 3.2.4.i with $Q := (\sigma \otimes \sigma)$, $A := \eta ; \pi_2^{\mathsf{T}}$, $B := \eta ; \rho_2^{\mathsf{T}}$, and the projections

$$\pi_2, \rho_2 : \mathbf{2}^{X' \times X''} \times \mathbf{2}^{X' \times X''} \longrightarrow \mathbf{2}^{X' \times X''}.$$

B: Here it has been used that η and $\mathfrak{M}_2$ commute somehow via the Kronecker-fork shuffle of Definition 3.3.7, which is a bijective map:

$$
\begin{aligned}
(\eta \otimes \eta) ; \mathfrak{M}_2 &= (\eta \otimes \eta) ; \mathtt{syq}((\varepsilon \olessthan \varepsilon), \varepsilon) \\
&= \mathtt{syq}((\varepsilon \olessthan \varepsilon) ; (\eta \otimes \eta)^{\mathsf{T}}, \varepsilon) \\
&= \mathtt{syq}((\varepsilon ; \eta^{\mathsf{T}} \olessthan \varepsilon ; \eta^{\mathsf{T}}), \varepsilon) \\
&= \mathtt{syq}(((\varepsilon' \otimes \varepsilon'') \olessthan (\varepsilon' \otimes \varepsilon'')), \varepsilon) \\
&= \mathtt{syq}(((\varepsilon' \olessthan \varepsilon') \otimes (\varepsilon'' \olessthan \varepsilon'')) ; \mathfrak{K}, \varepsilon) \quad \text{see Proposition 3.3.7} \\
&= \mathfrak{K}^{\mathsf{T}} ; \mathtt{syq}(((\varepsilon' \olessthan \varepsilon') \otimes (\varepsilon'' \olessthan \varepsilon'')), \varepsilon) \quad \text{since } \mathfrak{K} \text{ is a bijective map} \\
&= \mathfrak{K}^{\mathsf{T}} ; \mathtt{syq}((\varepsilon' ; \mathfrak{M}_2'^{\mathsf{T}} \otimes \varepsilon'' ; \mathfrak{M}_2''^{\mathsf{T}}), \varepsilon) \quad \text{Proposition 4.3.2.i} \\
&= \mathfrak{K}^{\mathsf{T}} ; \mathtt{syq}((\varepsilon' \otimes \varepsilon'') ; (\mathfrak{M}_2' \otimes \mathfrak{M}_2'')^{\mathsf{T}}, \varepsilon) \\
&= \mathfrak{K}^{\mathsf{T}} ; (\mathfrak{M}_2' \otimes \mathfrak{M}_2'') ; \mathtt{syq}((\varepsilon' \otimes \varepsilon''), \varepsilon) \\
&= \mathfrak{K}^{\mathsf{T}} ; (\mathfrak{M}_2' \otimes \mathfrak{M}_2'') ; \eta
\end{aligned}
$$

iv) The task is to prove $\mathcal{U} \subseteq \overline{\mathcal{U} ; \varepsilon^{\mathsf{T}} ; \overline{\mathcal{U}}}$, or in fully expanded form

$$(\mathcal{U}' \otimes \mathcal{U}'') ; \eta ; \Omega \subseteq \overline{(\mathcal{U}' \otimes \mathcal{U}'') ; \eta ; \Omega ; \varepsilon^{\mathsf{T}} ; \overline{(\mathcal{U}' \otimes \mathcal{U}'') ; \eta ; \Omega}}.$$

Two observations facilitate this task considerably; it holds namely

$$\Omega ; \overline{\varepsilon^{\mathsf{T}} ; X} = \overline{\varepsilon^{\mathsf{T}} ; X} \quad \text{and} \quad \overline{Y ; \overline{Z ; \Omega}} = \overline{Y ; \overline{Z ; \Omega} ; \Omega^{\mathsf{T}}} ; \Omega,$$

which is easily evaluated using $\mathbb{I} \subseteq \Omega$ and the Schröder rule. Therefore the formula to prove is equivalent with

$$(\mathcal{U}' \otimes \mathcal{U}'') ; \eta ; \Omega \subseteq \overline{(\mathcal{U}' \otimes \mathcal{U}'') ; \eta ; \varepsilon^{\mathsf{T}} ; \overline{(\mathcal{U}' \otimes \mathcal{U}'') ; \eta ; \Omega} ; \Omega^{\mathsf{T}}} ; \Omega,$$

allowing us to drop the factor Ω on the smaller side, so that it suffices to show

$$(\mathcal{U}' \otimes \mathcal{U}'') ; \eta \subseteq \overline{(\mathcal{U}' \otimes \mathcal{U}'') ; \eta ; \varepsilon^{\mathsf{T}} ; \overline{(\mathcal{U}' \otimes \mathcal{U}'') ; \eta}}.$$

Shunting η and applying the formula for $\eta ; \varepsilon^{\mathsf{T}}$ reduces the task to

$$(\mathcal{U}' \otimes \mathcal{U}'') \subseteq \overline{(\mathcal{U}' \otimes \mathcal{U}'') ; (\varepsilon' \otimes \varepsilon'')^{\mathsf{T}} ; \overline{(\mathcal{U}' \otimes \mathcal{U}'')}},$$

where in addition $\eta ; \eta^{\mathsf{T}}$ has been dropped since $\mathbb{I} \subseteq \eta ; \eta^{\mathsf{T}}$ instead suffices.

Now we may proceed with Proposition 3.1.9:

$$
\begin{aligned}
&\overline{(\varepsilon' \otimes \varepsilon'')^{\mathsf{T}} ; \overline{(\mathcal{U}' \otimes \mathcal{U}'')}} = \\
&= (\overline{\varepsilon'^{\mathsf{T}} ; \overline{\mathcal{U}'}} \otimes \overline{\varepsilon''^{\mathsf{T}} ; \overline{\mathcal{U}''}}) \cup \pi_{\times} ; \overline{\varepsilon'^{\mathsf{T}} ; \mathbb{T}} \cup \rho_{\times} ; \overline{\varepsilon''^{\mathsf{T}} ; \mathbb{T}}
\end{aligned}
$$

Using this

$$
\begin{aligned}
&(\mathcal{U}' \otimes \mathcal{U}'') ; \overline{(\varepsilon' \otimes \varepsilon'')^{\mathsf{T}} ; \overline{(\mathcal{U}' \otimes \mathcal{U}'')}} \\
=\ & (\mathcal{U}' \otimes \mathcal{U}'') ; (\overline{\varepsilon'^{\mathsf{T}} ; \overline{\mathcal{U}'}} \otimes \overline{\varepsilon''^{\mathsf{T}} ; \overline{\mathcal{U}''}}) \cup (\mathcal{U}' \otimes \mathcal{U}'') ; \left[\pi_\times ; \overline{\varepsilon'^{\mathsf{T}} ; \mathbb{T}} \cup \rho_\times ; \overline{\varepsilon''^{\mathsf{T}} ; \mathbb{T}}\right] \\
=\ & (\mathcal{U}' \otimes \mathcal{U}'') ; (\overline{\varepsilon'^{\mathsf{T}} ; \overline{\mathcal{U}'}} \otimes \overline{\varepsilon''^{\mathsf{T}} ; \overline{\mathcal{U}''}}) \cup (\pi ; \mathcal{U}' \cap \rho ; \mathcal{U}'' ; \mathbb{T}) ; \overline{\varepsilon'^{\mathsf{T}} ; \mathbb{T}} \cup \dots \\
=\ & (\mathcal{U}' \otimes \mathcal{U}'') ; (\overline{\varepsilon'^{\mathsf{T}} ; \overline{\mathcal{U}'}} \otimes \overline{\varepsilon''^{\mathsf{T}} ; \overline{\mathcal{U}''}}) \cup \perp\!\!\!\perp \cup \perp\!\!\!\perp \\
=\ & (\mathcal{U}' \otimes \mathcal{U}'') ; (\overline{\varepsilon'^{\mathsf{T}} ; \overline{\mathcal{U}'}} \otimes \overline{\varepsilon''^{\mathsf{T}} ; \overline{\mathcal{U}''}}) \\
=\ & (\mathcal{U}' ; \overline{\varepsilon'^{\mathsf{T}} ; \overline{\mathcal{U}'}} \otimes \mathcal{U}'' ; \overline{\varepsilon''^{\mathsf{T}} ; \overline{\mathcal{U}''}}) \qquad \text{see remark below} \\
\supseteq\ & (\mathcal{U}' \otimes \mathcal{U}'')
\end{aligned}
$$

Again, we apply Corollary 3.2.4.i; however, this time setting $Q := \sigma''$ and

$$
P := \mathcal{U}',\ S := \mathcal{U}'',\ A := \overline{\varepsilon'^{\mathsf{T}} ; \overline{\mathcal{U}'}} ; \pi_\times^{\mathsf{T}}, \quad \text{and} \quad B := \overline{\varepsilon''^{\mathsf{T}} ; \overline{\mathcal{U}''}} ; \rho_\times^{\mathsf{T}}. \qquad \square
$$

Proposition 6.3.3 *The projections π, ρ of a direct product to its constituent factors (see Fig. 6.5) are continuous.*

Proof We will show the case π only and start rather willful:

$$
\begin{aligned}
&\pi ; \mathcal{U}' = \pi ; \mathcal{U}' ; \Omega' = [\pi ; \mathcal{U}' \cap \mathbb{T}] ; \Omega' \\
&= [\pi ; \mathcal{U}' \cap \rho ; \mathcal{U}'' ; \mathbb{T}] ; \Omega' \cup \perp\!\!\!\perp \\
&= [\pi ; \mathcal{U}' \cap \rho ; \mathcal{U}'' ; \mathbb{T}] ; \Omega' \cup (\pi ; \mathcal{U}' ; \mathbb{T} \cap \rho ; \mathcal{U}'') ; \overline{\varepsilon''^{\mathsf{T}} ; \mathbb{T}} \quad \text{since } \mathcal{U}'' ; \overline{\varepsilon''^{\mathsf{T}} ; \mathbb{T}} \subseteq \perp\!\!\!\perp \\
&= (\mathcal{U}' \otimes \mathcal{U}'') ; \pi_\times ; \Omega' \cup (\mathcal{U}' \otimes \mathcal{U}'') ; \rho_\times ; \overline{\varepsilon''^{\mathsf{T}} ; \mathbb{T}} \quad \text{Proposition 3.1.3 twice} \\
&= (\mathcal{U}' \otimes \mathcal{U}'') ; \left[\pi_\times ; \Omega' \cup \rho_\times ; \overline{\varepsilon''^{\mathsf{T}} ; \mathbb{T}}\right] \\
&= (\mathcal{U}' \otimes \mathcal{U}'') ; \eta ; \Omega ; \vartheta_{\pi^{\mathsf{T}}}^{\mathsf{T}} \quad \text{see below} \\
&= \mathcal{U} ; \vartheta_{\pi^{\mathsf{T}}}^{\mathsf{T}} \quad \text{definition of } \mathcal{U}
\end{aligned}
$$

The remaining part:

$$
\begin{aligned}
\overline{\pi_\times ; \varepsilon'^{\mathsf{T}} ; \overline{\varepsilon'}} \cup \overline{\rho_\times ; \varepsilon''^{\mathsf{T}} ; \mathbb{T}} &= \overline{\pi_\times ; \varepsilon'^{\mathsf{T}} ; \overline{\varepsilon'} \cap \rho_\times ; \varepsilon''^{\mathsf{T}} ; \mathbb{T}} \\
&= \overline{(\pi_\times ; \varepsilon'^{\mathsf{T}} \cap \rho_\times ; \varepsilon''^{\mathsf{T}} ; \mathbb{T}) ; \overline{\varepsilon'}} \quad \text{masking} \\
&= \overline{(\pi_\times ; \varepsilon'^{\mathsf{T}} ; \pi^{\mathsf{T}} \cap \rho_\times ; \varepsilon''^{\mathsf{T}} ; \rho^{\mathsf{T}}) ; \pi ; \overline{\varepsilon'}} \quad \text{destroy and append rule} \\
&= \overline{(\varepsilon' \otimes \varepsilon'')^{\mathsf{T}} ; \overline{\pi ; \varepsilon'}} \\
&= \overline{\eta ; \varepsilon^{\mathsf{T}} ; \overline{\pi ; \varepsilon'}} \quad \text{Proposition 6.3.1.ii} \\
&= \overline{\eta ; \varepsilon^{\mathsf{T}} ; \overline{\varepsilon ; \vartheta_{\pi^{\mathsf{T}}}^{\mathsf{T}}}} \quad \text{standard rule for existential image} \\
&= \eta ; \overline{\varepsilon^{\mathsf{T}} ; \overline{\varepsilon}} ; \vartheta_{\pi^{\mathsf{T}}}^{\mathsf{T}} = \eta ; \Omega ; \vartheta_{\pi^{\mathsf{T}}}^{\mathsf{T}} \quad \text{since } \vartheta_{\pi^{\mathsf{T}}} \text{ is a map} \qquad \square
\end{aligned}
$$

There seem to be not many other broadly applicable constructions of new topologies out of given ones. In studying these three, product, subset or relative, and quotient topology, we had the opportunity to work with the universal constructions of relational mathematics mentioned in Sect. 3.1. Then we have demonstrated that the Kronecker, fork, and join operators indeed suffice to achieve such proofs.

There exist, however, descriptional ways of talking about the initial, respectively final, object in a category—here topologies together with their continuous maps—with respect to a not necessarily finite set of constituents.

The characteristic mappings occurring as projections in connection with the direct product, as injection for the relative topology or as the natural projection for quotient forming all turned out to be continuous. So the universal relational constructions subsume under the categorical ones.

Finally we should mention that in the cases of relative topology and product topology, sharp factorization was necessary: It had been possible because there had been membership and singleton injection available so as to allow to apply the respective propositions.

Exercises

Exercise 6.1 In the setting of relative topology in Fig. 6.3, assume an arbitrary continuous map $\psi : Z \longrightarrow Y$. Prove that ψ is continuous if and only if $\psi ; \varphi$ is.

Exercise 6.2 In the setting of quotient topology in Fig. 6.1, assume an arbitrary continuous map $\psi : X_{\Xi} \longrightarrow Z$. Prove that ψ is continuous if and only if $\xi ; \psi$ is.

Chapter 7
Closures and Their Aumann Contacts

Topology has been shown to be definable in several cryptomorphically equivalent ways: by a neighborhood system, by a collection of open sets (be these given as a vector along the powerset or as a partial diagonal on it), by a collection of closed sets, or by a mapping to open kernels. It is not commonly known that also certain Aumann contact relations as originating from [Aum70, Aum74] give rise to topologies. These in turn always lead to contact relations.

7.1 Aumann Contact Related to Topology

Aumanns contacts have been defined as relations $C : X \longrightarrow \mathbf{2}^X$ above the membership ε, but never in relation to the empty set, satisfying this law

$$\forall x \in X : \forall u \in \mathbf{2}^X : \big[\exists v \in \mathbf{2}^X : C_{xv} \wedge \big(\forall y \in X : \varepsilon_{yv} \rightarrow C_{yu}\big)\big] \rightarrow C_{xu}.$$

In plain words it says that x is in contact with a subset u when there exists some subset v to which x is in contact of which all the elements are already in contact with u. We will see how this occurs in connection with closures. However, to work with the property, we better lift it to the relational level. The following is, thus, some sort of a free re-interpretation of Aumanns concept in a quantifier-free style.

Definition 7.1.1 We consider a set related to its powerset, with a membership relation $\varepsilon : X \longrightarrow \mathbf{2}^X$. Then a relation $C : X \longrightarrow \mathbf{2}^X$ is called an **Aumann**[1] **contact**

[1]Georg Aumann (1906–1980) was a professor at TU München since 1960. Already in 1934/35 he visited the Institute for Advanced Studies in Princeton as a Rockefeller scholar. Some consider him as one of the more significant mathematicians of the first half of the twentieth century, not least because of his book *Reelle Funktionen*, [Aum69]. The first author has in 1968 been with him

G. Schmidt, M. Winter, *Relational Topology*, Lecture Notes
in Mathematics 2208, https://doi.org/10.1007/978-3-319-74451-3_7

C

	{}	{a}	{b}	{a,b}	{c}	{a,c}	{b,c}	{a,b,c}	{d}	{a,d}	{b,d}	{a,b,d}	{c,d}	{a,c,d}	{b,c,d}	{a,b,c,d}
a	0	1	0	1	0	1	0	1	0	1	0	1	0	1	0	1
b	0	0	1	1	0	1	1	1	0	1	1	1	0	1	1	1
c	0	0	1	1	1	1	1	1	1	1	1	1	1	1	1	1
d	0	0	0	1	0	1	0	1	1	1	1	1	1	1	1	1

C'

	{}	{a}	{b}	{a,b}	{c}	{a,c}	{b,c}	{a,b,c}	{d}	{a,d}	{b,d}	{a,b,d}	{c,d}	{a,c,d}	{b,c,d}	{a,b,c,d}
a	0	1	0	1	0	1	0	1	0	1	0	1	0	1	0	1
b	0	0	1	1	0	1	1	1	0	1	1	1	0	1	1	1
c	0	0	**O**	1	1	1	1	1	**O**	1	1	1	1	1	1	1
d	0	0	0	1	0	1	0	1	1	1	1	1	1	1	1	1

Fig. 7.1 Aumann contact with or without modification of Proposition 7.1.3: the **O** 's

relation, provided

i) $\varepsilon \subseteq C \subseteq \mathbb{T};\varepsilon$

ii) $C;\overline{\varepsilon^{\mathsf T};\overline{C}} \subseteq C$, or equivalently, $C^{\mathsf T};\overline{C} \subseteq \varepsilon^{\mathsf T};\overline{C}$.

We call C a **topological Aumann contact relation**, when in addition

$$(\overline{C} \circledless \overline{C}) \subseteq \overline{C};\mathfrak{J}_2^{\mathsf T}. \qquad \square$$

This definition is slightly more restrictive than that of [Sch11, Def. 11.18] in as far as contact with the empty set is concerned; e.g., the first column of C in Fig. 7.1 is demanded to be a **0**-column. Aumann contacts may always be generated from an arbitrary relation $R : X \longrightarrow Y$ using the membership relation $\varepsilon : X \longrightarrow \mathbf{2}^X$ as $C := \mathtt{lbd}_R(\mathtt{ubd}_R(\varepsilon)) = \overline{\overline{R};\overline{\overline{R}^{\mathsf T};\varepsilon}} = R/(\varepsilon\backslash R)$. In particular, every contact generates itself, i.e., $C = \overline{\overline{C};\overline{\overline{C}^{\mathsf T};\varepsilon}} = C/(\varepsilon\backslash C)$. This may be shown remembering the upper and lower bound functionals:

$$\begin{aligned} \overline{\overline{C};\overline{\overline{C}^{\mathsf T};\varepsilon}} &= \overline{\overline{C};\overline{\overline{C}^{\mathsf T};C}} && \text{anticipating Proposition 7.1.2.i} \\ &= \overline{\overline{C};\overline{\overline{C}^{\mathsf T};\overline{\overline{C};\mathbb{I}}}} = \overline{\overline{C};\mathbb{I}} = C && \text{since } \mathtt{lbd}_C(\mathtt{ubd}_C(\mathtt{lbd}_C(\mathbb{I}))) = \mathtt{lbd}_C(\mathbb{I}) \end{aligned}$$

The preceding proof might—of course—be reformulated via residuals[2] and would then read

$$C/(\varepsilon\backslash C) = C/(C\backslash C) = C/(C\backslash(C/\mathbb{I})) = C/\mathbb{I} = C.$$

A few additional facts concerning Aumann contacts are now summarized.

among those who formally founded the Mathematics unit of TUM—terminating its existence as an informal substructure of the old faculty of 'Allgemeine Wissenschaften'.

[2]One has a rather firm feeling for negation; e.g. monotony when doubly negated. Do we have a corresponding feeling for "/" and "\" and how they operate together? Earlier denotations "$\overset{+}{;}$" (once designed contrasting to ";"), "·.", and ".·" (in diverging intention!) have provided some confusion as it has been reported already in [SS89, SS93].

Proposition 7.1.2 *The following formulae hold for Aumann contacts.*

i) $C;\overline{\varepsilon^{\mathsf{T}};\overline{C}} = C \qquad C^{\mathsf{T}};\overline{C} = \varepsilon^{\mathsf{T}};\overline{C} \qquad C\backslash C = \varepsilon\backslash C$

ii) C *is up-closed, i.e.* $C;\Omega = C$.

iii) The construct $\rho' := \mathtt{syq}(C,\varepsilon)$ *is a closure mapping with respect to the powerset ordering* Ω *that satisfies* $\rho';\varepsilon^{\mathsf{T}} = C^{\mathsf{T}}$.

iv) For a topological Aumann contact always $(\overline{C} \olessthan \overline{C}) = \overline{C};\mathfrak{J}_2^{\mathsf{T}}$.

v) Any closure map ρ_1 *wrt.* Ω *that sends the empty set to itself leads to an Aumann contact* $C_1 := \varepsilon;\rho_1^{\mathsf{T}}$.

Proof

i) In view of Definition 7.1.1.ii, only "$\supseteq$" has to be shown:

$$\varepsilon;\mathbb{I} \subseteq C \iff \varepsilon^{\mathsf{T}};\overline{C} \subseteq \overline{\mathbb{I}} \iff \overline{\varepsilon^{\mathsf{T}};\overline{C}} \supseteq \mathbb{I}$$

While "$\supseteq$" of the second claim is obvious looking at Definition 7.1.1.i, "$\subseteq$" follows with Definition 7.1.1.ii. The third formula is just a transcription of the second.

ii) Direction "$\supseteq$" follows since Ω is reflexive. With Definition 7.1.1.i,ii we obtain

$$C;\Omega = C;\overline{\varepsilon^{\mathsf{T}};\overline{\varepsilon}} \subseteq C;\overline{\varepsilon^{\mathsf{T}};\overline{C}} \subseteq C.$$

iii) An immediate result is $\rho';\varepsilon^{\mathsf{T}} = \mathtt{syq}(C,\varepsilon);\varepsilon^{\mathsf{T}} = \left[\varepsilon;\mathtt{syq}(\varepsilon,C)\right]^{\mathsf{T}} = C^{\mathsf{T}}$. The mapping property of ρ' follows since the membership relation ε is positioned on the right side of the symmetric quotient. ρ' is idempotent since via shunting

$$\rho';\rho' \subseteq \rho' \iff \rho' \subseteq \rho';\rho'^{\mathsf{T}}$$

and

$$\begin{array}{lll} \rho';\rho'^{\mathsf{T}} & = \rho';\mathtt{syq}(C,\varepsilon)^{\mathsf{T}} = \rho';\mathtt{syq}(\varepsilon,C) & = \mathtt{syq}(\varepsilon;\rho'^{\mathsf{T}},C) \\ & = \mathtt{syq}(C,C) & \text{see above} \\ & = \overline{\overline{C}^{\mathsf{T}};C} \cap \overline{C^{\mathsf{T}};\overline{C}} & \text{expanded} \\ & = \overline{\overline{C}^{\mathsf{T}};\varepsilon} \cap \overline{C^{\mathsf{T}};\overline{C}} & \text{following (i)} \\ & \supseteq \overline{\overline{C}^{\mathsf{T}};\varepsilon} \cap \overline{C^{\mathsf{T}};\overline{\varepsilon}} = \mathtt{syq}(C,\varepsilon) = \rho' & \text{monotony and definition of } \rho' \end{array}$$

The last line shows in addition that $\rho' \subseteq \overline{C^{\mathsf{T}};\overline{\varepsilon}} \subseteq \overline{\varepsilon^{\mathsf{T}};\overline{\varepsilon}} \subseteq \Omega$. Finally we prove via shunting that ρ' is monotonic:

$$\Omega;\rho' \subseteq \rho';\Omega \iff \Omega \subseteq \rho';\Omega;\rho'^{\mathsf{T}},$$

where the latter is indeed satisfied:

$$\rho' ; \Omega ; \rho'^{\top} = \rho' ; \overline{\varepsilon^{\top} ; \overline{\varepsilon}} ; \rho'^{\top} = \overline{\rho' ; \varepsilon^{\top} ; \overline{\overline{\varepsilon} ; \rho'^{\top}}} = \overline{C^{\top} ; \overline{C}} = \overline{\varepsilon^{\top} ; \overline{C}} \quad \text{using (i) again}$$
$$\supseteq \overline{\varepsilon^{\top} ; \overline{\varepsilon}} = \Omega$$

iv) In proving $\overline{C} ; \mathfrak{J}_2^{\top} \subseteq (\overline{C} \circledless \overline{C}) = \overline{C} ; \pi^{\top} \cap \overline{C} \rho^{\top}$, we restrict to $\overline{C} ; \mathfrak{J}_2^{\top} \subseteq \overline{C} ; \pi^{\top}$ or equivalently $\overline{C} ; \mathfrak{J}_2^{\top} ; \pi \subseteq \overline{C}$ obtained by shunting π. From Proposition 4.2.2.vii we have that $\mathfrak{J}_2^{\top} ; \pi = \Omega^{\top}$ and from (ii) that $\overline{C} ; \Omega^{\top} \subseteq \overline{C}$.

v) The condition $\varepsilon \subseteq C_1$ follows via shunting from $\varepsilon ; \rho_1 \subseteq \varepsilon ; \Omega = \varepsilon$ since ρ_1 is expanding. The construct $z := \overline{\mathbb{T} ; \varepsilon}^{\top}$ is a point representing the empty subset. Using the condition $\rho_1^{\top} ; z = z$, namely that ρ_1 sends the empty set to itself, implies $\mathbb{T} = \overline{\varepsilon} ; z = \overline{\varepsilon} ; \rho_1^{\top} ; z$. Shunted we obtain $\mathbb{T} ; z^{\top} \subseteq \varepsilon ; \rho_1^{\top} = \overline{\overline{\varepsilon} ; \rho_1^{\top}} = \overline{C_1}$ and negated $C_1 \subseteq \overline{\mathbb{T} ; z^{\top}} = \overline{\mathbb{T} ; \overline{\mathbb{T} ; \varepsilon}} = \mathbb{T} ; \varepsilon$.

It remains to show the second condition that now reads $\rho_1 ; \varepsilon^{\top} \overline{\overline{\varepsilon} ; \rho_1^{\top}} \subseteq \varepsilon^{\top} \overline{\overline{\varepsilon} ; \rho_1^{\top}}$. An equivalent form where ρ_1 slipped out/in negation in $\Omega = \overline{\varepsilon^{\top} ; \overline{\varepsilon}}$, and shunted is

$$\overline{\rho_1 ; \Omega} ; \rho_1^{\top} ; \rho_1 \subseteq \overline{\Omega}.$$

This can indeed be shown employing Schröder rule, idempotency $\rho_1 ; \rho_1 = \rho_1$, univalency $\rho_1^{\top} ; \rho_1 \subseteq \mathbb{I}$, and monotony $\Omega ; \rho_1 \subseteq \rho_1 ; \Omega$ in

$$\Omega ; \rho_1^{\top} ; \rho_1 = \Omega ; \rho_1^{\top} ; \rho_1 ; \rho_1 \subseteq \Omega ; \rho_1 \subseteq \rho_1 ; \Omega. \qquad \square$$

With the following proposition we exhibit a slightly specialized version of an Aumann contact.

Proposition 7.1.3 *Whenever C is an Aumann contact, then so is the possibly smaller relation $C' := C \cap \overline{\overline{\mathbb{I}} ; \sigma}$ with $\sigma := \mathrm{syq}(\mathbb{I}, \varepsilon)$ the singleton injection.*

Proof

i) From Proposition 2.2.2.i follows $\varepsilon ; \sigma^{\top} \subseteq \mathbb{I}$, so that $\varepsilon \subseteq \overline{\overline{\mathbb{I}} ; \sigma}$, and obviously $C' \subseteq C \subseteq \mathbb{T} ; \varepsilon$.

ii) We have to prove $(C \cap \overline{\overline{\mathbb{I}} ; \sigma})^{\top} ; (\overline{C} \cup \overline{\mathbb{I}} ; \sigma) \subseteq \varepsilon^{\top} ; (\overline{C} \cup \overline{\mathbb{I}} ; \sigma)$, from which the product with $\overline{C}$ is trivial since C is an Aumann contact by assumption. It suffices then to show that $(C \cap \overline{\overline{\mathbb{I}};\sigma})^{\mathsf{T}};\overline{\mathbb{I}};\sigma \subseteq \varepsilon^{\mathsf{T}};\overline{\mathbb{I}};\sigma$:

$$
\begin{array}{lll}
\Longleftarrow & (C \cap \overline{\overline{\mathbb{I}};\sigma})^{\mathsf{T}};\overline{\mathbb{I}} \subseteq \varepsilon^{\mathsf{T}};\overline{\mathbb{I}} & \\
\Longleftarrow & (\mathbb{T};\varepsilon \cap \overline{\overline{\mathbb{I}};\sigma})^{\mathsf{T}};\overline{\mathbb{I}} \subseteq \varepsilon^{\mathsf{T}};\overline{\mathbb{I}} & \text{because } C \subseteq \mathbb{T};\varepsilon \\
\Longleftrightarrow & \overline{\mathbb{I}};(\mathbb{T};\varepsilon \cap \overline{\overline{\mathbb{I}};\sigma}) \subseteq \overline{\mathbb{I}};\varepsilon & \text{transposed} \\
\Longleftrightarrow & \overline{\mathbb{I}};(\varepsilon \cap \overline{\overline{\mathbb{I}};\sigma}) \subseteq \overline{\mathbb{I}};\varepsilon \ \text{ and } \ \overline{\mathbb{I}};(\overline{\mathbb{I}};\varepsilon \cap \overline{\overline{\mathbb{I}};\sigma}) \subseteq \overline{\mathbb{I}};\varepsilon, & \\
 & & \text{splitted } \mathbb{T} = \mathbb{I} \cup \overline{\mathbb{I}} \\
\Longleftarrow & \overline{\mathbb{I}};(\overline{\mathbb{I}};\varepsilon \cap \overline{\overline{\mathbb{I}};\sigma}) \subseteq \overline{\mathbb{I}};\varepsilon & \text{since the first one is trivial} \\
\Longleftarrow & \overline{\mathbb{I}};(\overline{\mathbb{I}};\varepsilon \cap \overline{\overline{\mathbb{I}};\sigma}) = \overline{\mathbb{I}};(\mathbb{T};\varepsilon \cap \overline{\mathbb{T};\sigma}) & \text{see below} \\
 & \quad = \mathbb{T};\varepsilon \cap \overline{\mathbb{I}};\overline{\mathbb{T};\sigma} & \text{masking} \\
 & \quad \subseteq \mathbb{T};\varepsilon \cap \overline{\mathbb{T};\sigma} & \\
 & \quad = \overline{\mathbb{I}};\varepsilon \cap \overline{\overline{\mathbb{I}};\sigma} & \text{again as shown below} \\
 & \quad \subseteq \overline{\mathbb{I}};\varepsilon &
\end{array}
$$

Here, we had been allowed to replace $\overline{\mathbb{I}}$ by $\mathbb{T}$; which is trivial—when interpreted in matrices:

$$
\begin{aligned}
\mathbb{T};\varepsilon \cap \overline{\mathbb{T};\sigma} &= (\mathbb{I};\varepsilon \cup \overline{\mathbb{I}};\varepsilon) \cap \overline{\mathbb{I};\sigma \cup \overline{\mathbb{I}};\sigma} = (\varepsilon \cup \overline{\mathbb{I}};\varepsilon) \cap \overline{\sigma} \cap \overline{\overline{\mathbb{I}};\sigma} \\
&= (\varepsilon \cup \overline{\mathbb{I}};\varepsilon) \cap (\overline{\mathbb{I}};\varepsilon \cup \overline{\varepsilon}) \cap \overline{\overline{\mathbb{I}};\sigma} = \big[\overline{\mathbb{I}};\varepsilon \cup (\varepsilon \cap \overline{\varepsilon})\big] \cap \overline{\overline{\mathbb{I}};\sigma} = \overline{\mathbb{I}};\varepsilon \cap \overline{\overline{\mathbb{I}};\sigma}
\end{aligned}
$$

□

In C', it is no longer allowed that an element is in contact to a singleton set it is not contained in.

From such contact relation, we got the closure operation $\rho' := \mathtt{syq}(C, \varepsilon)$ in Proposition 7.1.2.iii, from which in turn a topology may be derived. One should keep in mind that then $\rho';\varepsilon^{\mathsf{T}} = C^{\mathsf{T}}$.

Proposition 7.1.4 *Given an arbitrary Aumann contact relation C, the construct*

$$\mathcal{U} := C;\rho';\Omega = C;\overline{C^{\mathsf{T}};\overline{\varepsilon}} = C;(C\backslash\varepsilon)$$

is indeed a neighborhood topology as defined in Definition 5.2.1.

Proof At the beginning, we prove equivalence of the definition variants

$$\rho' ; \Omega = \rho' ; \overline{\varepsilon^{\mathsf{T}} ; \overline{\varepsilon}} = \overline{\rho' ; \varepsilon^{\mathsf{T}} ; \overline{\varepsilon}} = \overline{C^{\mathsf{T}} ; \overline{\varepsilon}} = C \backslash \varepsilon.$$

Then we follow the numbering scheme of Definition 5.2.1.

i) The relation $\mathcal{U}$ is total since $C ; \rho' ; \Omega ; \mathbb{T} = C ; \rho' ; \mathbb{T} = C ; \mathbb{T} \supseteq \varepsilon ; \mathbb{T} = \mathbb{T}$.

$$C ; \rho' ; \Omega \subseteq \varepsilon \quad \Longleftrightarrow \quad \overline{\varepsilon} = \overline{\varepsilon} ; \Omega^{\mathsf{T}} \subseteq \overline{C ; \rho'} \quad \Longleftrightarrow \quad C ; \rho' \subseteq \varepsilon \quad \Longleftrightarrow \quad C \subseteq \varepsilon ; \rho'^{\mathsf{T}}.$$

ii) is trivial.

iii)

$$\begin{aligned}
&(\mathcal{U} \textcircled{<} \mathcal{U}) ; \mathfrak{M}_2 = (C ; \rho' ; \Omega \textcircled{<} C ; \rho' ; \Omega) ; \mathfrak{M}_2 && \text{by definition}\\
&= (C \textcircled{<} C) ; (\rho' \otimes \rho') ; (\Omega \otimes \Omega) ; \mathfrak{M}_2 && \text{Corollary 3.2.3}\\
&= (C \textcircled{<} C) ; (\rho' \otimes \rho') ; \mathfrak{M}_2 ; \Omega && \text{Proposition 4.3.8.iv}\\
&= (C \textcircled{<} C) ; (\rho' \otimes \rho') ; \mathtt{syq}((\varepsilon \textcircled{<} \varepsilon), \varepsilon) ; \Omega && \text{by definition}\\
&= (C \textcircled{<} C) ; \mathtt{syq}((\varepsilon \textcircled{<} \varepsilon) ; (\rho' \otimes \rho')^{\mathsf{T}}, \varepsilon) ; \Omega && \text{Proposition 2.1.4.i}\\
&= (C \textcircled{<} C) ; \mathtt{syq}((\varepsilon ; \rho'^{\mathsf{T}} \textcircled{<} \varepsilon ; \rho'^{\mathsf{T}}), \varepsilon) ; \Omega &&\\
&= (C \textcircled{<} C) ; \mathtt{syq}((C \textcircled{<} C), \varepsilon) ; \Omega &&\\
&= \left[\varepsilon \cap \mathbb{T} ; \mathtt{syq}((C \textcircled{<} C), \varepsilon)\right] ; \Omega && \text{Proposition 2.1.1}\\
&\subseteq \left[\varepsilon \cap \mathbb{T} ; \mathtt{syq}(C, \varepsilon)\right] ; \Omega && \text{see below}\\
&= C ; \mathtt{syq}(C, \varepsilon) ; \Omega && \text{Proposition 2.1.1 again}\\
&= C ; \rho' ; \Omega = \mathcal{U} &&
\end{aligned}$$

It remains to prove $\mathbb{T} ; \mathtt{syq}((C \textcircled{<} C), \varepsilon) \subseteq \mathbb{T} ; \mathtt{syq}(C, \varepsilon)$ which follows from

$$\begin{aligned}
\mathbb{T} ; \mathtt{syq}((C \textcircled{<} C), \varepsilon) &= \mathbb{T} ; \mathtt{syq}((\varepsilon ; \rho'^{\mathsf{T}} \textcircled{<} \varepsilon ; \rho'^{\mathsf{T}}), \varepsilon)\\
&= \mathbb{T} ; (\rho' \otimes \rho') ; \mathtt{syq}((\varepsilon \textcircled{<} \varepsilon), \varepsilon) = \mathbb{T} ; (\rho' \otimes \rho') ; \mathfrak{M}_2\\
&\subseteq \mathbb{T} ; \rho' = \mathbb{T} ; \mathtt{syq}(C, \varepsilon),
\end{aligned}$$

when we manage to show

$$\mathbb{T} ; (\rho' \otimes \rho') ; \mathfrak{M}_2 \subseteq \mathbb{T} ; \rho'$$

or equivalently after shunting twice

$$\mathbb{T} \subseteq \mathbb{T} ; \rho' ; \mathfrak{M}_2^{\mathsf{T}} ; (\rho' \otimes \rho')^{\mathsf{T}}.$$

The term considered indeed evaluates to an obviously surjective relation, namely

$$\begin{aligned}
\rho' ; \mathfrak{M}_2^{\mathsf{T}} ; (\rho' \otimes \rho')^{\mathsf{T}} &= \rho' ; \mathtt{syq}(\varepsilon, (\varepsilon \circledless \varepsilon)) ; (\rho' \otimes \rho')^{\mathsf{T}} \\
&= \mathtt{syq}(\varepsilon ; \rho'^{\mathsf{T}}, (\varepsilon \circledless \varepsilon) ; (\rho' \otimes \rho')^{\mathsf{T}}) \\
&= \mathtt{syq}(C, (C \circledless C)) = \overline{\overline{C}^{\mathsf{T}} ; (C \circledless C)} \cap \overline{C^{\mathsf{T}} ; \overline{(C \circledless C)}} \\
&\supseteq \overline{\overline{\varepsilon}^{\mathsf{T}} ; (C \circledless C)} \cap \overline{C^{\mathsf{T}} ; \left[\overline{C} ; \pi^{\mathsf{T}} \cup \overline{C} ; \rho^{\mathsf{T}}\right]} \\
&= \overline{\overline{\varepsilon}^{\mathsf{T}} ; (C \circledless C)} \cap \overline{\varepsilon^{\mathsf{T}} ; \overline{C} ; \pi^{\mathsf{T}} \cup \varepsilon^{\mathsf{T}} ; \overline{C} ; \rho^{\mathsf{T}}} \quad \text{Proposition 7.1.2.i} \\
&= \overline{\overline{\varepsilon}^{\mathsf{T}} ; (C \circledless C)} \cap \overline{\varepsilon^{\mathsf{T}} ; \overline{(C \circledless C)}} = \mathtt{syq}(\varepsilon, (C \circledless C))
\end{aligned}$$

iv)

$$\begin{aligned}
\mathcal{U} = C ; \rho' ; \Omega &= C ; \overline{C^{\mathsf{T}} ; \overline{C ; \rho' ; \Omega}} = C ; \overline{\rho' ; \varepsilon^{\mathsf{T}} ; \overline{C ; \rho' ; \Omega}} \\
&= C ; \rho' ; \overline{\varepsilon^{\mathsf{T}} ; \overline{C ; \rho' ; \Omega}} \subseteq C ; \rho' ; \Omega ; \overline{\varepsilon^{\mathsf{T}} ; \overline{C ; \rho' ; \Omega}} = \mathcal{U} ; \overline{\varepsilon^{\mathsf{T}} ; \overline{\mathcal{U}}}
\end{aligned}$$

□

But also the other way round: Then one obtains, however, always a topological Aumann contact. The concept of an Aumann contact is, thus, the more general one.

Proposition 7.1.5 *Given a neighborhood topology $\mathcal{U}$, we will always obtain a topological Aumann contact with the construct $C := \overline{\mathcal{U}} ; \mathcal{N}$.*

Proof

i) $\varepsilon \subseteq \overline{\mathcal{U}} ; \mathcal{N} \iff \overline{\varepsilon} = \varepsilon ; \mathcal{N} \subseteq \overline{\mathcal{U}} \iff \mathcal{U} \subseteq \varepsilon$ holds by definition.

We consider $g := \mathtt{syq}(\mathbb{T}, \varepsilon) = \overline{\mathbb{T} ; \overline{\varepsilon}}$, the mapping that sends every element to the powerset element corresponding to the greatest subset. Then obviously

$$\Omega ; g^{\mathsf{T}} = \overline{\varepsilon^{\mathsf{T}} ; \overline{\varepsilon}} ; g^{\mathsf{T}} = \overline{\varepsilon^{\mathsf{T}} ; \overline{\varepsilon ; g^{\mathsf{T}}}} = \overline{\varepsilon^{\mathsf{T}} ; \overline{\varepsilon ; \mathtt{syq}(\varepsilon, \mathbb{T})}} = \overline{\varepsilon^{\mathsf{T}} ; \overline{\mathbb{T}}} = \overline{\varepsilon^{\mathsf{T}} ; \mathbb{L}} = \mathbb{T}.$$

This allows us to proceed as follows

$$\mathbb{T} = \mathcal{U} ; \mathbb{T} = \mathcal{U} ; \Omega ; g^{\mathsf{T}} = \mathcal{U} ; g^{\mathsf{T}} \quad \Longrightarrow \quad \mathbb{T} ; g \subseteq \mathcal{U} \quad \text{via shunting}$$

in order to finally arrive at

$$C = \overline{\mathcal{U}} ; \mathcal{N} \subseteq \mathbb{T} ; \varepsilon \quad \iff \quad \overline{\mathcal{U}} \subseteq \mathbb{T} ; \varepsilon ; \mathcal{N} = \mathbb{T} ; \overline{\varepsilon} = \overline{\mathbb{T} ; g}.$$

ii)

$$C;\overline{\varepsilon^{\mathsf{T}};\overline{C}} = \overline{\mathcal{U}};\mathcal{N};\overline{\varepsilon^{\mathsf{T}};\overline{\overline{\mathcal{U}};\mathcal{N}}} \subseteq \overline{\mathcal{U}};\mathcal{N} = C \quad \Longleftarrow \quad \overline{\mathcal{U}};\mathcal{N};\overline{\varepsilon^{\mathsf{T}};\overline{\overline{\overline{\mathcal{U}}}}} \subseteq \overline{\mathcal{U}}$$
$$\Longleftrightarrow \quad \overline{\mathcal{U}};\overline{\overline{\varepsilon}^{\mathsf{T}};\mathcal{U}} \subseteq \overline{\mathcal{U}} \quad \Longleftrightarrow \quad \mathcal{U};\overline{\mathcal{U}^{\mathsf{T}};\overline{\varepsilon}} \subseteq \mathcal{U} \quad \Longleftarrow \quad \text{Proposition 5.2.2.v}$$

In addition, we obtain that the Aumann contact is a topological one:

$$(\overline{C} \otimes\!\!< \overline{C});\mathfrak{J}_2 = (\mathcal{U};\mathcal{N} \otimes\!\!< \mathcal{U};\mathcal{N});\mathfrak{J}_2 = (\mathcal{U} \otimes\!\!< \mathcal{U});(\mathcal{N} \otimes \mathcal{N});\mathfrak{J}_2$$
$$= (\mathcal{U} \otimes\!\!< \mathcal{U});\mathfrak{M}_2;\mathcal{N} \subseteq \mathcal{U};\mathcal{N} = \overline{C} \quad \text{using Proposition 4.3.1.ii} \qquad \square$$

Topological Aumann contacts are, as we are about to show, in direct correspondence with neighborhood topologies when we consider a transition to topology that is different from Proposition 7.1.4.

Proposition 7.1.6 *Given a topological Aumann contact C, the construct*

$$\mathcal{U} := \overline{C};\mathcal{N}$$

is indeed a neighborhood topology as defined in Definition 5.2.1.

Proof We follow the numbering scheme of Definition 5.2.1.

i)

$$\mathcal{U};\mathbb{T} = \overline{C};\mathcal{N};\mathbb{T} = \overline{C};\mathbb{T} \supseteq \overline{\mathbb{T};\varepsilon};\mathbb{T} \supseteq \mathtt{syq}(\mathbb{L}, \varepsilon);\mathbb{T} = \mathbb{T}$$

$\mathcal{U} \subseteq \varepsilon$ means $\overline{C};\mathcal{N} \subseteq \varepsilon \iff \overline{C} \subseteq \varepsilon;\mathcal{N} = \overline{\varepsilon}$, which follows from $\varepsilon \subseteq C$.

ii) From Proposition 7.1.2.ii, we know that $C;\Omega \subseteq C$, which also means $\overline{C};\Omega^{\mathsf{T}} \subseteq \overline{C}$. Because obviously $\Omega^{\mathsf{T}} = \mathcal{N};\Omega;\mathcal{N}$, this gives the result when considering

$$\mathcal{U};\Omega = \overline{C};\mathcal{N};\Omega \subseteq \overline{C};\mathcal{N} = \mathcal{U}.$$

iii) We start from $(\overline{C} \otimes\!\!< \overline{C}) \subseteq \overline{C};\mathfrak{J}_2^{\mathsf{T}}$ for a topological Aumann contact and modify both sides:

$$(\overline{C} \otimes\!\!< \overline{C}) = (\overline{\overline{\mathcal{U}};\mathcal{N}} \otimes\!\!< \overline{\overline{\mathcal{U}};\mathcal{N}}) = (\mathcal{U};\mathcal{N} \otimes\!\!< \mathcal{U};\mathcal{N})$$
$$= (\mathcal{U} \otimes\!\!< \mathcal{U});(\mathcal{N} \otimes \mathcal{N}) \subseteq \mathcal{U};\mathfrak{M}_2^{\mathsf{T}};(\mathcal{N} \otimes \mathcal{N})$$
$$\overline{C};\mathfrak{J}_2^{\mathsf{T}} = \overline{\overline{\mathcal{U}};\mathcal{N}};\mathfrak{J}_2^{\mathsf{T}} = \mathcal{U};\mathcal{N};\mathfrak{J}_2^{\mathsf{T}}$$

after which procedure we may apply Proposition 4.3.1.ii.

iv) $\mathcal{U} \subseteq \mathcal{U};\overline{\varepsilon^\mathsf{T};\overline{\mathcal{U}}}$ is to be shown, i.e. $\overline{C};\mathcal{N} \subseteq \overline{C};\mathcal{N};\overline{\varepsilon^\mathsf{T};\overline{\overline{C};\mathcal{N}}}$ or else $\overline{C} \subseteq \overline{C};\overline{\overline{\varepsilon}^\mathsf{T};C}$.
It suffices to prove

$$\overline{C} \subseteq \overline{C};\mathtt{syq}(\varepsilon, C) = \overline{C};\rho'^\mathsf{T} \iff \overline{C};\rho' \subseteq \overline{C} \iff C;\rho'^\mathsf{T} \subseteq C.$$

This holds because with ρ' idempotent $C;\rho'^\mathsf{T} = \varepsilon;\rho'^\mathsf{T};\rho'^\mathsf{T} = \varepsilon;\rho'^\mathsf{T} = C$. □

The concepts underlying the Aumann contact have attracted further attention. In the voluminous *Theory of convex structures*, [vdV93], the concept of betweenness is defined in predicate logic form which we lift to point-free style without quantifiers as follows.

Definition 7.1.7 A relation $B : X \longrightarrow \mathbf{2}^X$ has been called **betweenness** provided it satisfies in combination with a membership relation $\varepsilon : X \longrightarrow \mathbf{2}^X$ the following:

i) $B \subseteq \mathbb{T};\varepsilon$, i.e., no point is 'between' the empty set,
ii) $\varepsilon \subseteq B$,
iii) $B^\mathsf{T};\overline{B} \subseteq \varepsilon^\mathsf{T};\overline{B}$. □

The comparison with Definition 7.1.1 makes it evident that this concept coincides with the earlier one of an Aumann contact. A detailed study of certain aspects of betweenness may also be found in [AN98].

7.2 Overview of Relationships

In total, we have the interrelationship of these topological concepts as shown in the following diagram. The result of Proposition 7.1.5 does not help in identifying the way back from $\mathcal{U}$ to C; it gives a different contact relation (Fig. 7.2).

To the lowest two we may go also directly from $\mathcal{U}$:

$$\mathcal{O}_D = \mathbb{I} \cap \overline{\varepsilon^\mathsf{T};\overline{\mathcal{U}}} \qquad \mathcal{O}_V = \overline{\varepsilon^\mathsf{T};\overline{\mathcal{U}}};\mathbb{T}.$$

This follows since

$$\mathcal{K}^\mathsf{T} = \mathtt{syq}(\varepsilon, \mathcal{U}) = \overline{\varepsilon^\mathsf{T};\overline{\mathcal{U}}} \cap \overline{\overline{\varepsilon}^\mathsf{T};\mathcal{U}} \supseteq \overline{\varepsilon^\mathsf{T};\overline{\mathcal{U}}} \cap \overline{\overline{\varepsilon}^\mathsf{T};\varepsilon} = \overline{\varepsilon^\mathsf{T};\overline{\mathcal{U}}} \cap \Omega^\mathsf{T},$$

but also $\overline{\varepsilon^\mathsf{T};\overline{\mathcal{U}}} \subseteq \overline{\varepsilon^\mathsf{T};\overline{\varepsilon}} = \Omega$.

The situation around $R, C, \mathcal{U}, C_2$ is further illustrated with Fig. 7.3.

The toggling between $\mathcal{U}$ and C_2 may simply be described by "negate and flip horizontally".

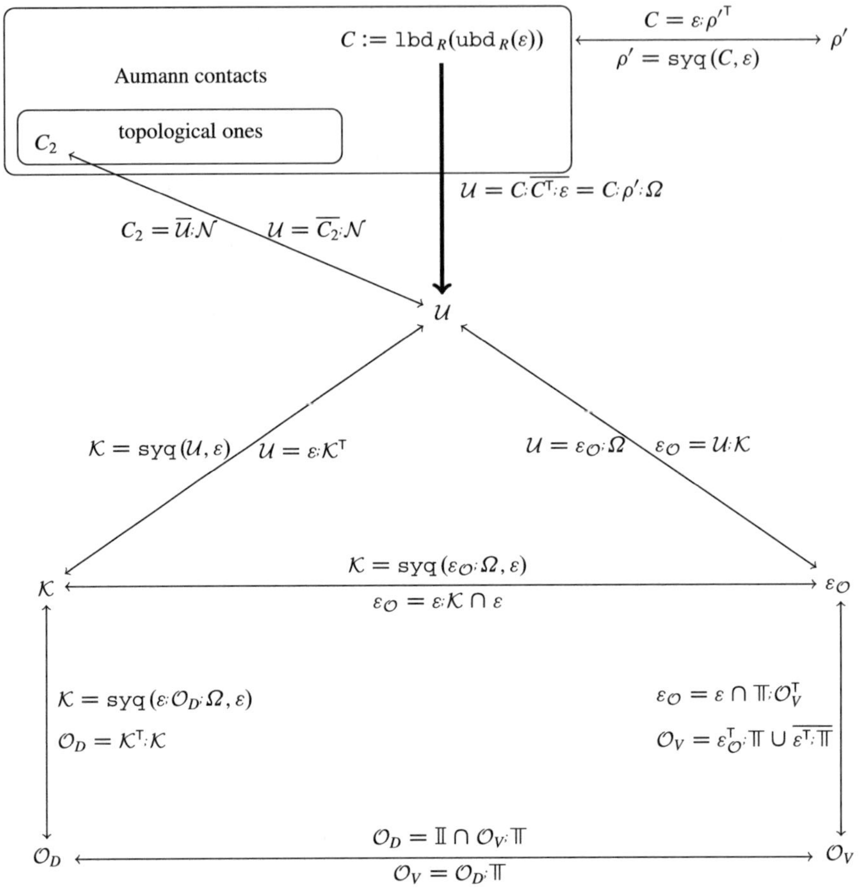

Fig. 7.2 Overview of topology definitions originating from an arbitrary R

There seems not to exist a natural way of getting back from $\mathcal{U}$ to C. The topological Aumann contact C_2 we obtain from $\mathcal{U}$ also leads to a closure

$$\begin{aligned}\rho_2' &= \mathtt{syq}(C_2,\varepsilon) = \mathtt{syq}(\overline{\mathcal{U}};\mathcal{N},\varepsilon) = \mathcal{N};\mathtt{syq}(\overline{\mathcal{U}},\varepsilon)\\ &= \mathcal{N};\mathtt{syq}(\mathcal{U},\overline{\varepsilon}) = \mathcal{N};\mathtt{syq}(\mathcal{U},\varepsilon;\mathcal{N}) = \mathcal{N};\mathtt{syq}(\mathcal{U},\varepsilon);\mathcal{N} = \mathcal{N};\mathcal{K};\mathcal{N},\end{aligned}$$

that—up to negations—is related with the kernel operation $\mathcal{K}$. Furthermore, C_2 coincides with ε^- from p. 91; see Fig. 7.4.

It should be mentioned that an arbitrary topology $\mathcal{U}$ always comes via $C_2 := \overline{\mathcal{U}};\mathcal{N}$ and $\mathcal{U}_2 := C_2;\overline{C_2^{\mathsf{T}};\overline{\varepsilon}} = \overline{\mathcal{U}};\overline{\overline{\mathcal{U}}^{\mathsf{T}};\overline{\varepsilon}}$ together with a second topology. This in turn has the Aumann contact $C_3 := \overline{\mathcal{U}_2};\mathcal{N}$ which is necessarily a topological one.

$R =$ given arbitrarily

	1	2	3	4	5	6
1	1	0	0	1	1	0
2	0	1	1	0	0	0
3	0	0	0	1	1	1
4	0	0	0	0	0	0
5	0	0	0	1	0	0

$C := \mathtt{lbd}_R(\mathtt{ubd}_R(\varepsilon)) =$

	{}	{1}	{2}	{1,2}	{3}	{1,3}	{2,3}	{1,2,3}	{4}	{1,4}	{2,4}	{1,2,4}	{3,4}	{1,3,4}	{2,3,4}	{1,2,3,4}	{5}	{1,5}	{2,5}	{1,2,5}	{3,5}	{1,3,5}	{2,3,5}	{1,2,3,5}	{4,5}	{1,4,5}	{2,4,5}	{1,2,4,5}	{3,4,5}	{1,3,4,5}	{2,3,4,5}	{1,2,3,4,5}
1	0	1	0	1	0	1	1	1	1	1	1	1	1	1	1	1	1	1	1	1	1	1	1	1	1	1	1	1	1	1	1	1
2	0	0	1	1	0	0	1	1	1	1	1	1	1	1	1	1	0	0	1	1	0	0	1	1	1	1	1	1	1	1	1	1
3	0	0	0	1	1	1	1	1	1	1	1	1	1	1	1	1	1	1	1	1	1	1	1	1	1	1	1	1	1	1	1	1
4	0	0	0	1	0	0	1	1	1	1	1	1	1	1	1	1	0	0	1	1	0	0	1	1	1	1	1	1	1	1	1	1
5	0	0	0	1	0	0	1	1	1	1	1	1	1	1	1	1	1	1	1	1	1	1	1	1	1	1	1	1	1	1	1	1

$\mathcal{U} := C \mathbin{;} \overline{C^{\mathsf{T}} \mathbin{;} \overline{\varepsilon}} =$

	{}	{1}	{2}	{1,2}	{3}	{1,3}	{2,3}	{1,2,3}	{4}	{1,4}	{2,4}	{1,2,4}	{3,4}	{1,3,4}	{2,3,4}	{1,2,3,4}	{5}	{1,5}	{2,5}	{1,2,5}	{3,5}	{1,3,5}	{2,3,5}	{1,2,3,5}	{4,5}	{1,4,5}	{2,4,5}	{1,2,4,5}	{3,4,5}	{1,3,4,5}	{2,3,4,5}	{1,2,3,4,5}
1	0	1	0	1	0	1	0	1	0	1	0	1	0	1	0	1	0	1	0	1	0	1	0	1	0	1	0	1	0	1	0	1
2	0	0	1	1	0	0	1	1	0	0	1	1	0	0	1	1	0	0	1	1	0	0	1	1	0	0	1	1	0	0	1	1
3	0	0	0	0	1	1	1	1	0	0	0	0	1	1	1	1	0	0	0	0	1	1	1	1	0	0	0	0	1	1	1	1
4	0	0	0	0	0	0	0	0	0	0	0	0	0	0	0	0	0	0	0	0	0	0	0	0	0	0	0	0	0	0	0	1
5	0	0	0	0	0	0	0	0	0	0	0	0	0	0	0	0	0	0	0	0	0	1	0	1	0	0	0	0	0	1	0	1

$C_2 := \varepsilon \mathbin{;} \mathcal{H}^{\mathsf{T}} = \overline{\mathcal{U}} \mathbin{;} \mathcal{N} =$

	{}	{1}	{2}	{1,2}	{3}	{1,3}	{2,3}	{1,2,3}	{4}	{1,4}	{2,4}	{1,2,4}	{3,4}	{1,3,4}	{2,3,4}	{1,2,3,4}	{5}	{1,5}	{2,5}	{1,2,5}	{3,5}	{1,3,5}	{2,3,5}	{1,2,3,5}	{4,5}	{1,4,5}	{2,4,5}	{1,2,4,5}	{3,4,5}	{1,3,4,5}	{2,3,4,5}	{1,2,3,4,5}
1	0	1	0	1	0	1	0	1	0	1	0	1	0	1	0	1	0	1	0	1	0	1	0	1	0	1	0	1	0	1	0	1
2	0	0	1	1	0	0	1	1	0	0	1	1	0	0	1	1	0	0	1	1	0	0	1	1	0	0	1	1	0	0	1	1
3	0	0	0	0	1	1	1	1	0	0	0	0	1	1	1	1	0	0	0	0	1	1	1	1	0	0	0	0	1	1	1	1
4	0	1	1	1	1	1	1	1	1	1	1	1	1	1	1	1	1	1	1	1	1	1	1	1	1	1	1	1	1	1	1	1
5	0	1	0	1	1	1	1	1	0	1	0	1	1	1	1	1	1	1	1	1	1	1	1	1	1	1	1	1	1	1	1	1

Fig. 7.3 Arbitrary relation R leading to Aumann contact C, topology $\mathcal{U}$ and topological Aumann contact C_2

Such a behaviour is not often met in Mathematics. When we have a Galois correspondence, going forth and back immediately stabilizes after the second step. Here it is different. This fact seems to need a detailed study (not to be elaborated on here).

It was Georg Aumann who felt that something might be studied in addition to the more classical topology versions with neighborhood, open sets, kernel forming, etc. His early work on real functions having appeared in post-war Germany, with

	1	2	3	4	5
a	1	0	1	0	0
b	0	0	0	0	1
c	0	0	0	0	0
d	0	0	1	0	0

R

	{}	{a}	{b}	{a,b}	{c}	{a,c}	{b,c}	{a,b,c}	{d}	{a,d}	{b,d}	{a,b,d}	{c,d}	{a,c,d}	{b,c,d}	{a,b,c,d}
a	0	1	0	1	1	1	1	1	1	1	1	1	1	1	1	1
b	0	0	1	1	1	1	1	1	0	0	1	1	1	1	1	1
c	0	0	0	1	1	1	1	1	0	0	1	1	1	1	1	1
d	0	0	0	1	1	1	1	1	1	1	1	1	1	1	1	1

$C := \mathtt{lbd}_R(\mathtt{ubd}_R(\varepsilon))$

$C_2 := \overline{\mathcal{U}};\mathcal{N} = \overline{C;\overline{C^{\mathsf{T}};\overline{\varepsilon}}};\mathcal{N} =$

	{}	{a}	{b}	{a,b}	{c}	{a,c}	{b,c}	{a,b,c}	{d}	{a,d}	{b,d}	{a,b,d}	{c,d}	{a,c,d}	{b,c,d}	{a,b,c,d}
a	0	1	0	1	0	1	0	1	0	1	0	1	0	1	0	1
b	0	0	1	1	0	0	1	1	0	0	1	1	0	0	1	1
c	0	1	1	1	1	1	1	1	1	1	1	1	1	1	1	1
d	0	1	0	1	0	1	0	1	1	1	1	1	1	1	1	1

	{}	{a}	{b}	{a,b}	{c}	{a,c}	{b,c}	{a,b,c}	{d}	{a,d}	{b,d}	{a,b,d}	{c,d}	{a,c,d}	{b,c,d}	{a,b,c,d}
{}	1	0	0	0	0	0	0	0	0	0	0	0	0	0	0	0
{a}	0	1	0	0	0	0	0	0	0	0	0	0	0	0	0	0
{b}	0	0	1	0	0	0	0	0	0	0	0	0	0	0	0	0
{a,b}	0	0	0	0	0	0	0	0	0	0	0	0	0	0	0	1
{c}	0	0	0	0	0	0	0	0	0	0	0	0	0	0	0	1
{a,c}	0	0	0	0	0	0	0	0	0	0	0	0	0	0	0	1
{b,c}	0	0	0	0	0	0	0	0	0	0	0	0	0	0	0	1
{a,b,c}	0	0	0	0	0	0	0	0	0	0	0	0	0	0	0	1
{d}	0	0	0	0	0	0	0	0	0	1	0	0	0	0	0	0
{a,d}	0	0	0	0	0	0	0	0	0	1	0	0	0	0	0	0
{b,d}	0	0	0	0	0	0	0	0	0	0	0	0	0	0	0	1
{a,b,d}	0	0	0	0	0	0	0	0	0	0	0	0	0	0	0	1
{c,d}	0	0	0	0	0	0	0	0	0	0	0	0	0	0	0	1
{a,c,d}	0	0	0	0	0	0	0	0	0	0	0	0	0	0	0	1
{b,c,d}	0	0	0	0	0	0	0	0	0	0	0	0	0	0	0	1
{a,b,c,d}	0	0	0	0	0	0	0	0	0	0	0	0	0	0	0	1

$\rho' = \mathtt{syq}(C, \varepsilon)$

	{}	{a}	{b}	{a,b}	{c}	{a,c}	{b,c}	{a,b,c}	{d}	{a,d}	{b,d}	{a,b,d}	{c,d}	{a,c,d}	{b,c,d}	{a,b,c,d}
{}	1	0	0	0	0	0	0	0	0	0	0	0	0	0	0	0
{a}	0	0	0	0	0	0	0	0	0	0	0	0	0	1	0	0
{b}	0	0	0	0	0	0	1	0	0	0	0	0	0	0	0	0
{a,b}	0	0	0	0	0	0	0	0	0	0	0	0	0	0	0	1
{c}	0	0	0	0	1	0	0	0	0	0	0	0	0	0	0	0
{a,c}	0	0	0	0	0	0	0	0	0	0	0	0	0	1	0	0
{b,c}	0	0	0	0	0	0	1	0	0	0	0	0	0	0	0	0
{a,b,c}	0	0	0	0	0	0	0	0	0	0	0	0	0	0	0	1
{d}	0	0	0	0	0	0	0	0	0	0	0	0	1	0	0	0
{a,d}	0	0	0	0	0	0	0	0	0	0	0	0	0	1	0	0
{b,d}	0	0	0	0	0	0	0	0	0	0	0	0	0	0	1	0
{a,b,d}	0	0	0	0	0	0	0	0	0	0	0	0	0	0	0	1
{c,d}	0	0	0	0	0	0	0	0	0	0	0	0	1	0	0	0
{a,c,d}	0	0	0	0	0	0	0	0	0	0	0	0	0	1	0	0
{b,c,d}	0	0	0	0	0	0	0	0	0	0	0	0	0	0	1	0
{a,b,c,d}	0	0	0	0	0	0	0	0	0	0	0	0	0	0	0	1

$\rho'_2 = \mathtt{syq}(C_2, \varepsilon)$

Fig. 7.4 An Aumann contact C and a topological Aumann contact C_2, together with closures ρ', ρ'_2 originating from an arbitrary R

second edition as [Aum69], may have inspired him to do so. Being already retired, he tentatively began studying the contact relation of arbitrary closure mappings investigating it in various directions. The overview just given shows that indeed a new and more general concept has been introduced.

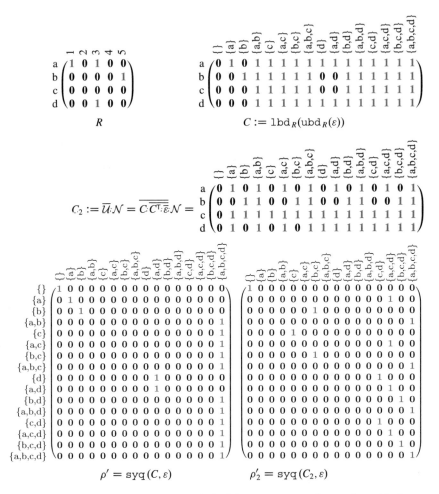

Chapter 8
Proximity and Nearness

Proximity is introduced when trying to axiomatize the concept of being in some sense "near" that may hold from a set to another set. Far better known are point-to-set notions that characterize being element of a neighborhood or of an open set. The first concept of proximity was described in 1908 by Frigyes Riesz and then ignored. Others to be mentioned for having worked on such ideas include V. A. Efremovič in 1934 and A. N. Wallace in 1940. More recently, we found some work in [NW70, VDDB02, BD07].

8.1 Proximity

The list of conditions for a relation δ to qualify as a proximity starts with several simple ones, namely being symmetric, to hold for nonempty sets only, to include nonempty intersection, and to be join-distributive. It ends with a complicated postulate yet to be discussed.

A proximity space (X, δ) is therefore a set X with a relation δ between subsets of X satisfying the following properties: For all subsets A, B, C and E of X

- $A\delta B \implies B\delta A$,
- $A\delta B \implies A \neq \emptyset$,
- $A \cap B \neq \emptyset \implies A\delta B$,
- $A\delta(B \cup C) \iff (A\delta B \text{ or } A\delta C)$,
- $\forall E, A\delta E \text{ or } B\delta(X\backslash E) \implies A\delta B$.

The first four items coincide with those required for a so-called *contact* in [BD07].

In [NW70], the last item of the above five axioms has been called the "strong" axiom—a name that indicates its importance. We will lift it over several steps to a point-free version that does no longer use quantifiers. If $A\delta B$, one says that "A is δ-near B" or "A and B are δ-proximal". It is not too easy to rephrase the intention

G. Schmidt, M. Winter, *Relational Topology*, Lecture Notes in Mathematics 2208, https://doi.org/10.1007/978-3-319-74451-3_8

of the last property above in plain words: Two arbitrary sets A, B aren't δ-near when the space can be split into two parts so that A is not near to the first part and B is not near to the second part.

The main properties of such a set neighborhood relation obviously ask for an alternative axiomatic characterization lifted to point-free form, thus avoiding quantifiers; it is provided with Definition 8.1.1. We restrict ourselves to justifying the lifting process for the most complicated of these laws, the strong one, in some more detail:

$\big[\forall E : A\delta E \vee B\delta(X\backslash E)\big] \longrightarrow A\delta B$

$\big[\neg\exists E : A\overline{\delta}E \wedge B(\overline{\delta};\mathcal{N})E\big] \longrightarrow A\delta B$

$A\overline{\overline{\delta};\mathcal{N};\overline{\delta}}B \longrightarrow A\delta B$

$\overline{\overline{\delta};\mathcal{N};\overline{\delta}} \subseteq \delta$

This leads us to define in a completely point- and quantifier-free form as follows:

Definition 8.1.1 We speak of a **pre-proximity relation** on a set X if in addition to membership $\varepsilon : X \longrightarrow \mathbf{2}^X$ and binary join $\mathfrak{J}_2 : \mathbf{2}^X \times \mathbf{2}^X \longrightarrow \mathbf{2}^X$ a relation $\Delta : \mathbf{2}^X \longrightarrow \mathbf{2}^X$ is given satisfying the following properties

i) $\Delta^{\mathsf{T}} \subseteq \Delta$,
ii) $\Delta;\mathbb{T} \subseteq \varepsilon^{\mathsf{T}};\mathbb{T}$,
iii) $\varepsilon^{\mathsf{T}};\varepsilon \subseteq \Delta$,
iv) $\mathfrak{J}_2;\Delta = (\pi \cup \rho);\Delta$ or, equivalently, $(\overline{\Delta}\otimes\overline{\Delta}) = \overline{\Delta};\mathfrak{J}_2^{\mathsf{T}}$.

A **proximity relation** is a pre-proximity satisfying the strong $\overline{\Delta} \subseteq \overline{\Delta};\mathcal{N};\overline{\Delta}$ in addition. □

Part (iv) allows the variant formulation closer to the Kronecker and fork calculus normally used here. The transition is obvious after negation and transposition.

The following remark identifies the coarsest proximity, so that there is no smaller one, and the biggest.

Remark 8.1.2 If any membership relation $\varepsilon : X \longrightarrow \mathbf{2}^X$ is given, the constructs $\Delta := \varepsilon^{\mathsf{T}};\mathbb{I};\varepsilon$, also called overlap, as well as $\Delta' := \varepsilon^{\mathsf{T}};\mathbb{T} \cap \mathbb{T};\varepsilon = \varepsilon^{\mathsf{T}};\mathbb{T};\varepsilon$ satisfy the requirements for a proximity.

Proof i,ii,iii) are trivial in both cases.
iv) is shown simultaneously for both cases using Proposition 4.3.2.i:

$$\varepsilon;\mathfrak{J}_2^{\mathsf{T}} = \varepsilon;\pi^{\mathsf{T}} \cup \varepsilon;\rho^{\mathsf{T}} = \varepsilon;(\pi^{\mathsf{T}} \cup \rho^{\mathsf{T}})$$

The additional "strong" property that makes these to proximities is also satisfied, which we show for the first variant with

$$\overline{\Delta} \subseteq \Omega;\overline{\Delta} = \overline{\varepsilon^{\mathsf{T}};\overline{\varepsilon}};\overline{\Delta} = \overline{\varepsilon^{\mathsf{T}};\varepsilon;\mathcal{N}};\overline{\Delta} = \overline{\Delta};\mathcal{N};\overline{\Delta}.$$

For the second, we recall that $l := \overline{\varepsilon^{\mathsf{T}}; \mathbb{T}}$ is a point, namely the least element in $\mathbf{2}^X$, and that

$$l \subseteq l; \mathcal{N}; l,$$

which is a consequence of shunting the point l

$$l \subseteq \overline{\varepsilon^{\mathsf{T}}; \mathbb{T}; \mathcal{N}; l} \quad \Longleftrightarrow \quad l; l^{\mathsf{T}} \subseteq \overline{\varepsilon^{\mathsf{T}}; \mathbb{T}; \mathcal{N}} = \overline{\varepsilon^{\mathsf{T}}; \mathbb{T}}; \mathcal{N} = \overline{\varepsilon^{\mathsf{T}}; \mathbb{T}}.$$

Therefore $\overline{\Delta'} = l \cup l^{\mathsf{T}} \subseteq (l \cup l^{\mathsf{T}}); \mathcal{N}; (l \cup l^{\mathsf{T}}) = \overline{\Delta'}; \mathcal{N}; \overline{\Delta'}$. □

Figure 8.1 shows the proximity $\Delta = \varepsilon^{\mathsf{T}}; \mathbb{I}; \varepsilon$ mentioned above and the one obtained from the topology of Fig. 8.2 following Proposition 8.1.3. With Fig. 8.1, it is easy to see that $\Delta' = \varepsilon^{\mathsf{T}}; \mathbb{T}; \varepsilon$ is the biggest conceivable proximity; biggest means: exactly first row and first column with **0** s.

	{}	{1}	{2}	{1,2}	{3}	{1,3}	{2,3}	{1,2,3}	{4}	{1,4}	{2,4}	{1,2,4}	{3,4}	{1,3,4}	{2,3,4}	{1,2,3,4}
{}	0	0	0	0	0	0	0	0	0	0	0	0	0	0	0	0
{1}	0	1	0	1	0	1	0	1	0	1	0	1	0	1	0	1
{2}	0	0	1	1	0	0	1	1	0	0	1	1	0	0	1	1
{1,2}	0	1	1	1	0	1	1	1	0	1	1	1	0	1	1	1
{3}	0	0	0	0	1	1	1	1	0	0	0	0	1	1	1	1
{1,3}	0	1	0	1	1	1	1	1	0	1	0	1	1	1	1	1
{2,3}	0	0	1	1	1	1	1	1	0	0	1	1	1	1	1	1
{1,2,3}	0	1	1	1	1	1	1	1	0	1	1	1	1	1	1	1
{4}	0	0	0	0	0	0	0	0	1	1	1	1	1	1	1	1
{1,4}	0	1	0	1	0	1	0	1	1	1	1	1	1	1	1	1
{2,4}	0	0	1	1	0	0	1	1	1	1	1	1	1	1	1	1
{1,2,4}	0	1	1	1	0	1	1	1	1	1	1	1	1	1	1	1
{3,4}	0	0	0	0	1	1	1	1	1	1	1	1	1	1	1	1
{1,3,4}	0	1	0	1	1	1	1	1	1	1	1	1	1	1	1	1
{2,3,4}	0	0	1	1	1	1	1	1	1	1	1	1	1	1	1	1
{1,2,3,4}	0	1	1	1	1	1	1	1	1	1	1	1	1	1	1	1

	{}	{1}	{2}	{1,2}	{3}	{1,3}	{2,3}	{1,2,3}	{4}	{1,4}	{2,4}	{1,2,4}	{3,4}	{1,3,4}	{2,3,4}	{1,2,3,4}
{}	0	0	0	0	0	0	0	0	0	0	0	0	0	0	0	0
{1}	0	1	1	1	0	1	1	1	0	1	1	1	0	1	1	1
{2}	0	1	1	1	0	1	1	1	0	1	1	1	0	1	1	1
{1,2}	0	1	1	1	0	1	1	1	0	1	1	1	0	1	1	1
{3}	0	0	0	0	1	1	1	1	1	1	1	1	1	1	1	1
{1,3}	0	1	1	1	1	1	1	1	1	1	1	1	1	1	1	1
{2,3}	0	1	1	1	1	1	1	1	1	1	1	1	1	1	1	1
{1,2,3}	0	1	1	1	1	1	1	1	1	1	1	1	1	1	1	1
{4}	0	0	0	0	1	1	1	1	1	1	1	1	1	1	1	1
{1,4}	0	1	1	1	1	1	1	1	1	1	1	1	1	1	1	1
{2,4}	0	1	1	1	1	1	1	1	1	1	1	1	1	1	1	1
{1,2,4}	0	1	1	1	1	1	1	1	1	1	1	1	1	1	1	1
{3,4}	0	0	0	0	1	1	1	1	1	1	1	1	1	1	1	1
{1,3,4}	0	1	1	1	1	1	1	1	1	1	1	1	1	1	1	1
{2,3,4}	0	1	1	1	1	1	1	1	1	1	1	1	1	1	1	1
{1,2,3,4}	0	1	1	1	1	1	1	1	1	1	1	1	1	1	1	1

Fig. 8.1 Coarsest and a bigger proximity

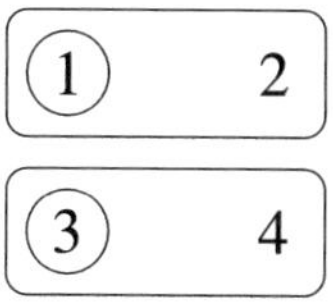

Fig. 8.2 The basis of open sets of the topology used for the right part of Fig. 8.1

Some interrelationships with topology seem obvious. The following proposition states that the relation between two points results in a pre-proximity when one takes their complements and finds a common point to which these are not neighborhoods.

Proposition 8.1.3 *From an arbitrary neighborhood topology determined by* $\mathcal{U}, \mathcal{K}, \mathcal{H}$*, one may obtain the pre-proximity relation*

$$\Delta := \mathcal{N} ; \overline{\mathcal{U}}^{\mathsf{T}} ; \overline{\mathcal{U}} ; \mathcal{N} = \mathcal{H} ; \overline{\Omega} ; \mathcal{K}^{\mathsf{T}} ; \mathcal{N}.$$

Proof Prior to the proof, we show equivalence of the variants:

$$\begin{aligned}\mathcal{N} ; \overline{\mathcal{U}}^{\mathsf{T}} ; \overline{\mathcal{U}} ; \mathcal{N} &= \mathcal{N} ; \overline{\mathcal{K} ; \varepsilon^{\mathsf{T}}} ; \overline{\varepsilon ; \mathcal{K}^{\mathsf{T}}} ; \mathcal{N} = \mathcal{N} ; \mathcal{K} ; \overline{\varepsilon}^{\mathsf{T}} ; \overline{\varepsilon} ; \mathcal{K}^{\mathsf{T}} ; \mathcal{N} \\ &= \mathcal{N} ; \mathcal{K} ; \mathcal{N} ; \varepsilon^{\mathsf{T}} ; \overline{\varepsilon} ; \mathcal{K}^{\mathsf{T}} ; \mathcal{N} = \mathcal{H} ; \overline{\Omega} ; \mathcal{K}^{\mathsf{T}} ; \mathcal{N}\end{aligned}$$

i) The pre-proximity Δ defined by the first variant is obviously symmetric by construction.

ii) We use that $\mathcal{U}$ is total, $\mathcal{U} ; \Omega = \mathcal{U}$ and $\overline{\Omega} = \varepsilon^{\mathsf{T}} ; \overline{\varepsilon}$ to show

$$\begin{aligned}&\mathbb{T} = \mathcal{U} ; \mathbb{T} = \mathcal{U} ; (\Omega \cup \overline{\Omega}) \subseteq \mathcal{U} \cup \mathbb{T} ; \overline{\varepsilon} \\ \iff \quad &\overline{\mathcal{U}}^{\mathsf{T}} \subseteq \overline{\varepsilon}^{\mathsf{T}} ; \mathbb{T} \quad \iff \quad \mathcal{N} ; \overline{\mathcal{U}}^{\mathsf{T}} \subseteq \varepsilon^{\mathsf{T}} ; \mathbb{T}.\end{aligned}$$

Now obviously $\Delta ; \mathbb{T} \subseteq \mathcal{N} ; \overline{\mathcal{U}}^{\mathsf{T}} ; \mathbb{T} \subseteq \varepsilon^{\mathsf{T}} ; \mathbb{T}$.

iii) $\varepsilon^{\mathsf{T}} ; \varepsilon \subseteq \mathcal{N} ; \overline{\mathcal{U}}^{\mathsf{T}} ; \overline{\mathcal{U}} ; \mathcal{N} \iff \mathcal{N} ; \varepsilon^{\mathsf{T}} ; \varepsilon ; \mathcal{N} = \overline{\varepsilon}^{\mathsf{T}} ; \overline{\varepsilon} \subseteq \overline{\mathcal{U}}^{\mathsf{T}} ; \overline{\mathcal{U}} \iff \mathcal{U} \subseteq \varepsilon$

iv)

$$\begin{aligned}\overline{\Delta} ; \mathfrak{J}_2^{\mathsf{T}} &= \mathcal{H} ; \Omega ; \mathcal{K}^{\mathsf{T}} ; \mathcal{N} ; \mathfrak{J}_2^{\mathsf{T}} && \text{maps slip below negation} \\ &= \mathcal{H} ; \Omega ; \mathcal{K}^{\mathsf{T}} ; \mathfrak{M}_2^{\mathsf{T}} ; (\mathcal{N} \otimes \mathcal{N}) && \text{De Morgan rule} \\ & && \text{Proposition 4.3.1.ii} \\ &= \mathcal{H} ; \Omega ; \mathfrak{M}_2^{\mathsf{T}} ; (\mathcal{K}^{\mathsf{T}} \otimes \mathcal{K}^{\mathsf{T}}) ; (\mathcal{N} \otimes \mathcal{N}) && \text{Proposition 5.2.3.iii,} \\ & && \mathfrak{M}_2, \mathcal{K} \text{ commute} \\ &= \mathcal{H} ; (\Omega \olessthan \Omega) ; (\mathcal{K}^{\mathsf{T}} \otimes \mathcal{K}^{\mathsf{T}}) ; (\mathcal{N} \otimes \mathcal{N}) && \text{Proposition 4.3.3.i} \\ &= (\mathcal{H} ; \Omega ; \mathcal{K}^{\mathsf{T}} ; \mathcal{N} \olessthan \mathcal{H} ; \Omega ; \mathcal{K}^{\mathsf{T}} ; \mathcal{N}) \\ &= (\overline{\Delta} \olessthan \overline{\Delta})\end{aligned}$$

□

We do not undertake here the proof of the additional "strong" axiom to establish even a proximity (and not just a pre-proximity). It would need to assume some separation such as by the Hausdorff-property.

Figure 8.3 shows that the topology of Fig. 5.2, transformed according to Proposition 8.1.3, does not result in a proximity; it violates the strong axiom and gives, thus, only a pre-proximity. The subsets $\{b\}$ and $\{d\}$ are related via $\overline{\Delta}$. However, the relation $\overline{\Delta} ; \mathcal{N} ; \overline{\Delta}$ relates $\{b\}$ with subsets $\{\}$ and $\{a\}$ only.

$\Delta =$

	{}	{a}	{b}	{a,b}	{c}	{a,c}	{b,c}	{a,b,c}	{d}	{a,d}	{b,d}	{a,b,d}	{c,d}	{a,c,d}	{b,c,d}	{a,b,c,d}
{}	0	0	0	0	0	0	0	0	0	0	0	0	0	0	0	0
{a}	0	1	0	1	0	1	0	1	0	1	0	1	0	1	0	1
{b}	0	0	1	1	1	1	1	1	0	0	1	1	1	1	1	1
{a,b}	0	1	1	1	1	1	1	1	0	1	1	1	1	1	1	1
{c}	0	0	1	1	1	1	1	1	1	1	1	1	1	1	1	1
{a,c}	0	1	1	1	1	1	1	1	1	1	1	1	1	1	1	1
{b,c}	0	0	1	1	1	1	1	1	1	1	1	1	1	1	1	1
{a,b,c}	0	1	1	1	1	1	1	1	1	1	1	1	1	1	1	1
{d}	0	0	0	0	1	1	1	1	1	1	1	1	1	1	1	1
{a,d}	0	1	0	1	1	1	1	1	1	1	1	1	1	1	1	1
{b,d}	0	0	1	1	1	1	1	1	1	1	1	1	1	1	1	1
{a,b,d}	0	1	1	1	1	1	1	1	1	1	1	1	1	1	1	1
{c,d}	0	0	1	1	1	1	1	1	1	1	1	1	1	1	1	1
{a,c,d}	0	1	1	1	1	1	1	1	1	1	1	1	1	1	1	1
{b,c,d}	0	0	1	1	1	1	1	1	1	1	1	1	1	1	1	1
{a,b,c,d}	0	1	1	1	1	1	1	1	1	1	1	1	1	1	1	1

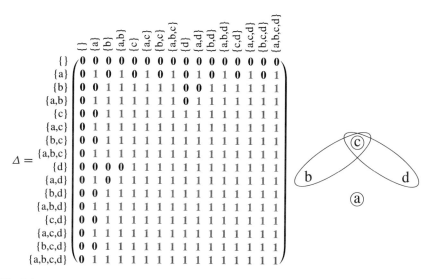

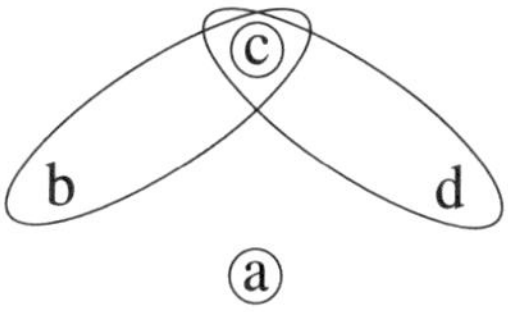

Fig. 8.3 The pre-proximity for the topology Fig. 5.2 is not a proximity because of $\{b\}, \{d\}$

8.2 Another Proximity Concept

A similar concept is provided in the following definition; see also [DV06, DL12]. For purposes of distinct notation, we will call the concept a DV-pre-proximity; others have sometimes termed it *contact*, which would be misleading in the present context. Such a DV-pre-proximity $\mathfrak{D} : B \longrightarrow B$ on a Boolean algebra B (with $0, \leq, -, \cup$) is given, provided the following properties hold:

- $x\mathfrak{D}y \quad \Longrightarrow \quad y\mathfrak{D}x$,
- $x\mathfrak{D}y \quad \Longrightarrow \quad x, y \neq 0$,
- $x \neq 0 \quad \Longrightarrow \quad x\mathfrak{D}x$,
- $x\mathfrak{D}(y \cup z) \quad \Longrightarrow \quad (x\mathfrak{D}y \text{ or } x\mathfrak{D}z)$,
- $x\mathfrak{D}y \text{ and } y \leq z \quad \Longrightarrow \quad x\mathfrak{D}z$.

These are the basic rules. Also here additional properties are often assumed to hold, such as:

- $\mathfrak{D}(x) = \mathfrak{D}(y)$ implies $x = y$,
- If $(\forall z)(x\mathfrak{D}z \text{ or } y\mathfrak{D} - z)$ then $x\mathfrak{D}y$.

The translation of this still partly predicate-logical version to a point-free form without quantifiers is immediate:

Definition 8.2.1 Given $\varepsilon, \Omega, \mathfrak{J}_2$ as usual, the relation $\mathfrak{D}$ is called a **DV-pre-proximity relation**, provided

i) $\mathfrak{D}^{\mathsf{T}} \subseteq \mathfrak{D}$,
ii) $\mathfrak{D} \subseteq \varepsilon^{\mathsf{T}} ; \mathbb{T} ; \varepsilon$,

iii) $\mathbb{I} \cap \varepsilon^{\mathsf{T}};\mathbb{T} \subseteq \mathfrak{D}$,
iv) $\mathfrak{D};\mathfrak{J}_2^{\mathsf{T}} \subseteq \mathfrak{D};(\pi \cup \rho)^{\mathsf{T}}$ or, equivalently, $(\overline{\mathfrak{D}} \otimes \overline{\mathfrak{D}}) \subseteq \overline{\mathfrak{D}};\mathfrak{J}_2^{\mathsf{T}}$,
v) $\mathfrak{D};\Omega \subseteq \mathfrak{D}$. □

Again, we have a variant in (iv) employing the fork operator. It is obtained via negation. Of course, the requirements of Definitions 8.1.1 and 8.2.1 appear to be somehow similar. We prove that they are indeed equivalent.

Proposition 8.2.2 *Every DV-pre-proximity is a pre-proximity and vice versa.*

Proof First we prove "Definition 8.1.1 satisfied $\Longrightarrow$ Definition 8.2.1 satisfied". Properties (i,iv) are obvious.

ii) From Definition 8.1.1.ii, we get $\Delta \subseteq \Delta;\mathbb{T} \subseteq \varepsilon^{\mathsf{T}};\mathbb{T}$ as well as by symmetry $\Delta \subseteq \mathbb{T};\varepsilon$, so that with masking

$$\Delta \subseteq \varepsilon^{\mathsf{T}};\mathbb{T} \cap \mathbb{T};\varepsilon = (\mathbb{T} \cap \varepsilon^{\mathsf{T}};\mathbb{T});\varepsilon = \varepsilon^{\mathsf{T}};\mathbb{T};\varepsilon.$$

iii) From Definition 8.1.1.iii, we get with the Dedekind rule

$$\varepsilon^{\mathsf{T}};\mathbb{T} \cap \mathbb{I} \subseteq (\varepsilon^{\mathsf{T}} \cap \mathbb{I};\mathbb{T});(\mathbb{T} \cap \varepsilon;\mathbb{I}) = \varepsilon^{\mathsf{T}};\varepsilon \subseteq \Delta.$$

v)

$$\begin{aligned}\Delta;\Omega &= \Delta;\pi^{\mathsf{T}};\mathfrak{J}_2 && \text{Proposition 4.2.2.vii}\\ &\subseteq \Delta;(\pi^{\mathsf{T}} \cup \rho^{\mathsf{T}});\mathfrak{J}_2 &&\\ &= \Delta;\mathfrak{J}_2^{\mathsf{T}};\mathfrak{J}_2 && \text{Definition 8.1.1.iv}\\ &\subseteq \Delta && \text{since } \mathfrak{J}_2 \text{ is univalent}\end{aligned}$$

Now we switch to proving "Definition 8.1.1 satisfied $\Longleftarrow$ Definition 8.2.1 satisfied":

i) follows from Definition 8.2.1.i.
ii) $\mathfrak{D};\mathbb{T} \subseteq \varepsilon^{\mathsf{T}};\mathbb{T}$ follows from Definition 8.2.1.ii because $\mathbb{T};\varepsilon;\mathbb{T} \subseteq \mathbb{T}$.
iii) From Definition 8.2.1.i,v, we get $\Omega^{\mathsf{T}};\mathfrak{D};\Omega \subseteq \mathfrak{D}$. Furthermore

$$\sigma;(\Omega \cap \varepsilon^{\mathsf{T}};\mathbb{T}) = \sigma;\Omega \cap \sigma;\varepsilon^{\mathsf{T}};\mathbb{T} = \sigma;\overline{\varepsilon^{\mathsf{T}};\overline{\varepsilon}} \cap \mathbb{I};\mathbb{T} = \overline{\sigma;\varepsilon^{\mathsf{T}};\overline{\varepsilon}} = \overline{\mathbb{I};\overline{\varepsilon}} = \varepsilon.$$

Applying both together with (iii) produces

$$\begin{aligned}\varepsilon^{\mathsf{T}};\varepsilon &= (\Omega^{\mathsf{T}} \cap \mathbb{T};\varepsilon);\sigma^{\mathsf{T}};\sigma;(\Omega \cap \varepsilon^{\mathsf{T}};\mathbb{T}) \subseteq (\Omega^{\mathsf{T}} \cap \mathbb{T};\varepsilon);\mathbb{I};(\Omega \cap \varepsilon^{\mathsf{T}};\mathbb{T})\\ &= \Omega^{\mathsf{T}};(\mathbb{I} \cap \varepsilon^{\mathsf{T}};\mathbb{T});\Omega \subseteq \Omega^{\mathsf{T}};\mathfrak{D};\Omega \subseteq \mathfrak{D}.\end{aligned}$$

iv) follows from Definition 8.2.1.iv concerning "$\subseteq$". Regarding "$\supseteq$", we prove, e.g.,

$$\mathfrak{D};\pi^{\mathsf{T}} \subseteq \mathfrak{D};\mathfrak{J}_2^{\mathsf{T}} \iff \mathfrak{D};\pi^{\mathsf{T}};\mathfrak{J}_2 \subseteq \mathfrak{D} \iff \mathfrak{D};\Omega \subseteq \mathfrak{D}$$

shunting, and using Proposition 4.2.2.vii as well as the present property of Definition 8.2.1.v. □

As this has now been proved, we may use either of these definitions together with the strong axiom as defined in Definition 8.1.1. With the following proposition, we see that a pre-proximity may arise from fairly trivial sources.

Proposition 8.2.3 *Given a reflexive and symmetric relation* $R : X \longrightarrow X$ *together with the corresponding membership* $\varepsilon : X \longrightarrow \mathbf{2}^X$*, the construct* $\mathfrak{D} := \varepsilon^\mathsf{T}{:}R{:}\varepsilon$ *is a DV-pre-proximity relation.*

Proof i) and (ii) are trivial for symmetric R. (v) follows from $\varepsilon{:}\Omega = \varepsilon$.
For (iii), we have obviously $\varepsilon{:}\mathbb{I} \subseteq R{:}\varepsilon$ when R is reflexive; therefore

$$\varepsilon^\mathsf{T}{:}\overline{R{:}\varepsilon} \subseteq \overline{\mathbb{I}}, \qquad \mathbb{I} \cap \varepsilon^\mathsf{T}{:}\overline{R{:}\varepsilon} \subseteq \perp\!\!\!\perp,$$

so that, splitting $\mathbb{T} = \left[\overline{R{:}\varepsilon} \cup R{:}\varepsilon\right]$,

$$\begin{aligned}\mathbb{I} \cap \varepsilon^\mathsf{T}{:}\mathbb{T} &= \mathbb{I} \cap \varepsilon^\mathsf{T}{:}\left[\overline{R{:}\varepsilon} \cup R{:}\varepsilon\right] \\ &= \mathbb{I} \cap \left[\varepsilon^\mathsf{T}{:}\overline{R{:}\varepsilon} \cup \varepsilon^\mathsf{T}{:}R{:}\varepsilon\right] \subseteq \perp\!\!\!\perp \cup\, \varepsilon^\mathsf{T}{:}R{:}\varepsilon = \mathfrak{D}.\end{aligned}$$

iv) We recall Proposition 4.3.2.ii, namely $\overline{\varepsilon}{:}\mathfrak{J}_2^\mathsf{T} = (\overline{\varepsilon} \circledless \overline{\varepsilon})$, implying

$$\begin{aligned}\overline{\mathfrak{D}}{:}\mathfrak{J}_2^\mathsf{T} &= \overline{\mathfrak{D}{:}\mathfrak{J}_2^\mathsf{T}} = \overline{\varepsilon^\mathsf{T}{:}R{:}\varepsilon{:}\mathfrak{J}_2^\mathsf{T}} = \overline{\varepsilon^\mathsf{T}{:}R{:}\overline{(\overline{\varepsilon} \circledless \overline{\varepsilon})}} = \overline{\varepsilon^\mathsf{T}{:}R{:}(\varepsilon{:}\pi^\mathsf{T} \cup \varepsilon{:}\rho^\mathsf{T})} \\ &= \overline{\varepsilon^\mathsf{T}{:}R{:}\varepsilon{:}\pi^\mathsf{T}} \cap \overline{\varepsilon^\mathsf{T}{:}R{:}\varepsilon{:}\rho^\mathsf{T}} = \overline{\varepsilon^\mathsf{T}{:}R{:}\varepsilon}{:}\pi^\mathsf{T} \cap \overline{\varepsilon^\mathsf{T}{:}R{:}\varepsilon}{:}\rho^\mathsf{T} = (\overline{\mathfrak{D}} \circledless \overline{\mathfrak{D}})\end{aligned}$$ □

It shall now even be shown that there is a one-to-one correspondence between reflexive and symmetric relations R and DV-pre-proximity relations $\mathfrak{D}$. In the following, we prepare this result by recalling the folklore properties of a Galois correspondence.

Proposition 8.2.4 *Let be given relations* $A : X \longrightarrow Y$ *and* $B : U \longrightarrow V$.

i) Then there holds a Galois correspondence, i.e.

$$R \subseteq \pi(C) \quad \Longleftrightarrow \quad C \supseteq \zeta(R),$$

between relations $R : X \longrightarrow U$ *and* $C : Y \longrightarrow V$ *when defining*

$$\pi(C) := \overline{A{:}\overline{C}{:}B^\mathsf{T}}, \qquad \zeta(R) := A^\mathsf{T}{:}R{:}B.$$

ii) Specializing to $A := B := \varepsilon : X \longrightarrow \mathbf{2}^X$, *the transition* $\zeta(R) := \varepsilon^\mathsf{T}{:}R{:}\varepsilon$ *turns out to be injective.*

Proof

i) We use the Schröder rule to obtain

$$\begin{aligned} R \subseteq \overline{A ; \overline{C} ; B^{\mathsf{T}}} &\iff A ; \overline{C} ; B^{\mathsf{T}} \subseteq \overline{R} \iff R ; B \subseteq \overline{A ; \overline{C}} \\ &\iff A ; \overline{C} \subseteq \overline{R ; B} \iff A^{\mathsf{T}} ; R ; B \subseteq C. \end{aligned}$$

ii) With two times membership deletion, Proposition 2.2.3, we get

$$\pi(\zeta(R)) = \overline{\varepsilon ; \overline{\varepsilon^{\mathsf{T}} ; \overline{R} ; \varepsilon} ; \varepsilon^{\mathsf{T}}} = \overline{\overline{\overline{R}} ; \varepsilon ; \varepsilon^{\mathsf{T}}} = \overline{\overline{R}} = R,$$

so that ζ must be injective, making this an embedding. □

We now focus on the special case of R being reflexive and symmetric and we see what it means in Proposition 8.2.3. It is obvious that symmetry propagates from R to $\mathfrak{D}$ and vice versa. Starting from $\mathfrak{D}$, we are in a position to prove that R is reflexive, mainly by membership relation deletion

$$R = \overline{\varepsilon ; \overline{\mathfrak{D}} ; \varepsilon^{\mathsf{T}}} \supseteq \overline{\varepsilon ; \overline{\varepsilon^{\mathsf{T}} ; \varepsilon} ; \varepsilon^{\mathsf{T}}} = \overline{\overline{\varepsilon} ; \varepsilon^{\mathsf{T}}} = \overline{\overline{\mathbb{I}} ; \varepsilon ; \varepsilon^{\mathsf{T}}} = \overline{\overline{\mathbb{I}}} = \mathbb{I}.$$

Nevertheless, it is possible as before to start from an arbitrary R and obtain the Aumann contact C with closure forming ρ (Fig. 8.4).

$$R = \begin{array}{c} \\ a \\ b \\ c \\ d \end{array} \begin{array}{cccc} a & b & c & d \\ 1 & 1 & 0 & 1 \\ 1 & 0 & 1 & 0 \\ 0 & 1 & 1 & 1 \\ 1 & 0 & 0 & 0 \end{array}$$

$$C = \begin{array}{c|cccccccccccccccc} & \{\} & \{a\} & \{b\} & \{a,b\} & \{c\} & \{a,c\} & \{b,c\} & \{a,b,c\} & \{d\} & \{a,d\} & \{b,d\} & \{a,b,d\} & \{c,d\} & \{a,c,d\} & \{b,c,d\} & \{a,b,c,d\} \\ \hline a & 0 & 1 & 0 & 1 & 0 & 1 & 0 & 1 & 1 & 1 & 1 & 1 & 1 & 1 & 1 & 1 \\ b & 0 & 0 & 1 & 1 & 0 & 0 & 1 & 1 & 1 & 1 & 1 & 1 & 1 & 1 & 1 & 1 \\ c & 0 & 0 & 0 & 0 & 1 & 1 & 1 & 1 & 0 & 0 & 0 & 0 & 1 & 1 & 1 & 1 \\ d & 0 & 0 & 0 & 1 & 0 & 0 & 0 & 1 & 1 & 1 & 1 & 1 & 1 & 1 & 1 & 1 \end{array}$$

$$\rho = \mathtt{syq}(C, \varepsilon) = \begin{array}{c|cccccccccccccccc} \{\} & 1 & 0 & 0 & 0 & 0 & 0 & 0 & 0 & 0 & 0 & 0 & 0 & 0 & 0 & 0 & 0 \\ \{a\} & 0 & 1 & 0 & 0 & 0 & 0 & 0 & 0 & 0 & 0 & 0 & 0 & 0 & 0 & 0 & 0 \\ \{b\} & 0 & 0 & 1 & 0 & 0 & 0 & 0 & 0 & 0 & 0 & 0 & 0 & 0 & 0 & 0 & 0 \\ \{a,b\} & 0 & 0 & 0 & 0 & 0 & 0 & 0 & 0 & 0 & 0 & 0 & 1 & 0 & 0 & 0 & 0 \\ \{c\} & 0 & 0 & 0 & 0 & 1 & 0 & 0 & 0 & 0 & 0 & 0 & 0 & 0 & 0 & 0 & 0 \\ \{a,c\} & 0 & 0 & 0 & 0 & 0 & 1 & 0 & 0 & 0 & 0 & 0 & 0 & 0 & 0 & 0 & 0 \\ \{b,c\} & 0 & 0 & 0 & 0 & 0 & 0 & 1 & 0 & 0 & 0 & 0 & 0 & 0 & 0 & 0 & 0 \\ \{a,b,c\} & 0 & 0 & 0 & 0 & 0 & 0 & 0 & 0 & 0 & 0 & 0 & 0 & 0 & 0 & 0 & 1 \\ \{d\} & 0 & 0 & 0 & 0 & 0 & 0 & 0 & 0 & 0 & 0 & 0 & 1 & 0 & 0 & 0 & 0 \\ \{a,d\} & 0 & 0 & 0 & 0 & 0 & 0 & 0 & 0 & 0 & 0 & 0 & 1 & 0 & 0 & 0 & 0 \\ \{b,d\} & 0 & 0 & 0 & 0 & 0 & 0 & 0 & 0 & 0 & 0 & 0 & 1 & 0 & 0 & 0 & 0 \\ \{a,b,d\} & 0 & 0 & 0 & 0 & 0 & 0 & 0 & 0 & 0 & 0 & 0 & 1 & 0 & 0 & 0 & 0 \\ \{c,d\} & 0 & 0 & 0 & 0 & 0 & 0 & 0 & 0 & 0 & 0 & 0 & 0 & 0 & 0 & 0 & 1 \\ \{a,c,d\} & 0 & 0 & 0 & 0 & 0 & 0 & 0 & 0 & 0 & 0 & 0 & 0 & 0 & 0 & 0 & 1 \\ \{b,c,d\} & 0 & 0 & 0 & 0 & 0 & 0 & 0 & 0 & 0 & 0 & 0 & 0 & 0 & 0 & 0 & 1 \\ \{a,b,c,d\} & 0 & 0 & 0 & 0 & 0 & 0 & 0 & 0 & 0 & 0 & 0 & 0 & 0 & 0 & 0 & 1 \end{array}$$

Fig. 8.4 Non-symmetric, non-reflexive R with contact C and closure mapping ρ

We are by Proposition 8.2.2 entitled to use properties of Definitions 8.1.1 and 8.2.1 jointly when we show that proximities give rise to topologies.

Proposition 8.2.5 *From an arbitrary proximity relation Δ (according to Definition 8.1.1 or else to Definition 8.2.1 with the strong axiom added), one may obtain the neighborhood topology $\mathcal{U} := \sigma;\overline{\Delta};\mathcal{N}$ (or equivalently $= \overline{\sigma;\Delta;\mathcal{N}}$).*

Proof We recall in advance

$$\begin{array}{ll} \varepsilon = \mathbb{I};\varepsilon = \sigma;\varepsilon^{\mathsf{T}};\varepsilon & \text{Lemma 2.2.2.i} \\ \subseteq \sigma;\Delta & \text{Definition 8.1.1.iii} \\ = \sigma;\Delta^{\mathsf{T}} \subseteq \sigma;\mathbb{T};\Delta^{\mathsf{T}} \subseteq \sigma;\mathbb{T};\varepsilon = \mathbb{T};\varepsilon & \text{Definition 8.1.1.ii and } \sigma \text{ is a mapping} \end{array}$$

i)

$$\begin{array}{l} \mathcal{U} = \overline{\sigma;\Delta;\mathcal{N}} \subseteq \overline{\varepsilon;\mathcal{N}} = \overline{\overline{\varepsilon}} = \varepsilon, \quad \text{see above} \\ \mathcal{U} = \overline{\sigma;\Delta;\mathcal{N}} \supseteq \overline{\mathbb{T};\varepsilon;\mathcal{N}} = \overline{\mathbb{T};\overline{\varepsilon}} \supseteq \mathtt{syq}(\mathbb{T},\varepsilon) \end{array}$$

Thus $\mathcal{U}$ is total since the definition of a membership ε demands that every $\mathtt{syq}(\varepsilon, X)$ be surjective.

ii) We have rather obviously $\mathcal{N};\Omega^{\mathsf{T}} = \Omega;\mathcal{N}$ and $\Delta;\Omega \subseteq \Delta$ (see Definition 8.2.1.v), so that

$$\mathcal{U};\Omega = \overline{\sigma;\Delta;\mathcal{N};\Omega} \subseteq \overline{\sigma;\Delta;\mathcal{N}} = \mathcal{U} \iff \sigma;\Delta;\mathcal{N};\Omega^{\mathsf{T}} = \sigma;\Delta;\Omega;\mathcal{N} \subseteq \sigma;\Delta;\mathcal{N}.$$

iii)

$$\begin{array}{ll} (\mathcal{U} \circledless \mathcal{U});\mathfrak{M}_2 & \\ \quad = (\sigma;\overline{\Delta};\mathcal{N} \circledless \sigma;\overline{\Delta};\mathcal{N});\mathfrak{M}_2 & \text{by definition} \\ \quad = \sigma;(\overline{\Delta} \circledless \overline{\Delta});(\mathcal{N} \otimes \mathcal{N});\mathfrak{M}_2 & \\ \quad = \sigma;(\overline{\Delta} \circledless \overline{\Delta});\mathfrak{J}_2;\mathcal{N} & \text{De Morgan rule, Proposition 4.3.1.ii} \\ \quad \subseteq \sigma;\overline{\Delta};\mathcal{N} = \mathcal{U} & \text{Definition 8.1.1.iv in shunted form} \end{array}$$

iv) In order to show $\mathcal{U} \subseteq \mathcal{U};\overline{\varepsilon^{\mathsf{T}};\overline{\mathcal{U}}}$, we use that $\Delta;\Omega \subseteq \Delta$; see above. In addition, $\sigma;\varepsilon^{\mathsf{T}};\overline{\varepsilon} = \mathbb{I};\overline{\varepsilon} = \overline{\varepsilon}$, so that we may employ $\sigma^{\mathsf{T}};\varepsilon \subseteq \overline{\varepsilon^{\mathsf{T}};\overline{\varepsilon}} = \Omega$.

$$\begin{array}{ll} \mathcal{U} = \sigma;\overline{\Delta};\mathcal{N} \subseteq \sigma;\overline{\Delta};\mathcal{N};\overline{\Delta};\mathcal{N} & \text{using strong xiom for a proximity} \\ \quad \subseteq \sigma;\overline{\Delta};\mathcal{N};\overline{\Omega^{\mathsf{T}};\Delta;\mathcal{N}} & \text{since } \mathcal{N}^{\mathsf{T}} \text{is a mapping and } \Delta;\Omega \subseteq \Delta \\ \quad \subseteq \sigma;\overline{\Delta};\mathcal{N};\overline{\varepsilon^{\mathsf{T}};\sigma;\Delta;\mathcal{N}} & \text{see above} \\ \quad = \mathcal{U};\overline{\varepsilon^{\mathsf{T}};\overline{\mathcal{U}}} & \end{array}$$

□

That one may also go in the reverse direction, i.e., from $\mathcal{U}$ to Δ, has partly been shown with Proposition 8.1.3.

As for every mathematical structure, one has also defined a structure-preserving mapping f for proximity in [NW70]. The definition postulates

$$(A,B) \in \Delta_1 \longrightarrow (f(A),f(B)) \in \Delta_2$$

in a not yet lifted form. We derive therefrom the following lifted version.

Definition 8.2.6 Given proximities $\Delta_i : \mathbf{2}^{X_i} \longrightarrow \mathbf{2}^{X_i}$, $i = 1,2$, and a mapping $f : X_1 \longrightarrow X_2$ of the underlying sets, we call

$$f \text{ a } \textbf{proximity mapping} \quad :\Longleftrightarrow \quad \Delta_1 ; \vartheta_f \subseteq \vartheta_f ; \Delta_2. \qquad \square$$

We have had problems to apply the traditional homomorphism scheme to continuity, when we define "traditional" to mean

$$\textit{structure} \times \textit{mapping} \quad \subseteq \quad \textit{mapping} \times \textit{structure}.$$

For proximity mappings, we can say that their definition adheres more or less to the traditional form; there is only a slight deviation, because f is given, but the definition is based on its existential image ϑ_f.

In Proposition 8.2.5, we have identified a topology for every proximity. It is remarkable that proximity mappings lead to continuous mappings between such topologies.

Proposition 8.2.7 *Any surjective proximity mapping $f : X_1 \longrightarrow X_2$ is continuous with respect to the neighborhood topologies $\mathcal{U}_1, \mathcal{U}_2$ according to Proposition 8.2.5.*

Proof We have to prove $f ; \mathcal{U}_2 \subseteq \mathcal{U}_1 ; \vartheta_{f^\mathsf{T}}^\mathsf{T}$, which expands to

$$
\begin{aligned}
& f ; \overline{\sigma_2 ; \Delta_2 ; \mathcal{N}_2} \subseteq \overline{\sigma_1 ; \Delta_1 ; \mathcal{N}_1} ; \vartheta_{f^\mathsf{T}}^\mathsf{T} = \overline{\sigma_1 ; \Delta_1 ; \mathcal{N}_1 ; \vartheta_{f^\mathsf{T}}^\mathsf{T}}. \\
\Longleftrightarrow \quad & f^\mathsf{T} ; \sigma_1 ; \Delta_1 ; \mathcal{N}_1 ; \vartheta_{f^\mathsf{T}}^\mathsf{T} = f^\mathsf{T} ; \sigma_1 ; \Delta_1 ; \vartheta_{f^\mathsf{T}}^\mathsf{T} ; \mathcal{N}_2 \subseteq \sigma_2 ; \Delta_2 ; \mathcal{N}_2 \\
& \qquad\qquad \text{since } \mathcal{N}_1 ; \vartheta_{f^\mathsf{T}}^\mathsf{T} = \vartheta_{f^\mathsf{T}}^\mathsf{T} ; \mathcal{N}_2, \text{ Proposition 2.2.11} \\
\Longleftrightarrow \quad & f^\mathsf{T} ; \sigma_1 ; \Delta_1 ; \vartheta_{f^\mathsf{T}}^\mathsf{T} \subseteq \sigma_2 ; \Delta_2 \\
\Longleftrightarrow \quad & \sigma_1 ; \Delta_1 ; \vartheta_{f^\mathsf{T}}^\mathsf{T} \subseteq f ; \sigma_2 ; \Delta_2 \quad \text{shunting}
\end{aligned}
$$

This shall now be proved:

$$
\begin{aligned}
\sigma_1 ; \Delta_1 ; \vartheta_{f^\mathsf{T}}^\mathsf{T} & \subseteq \sigma_1 ; \Delta_1 ; \vartheta_f && \text{Proposition 2.2.8.i for surjective } f \\
& \subseteq \sigma_1 ; \vartheta_f ; \Delta_2 && \text{Definition 8.2.6} \\
& = f ; \sigma_2 ; \Delta_2 && \text{Proposition 2.2.6.ii} \qquad \square
\end{aligned}
$$

It seems to be an interesting task to study how the additional "strong" properties sometimes demanded for pre-proximity as well as for DV-pre-proximity relations are related with one another.

8.3 Nearness

Closely related with "proximity" is the concept of "nearness". We have said "A and B are δ-proximal" if $A\delta B$. Now we proceed to saying that "B is in a δ-neighborhood of A", written $A \ll B$ when $A\delta(X\backslash B)$ is false. This changes the axioms slightly. The so changed axioms will later provide an alternative axiomatic characterization for proximity.

The nearness concept for subsets of a set X is formulated quantifying over all subsets A, B, C, and D of the set X in question, postulating the following six axioms:

- $X \ll X$
- $A \ll B \implies A \subseteq B$
- $A \subseteq B \ll C \subseteq D \implies A \ll D$
- $(A \ll B \text{ and } A \ll C) \implies A \ll B \cap C$
- $A \ll B \implies X\backslash B \ll X\backslash A$
- $A \ll B \implies \exists E : A \ll E \ll B$

This is now lifted more or less directly to a point-free as well as quantifier-free version.

Definition 8.3.1 We call the relation $R : \mathbf{2}^X \longrightarrow \mathbf{2}^X$ a **nearness**, provided

i)

$$\overline{\varepsilon^{\mathsf{T}} ; \overline{\mathbb{T}}} \cup \overline{\mathbb{T} ; \overline{\varepsilon}} \subseteq R \quad \text{(or more intuitively} \quad \mathtt{syq}(\varepsilon, \mathbb{L}) \subseteq R, \quad \mathtt{syq}(\varepsilon, \mathbb{T}) \subseteq R)$$

ii)

$$R \subseteq \Omega$$

iii)

$$\Omega ; R ; \Omega \subseteq R$$

iv)

$$(R \circledleq R) \subseteq R ; \mathfrak{M}_2^{\mathsf{T}}, \quad \text{in fact an equality—see below}$$

v)

$$R ; \mathcal{N} \subseteq \mathcal{N} ; R^{\mathsf{T}}$$

vi)

$$R \subseteq R ; R, \quad \text{i.e., } R \text{ is dense} \qquad \square$$

Equality for (iv) need not be postulated; it requires that also

$$R:\mathfrak{M}_2^{\mathsf{T}} \subseteq (R \otimes R) = R:\pi^{\mathsf{T}} \cap R:\rho^{\mathsf{T}},$$

but the first containment e.g. will hold when $R:\mathfrak{M}_2^{\mathsf{T}} \subseteq R:\pi^{\mathsf{T}}$. This is via shunting equivalent with $R:\mathfrak{M}_2^{\mathsf{T}}:\pi \subseteq R$. In view of (iii) and Proposition 4.2.2.vi, this holds indeed.

Remark 8.3.2 Given any membership relation $\varepsilon : X \longrightarrow \mathbf{2}^X$, the powerset ordering Ω satisfies all the requirements for a nearness.

Proof Again, (i,ii,iii,vi) are trivial.

iv) follows from Proposition 4.3.3.i
v) $\Omega:\mathcal{N} = \overline{\varepsilon^{\mathsf{T}}:\overline{\varepsilon}}:\mathcal{N} = \overline{\varepsilon^{\mathsf{T}}:\varepsilon} = \mathcal{N}:\overline{\varepsilon}^{\mathsf{T}}:\varepsilon = \mathcal{N}:\Omega^{\mathsf{T}}$ □

The nearness Ω, as just observed in Remark 8.3.2, is the greatest among all possible nearnesses. We see in property (ii) that there cannot exist a greater one.

Proposition 8.3.3 *For any given proximity Δ, the relation $R := \overline{\Delta:\mathcal{N}}$ is a nearness.*

Proof

i) Using Proposition 8.1.1.ii, we have $\Delta:\mathbb{T} \subseteq \varepsilon^{\mathsf{T}}:\mathbb{T}$, implying for the first

$$\overline{\varepsilon^{\mathsf{T}}:\mathbb{T}} \subseteq \overline{\Delta:\mathbb{T}} = \overline{\Delta:\mathcal{N}:\mathbb{T}} = \overline{\overline{R}:\mathbb{T}} \subseteq R;$$

similarly for the second.

ii)

$$R \subseteq \Omega \iff \overline{\Delta:\mathcal{N}} \subseteq \overline{\varepsilon^{\mathsf{T}}:\overline{\varepsilon}} \iff \varepsilon^{\mathsf{T}}:\overline{\varepsilon} \subseteq \Delta:\mathcal{N} \iff \varepsilon^{\mathsf{T}}:\overline{\varepsilon}:\mathcal{N} = \varepsilon^{\mathsf{T}}:\varepsilon \subseteq \Delta$$

where the latter is guaranteed by Definition 8.1.1.iii.

iii) Following Proposition 8.2.2, we are now entitled to use Definition 8.2.1.v, viz. $\Delta:\Omega \subseteq \Delta$, and in transposed form also $\Omega^{\mathsf{T}}:\Delta \subseteq \Delta$.

$$\begin{aligned} \Omega:R:\Omega \subseteq R \quad &\iff \quad \Omega:\overline{\Delta:\mathcal{N}}:\Omega \subseteq \overline{\Delta:\mathcal{N}} \\ \iff \quad \Omega^{\mathsf{T}}:\Delta:\mathcal{N} \subseteq \overline{\overline{\Delta:\mathcal{N}}:\Omega} \quad &\Longleftarrow \quad \Delta:\mathcal{N} \subseteq \overline{\overline{\Delta:\mathcal{N}}:\Omega} \\ \iff \quad \overline{\Delta:\mathcal{N}}:\Omega \subseteq \overline{\Delta:\mathcal{N}} \quad &\iff \quad \Delta:\mathcal{N}:\Omega^{\mathsf{T}} \subseteq \Delta:\mathcal{N} \\ \iff \quad \Delta:\Omega = \Delta:\mathcal{N}:\Omega^{\mathsf{T}}:\mathcal{N} \subseteq \Delta & \end{aligned}$$

iv) We prove even equality:

$$\begin{aligned} (R \otimes R):\mathfrak{M}_2 &= (\overline{\Delta:\mathcal{N}} \otimes \overline{\Delta:\mathcal{N}}):\mathfrak{M}_2 = (\overline{\Delta} \otimes \overline{\Delta}):(\mathcal{N} \otimes \mathcal{N}):\mathfrak{M}_2 \\ &= (\overline{\Delta} \otimes \overline{\Delta}):\mathfrak{J}_2:\mathcal{N} \quad \text{point-free De Morgan rule Proposition 4.3.1.ii} \\ &= \overline{\Delta}:\mathcal{N} \quad \text{using Definition 8.1.1.iv} \\ &= \overline{\Delta:\mathcal{N}} = R \end{aligned}$$

v)

$$R{:}\mathcal{N} = \overline{\Delta{:}\mathcal{N}}{:}\mathcal{N} = \overline{\Delta}{:}\mathcal{N}{:}\mathcal{N} = \overline{\Delta} = \mathcal{N}{:}\mathcal{N}{:}\overline{\Delta} = \mathcal{N}{:}\overline{\Delta{:}\mathcal{N}}^{\mathsf{T}} = \mathcal{N}{:}R^{\mathsf{T}}$$

vi)

$$\overline{\Delta} \subseteq \overline{\Delta}{:}\mathcal{N}{:}\overline{\Delta} \iff \overline{\Delta{:}\mathcal{N}} \subseteq \overline{\Delta{:}\mathcal{N}}{:}\overline{\Delta{:}\mathcal{N}} \iff R \subseteq R{:}R \qquad \square$$

In Figs. 8.5 and 8.6, we show an example of proximity and nearness (Fig. 8.7).
Nearly the same as Proposition 8.3.3 is possible in the other direction.

Proposition 8.3.4 *Given any nearness R, the relation $\Delta : \mathbf{2}^X \longrightarrow \mathbf{2}^X$ defined as*

$$\Delta := \overline{R{:}\mathcal{N}},$$

will be a proximity.

Proof

i) $\Delta^{\mathsf{T}} \subseteq \Delta \iff \overline{R{:}\mathcal{N}}^{\mathsf{T}} \subseteq \overline{R{:}\mathcal{N}} \iff R{:}\mathcal{N} \subseteq \mathcal{N}{:}R^{\mathsf{T}}$; this follows from Definition 8.3.1.v

	{}	{a}	{b}	{a,b}	{c}	{a,c}	{b,c}	{a,b,c}	{d}	{a,d}	{b,d}	{a,b,d}	{c,d}	{a,c,d}	{b,c,d}	{a,b,c,d}
{}	0	0	0	0	0	0	0	0	0	0	0	0	0	0	0	0
{a}	0	1	0	1	0	1	0	1	1	1	1	1	1	1	1	1
{b}	0	0	1	1	1	1	1	1	0	0	1	1	1	1	1	1
{a,b}	0	1	1	1	1	1	1	1	1	1	1	1	1	1	1	1
{c}	0	0	1	1	1	1	1	1	0	0	1	1	1	1	1	1
{a,c}	0	1	1	1	1	1	1	1	1	1	1	1	1	1	1	1
{b,c}	0	0	1	1	1	1	1	1	0	0	1	1	1	1	1	1
{a,b,c}	0	1	1	1	1	1	1	1	1	1	1	1	1	1	1	1
{d}	0	1	0	1	0	1	0	1	1	1	1	1	1	1	1	1
{a,d}	0	1	0	1	0	1	0	1	1	1	1	1	1	1	1	1
{b,d}	0	1	1	1	1	1	1	1	1	1	1	1	1	1	1	1
{a,b,d}	0	1	1	1	1	1	1	1	1	1	1	1	1	1	1	1
{c,d}	0	1	1	1	1	1	1	1	1	1	1	1	1	1	1	1
{a,c,d}	0	1	1	1	1	1	1	1	1	1	1	1	1	1	1	1
{b,c,d}	0	1	1	1	1	1	1	1	1	1	1	1	1	1	1	1
all	0	1	1	1	1	1	1	1	1	1	1	1	1	1	1	1

	{}	{a}	{b}	{a,b}	{c}	{a,c}	{b,c}	{a,b,c}	{d}	{a,d}	{b,d}	{a,b,d}	{c,d}	{a,c,d}	{b,c,d}	{a,b,c,d}
{}	1	1	1	1	1	1	1	1	1	1	1	1	1	1	1	1
{a}	0	0	0	0	0	0	0	0	0	1	0	1	0	1	0	1
{b}	0	0	0	0	0	0	1	1	0	0	0	0	0	0	1	1
{a,b}	0	0	0	0	0	0	0	0	0	0	0	0	0	0	0	1
{c}	0	0	0	0	0	0	1	1	0	0	0	0	0	0	1	1
{a,c}	0	0	0	0	0	0	0	0	0	0	0	0	0	0	0	1
{b,c}	0	0	0	0	0	0	1	1	0	0	0	0	0	0	1	1
{a,b,c}	0	0	0	0	0	0	0	0	0	0	0	0	0	0	0	1
{d}	0	0	0	0	0	0	0	0	0	1	0	1	0	1	0	1
{a,d}	0	0	0	0	0	0	0	0	0	1	0	1	0	1	0	1
{b,d}	0	0	0	0	0	0	0	0	0	0	0	0	0	0	0	1
{a,b,d}	0	0	0	0	0	0	0	0	0	0	0	0	0	0	0	1
{c,d}	0	0	0	0	0	0	0	0	0	0	0	0	0	0	0	1
{a,c,d}	0	0	0	0	0	0	0	0	0	0	0	0	0	0	0	1
{b,c,d}	0	0	0	0	0	0	0	0	0	0	0	0	0	0	0	1
all	0	0	0	0	0	0	0	0	0	0	0	0	0	0	0	1

Fig. 8.5 A pair of proximity and nearness based on the open set basis of Fig. 8.6

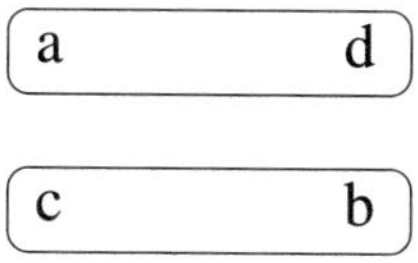

Fig. 8.6 The basis of open sets for Fig. 8.5

	{}	{1}	{2}	{1,2}	{3}	{1,3}	{2,3}	{1,2,3}	{4}	{1,4}	{2,4}	{1,2,4}	{3,4}	{1,3,4}	{2,3,4}	{1,2,3,4}
{}	1	1	1	1	1	1	1	1	1	1	1	1	1	1	1	1
{1}	0	1	0	1	0	1	0	1	0	1	0	1	0	1	0	1
{2}	0	0	1	1	0	0	1	1	0	0	1	1	0	0	1	1
{1,2}	0	0	0	1	0	0	0	1	0	0	0	1	0	0	0	1
{3}	0	0	0	0	1	1	1	1	0	0	0	0	1	1	1	1
{1,3}	0	0	0	0	0	1	0	1	0	0	0	0	0	1	0	1
{2,3}	0	0	0	0	0	0	1	1	0	0	0	0	0	0	1	1
{1,2,3}	0	0	0	0	0	0	0	1	0	0	0	0	0	0	0	1
{4}	0	0	0	0	0	0	0	0	1	1	1	1	1	1	1	1
{1,4}	0	0	0	0	0	0	0	0	0	1	0	1	0	1	0	1
{2,4}	0	0	0	0	0	0	0	0	0	0	1	1	0	0	1	1
{1,2,4}	0	0	0	0	0	0	0	0	0	0	0	1	0	0	0	1
{3,4}	0	0	0	0	0	0	0	0	0	0	0	0	1	1	1	1
{1,3,4}	0	0	0	0	0	0	0	0	0	0	0	0	0	1	0	1
{2,3,4}	0	0	0	0	0	0	0	0	0	0	0	0	0	0	1	1
all	0	0	0	0	0	0	0	0	0	0	0	0	0	0	0	1

	{}	{1}	{2}	{1,2}	{3}	{1,3}	{2,3}	{1,2,3}	{4}	{1,4}	{2,4}	{1,2,4}	{3,4}	{1,3,4}	{2,3,4}	{1,2,3,4}
{}	1	1	1	1	1	1	1	1	1	1	1	1	1	1	1	1
{1}	0	0	0	1	0	0	0	1	0	0	0	1	0	0	0	1
{2}	0	0	0	1	0	0	0	1	0	0	0	1	0	0	0	1
{1,2}	0	0	0	1	0	0	0	1	0	0	0	1	0	0	0	1
{3}	0	0	0	0	0	0	0	0	0	0	0	0	1	1	1	1
{1,3}	0	0	0	0	0	0	0	0	0	0	0	0	0	0	0	1
{2,3}	0	0	0	0	0	0	0	0	0	0	0	0	0	0	0	1
{1,2,3}	0	0	0	0	0	0	0	0	0	0	0	0	0	0	0	1
{4}	0	0	0	0	0	0	0	0	0	0	0	0	1	1	1	1
{1,4}	0	0	0	0	0	0	0	0	0	0	0	0	0	0	0	1
{2,4}	0	0	0	0	0	0	0	0	0	0	0	0	0	0	0	1
{1,2,4}	0	0	0	0	0	0	0	0	0	0	0	0	0	0	0	1
{3,4}	0	0	0	0	0	0	0	0	0	0	0	0	1	1	1	1
{1,3,4}	0	0	0	0	0	0	0	0	0	0	0	0	0	0	0	1
{2,3,4}	0	0	0	0	0	0	0	0	0	0	0	0	0	0	0	1
all	0	0	0	0	0	0	0	0	0	0	0	0	0	0	0	1

Fig. 8.7 Biggest and a smaller nearness corresponding to Fig. 8.1

ii) $\Delta;\mathbb{T} = \overline{R;\mathcal{N}};\mathbb{T} = \overline{R};\mathcal{N};\mathbb{T} = \overline{R};\mathbb{T} \subseteq \overline{\overline{\varepsilon^{\mathsf{T}};\mathbb{T}} \cup \overline{\mathbb{T};\overline{\varepsilon}}};\mathbb{T} = (\varepsilon^{\mathsf{T}};\mathbb{T} \cap \mathbb{T};\overline{\varepsilon});\mathbb{T} \subseteq \varepsilon^{\mathsf{T}};\mathbb{T}$,
i.e. with Definition 8.3.1.i

iii) $\varepsilon^{\mathsf{T}};\varepsilon \subseteq \Delta = \overline{R;\mathcal{N}} \iff \varepsilon^{\mathsf{T}};\overline{\varepsilon} = \varepsilon^{\mathsf{T}};\varepsilon;\mathcal{N} \subseteq \overline{R} \iff R \subseteq \Omega$,
i.e. Definition 8.3.1.ii.

iv)

$$\begin{aligned}
(\overline{\Delta} \olessthan \overline{\Delta});\mathfrak{J}_2 &= (R;\mathcal{N} \olessthan R;\mathcal{N});\mathfrak{J}_2 = (R \olessthan R);(\mathcal{N} \otimes \mathcal{N});\mathfrak{J}_2 \\
&= (R \olessthan R);\mathfrak{M}_2;\mathcal{N} && \text{point-free De Morgan rule} \\
&= R;\mathcal{N} && \text{Definition 8.3.1.iv} \\
&= \overline{\Delta} && \text{by definition}
\end{aligned}$$

v) We start from Definition 8.3.1.vi to obtain

$$R \subseteq R;R \iff R;\mathcal{N} \subseteq R;R;\mathcal{N} = R;\mathcal{N};\mathcal{N};R;\mathcal{N} \iff \overline{\Delta} \subseteq \overline{\Delta};\mathcal{N};\overline{\Delta}$$

□

With Propositions 8.3.3 and 8.3.4, we have thus shown that proximity and nearness are cryptomorphic concepts.

8.4 Apartness and Connection Algebra

The concept of being near (or being proximal) has as contrast for a point that of being in some sense "apart" from a set of points. This is a concept formulated in the environment of constructive mathematics, where a sophisticated distinction is made

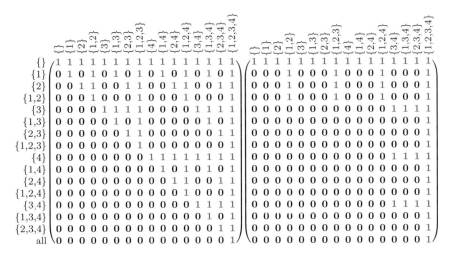

between logical complements "¬", complements "−", and an apartness complement "∼"; see e.g. [BStV01]. We can here, of course, not dive into such details.

When we, however, approach parts of this definition naively, neglecting complicated details and even one of the axioms, it might read as follows: Assume a set X and a relation **apart** $: X \longrightarrow \mathbf{2}^X$, intended to express that a point x *is apart from a subset* u, that satisfies

- $x \neq y \quad\Longrightarrow\quad \mathbf{apart}(x, \{y\})$,
- $\mathbf{apart}(x, u) \quad\Longrightarrow\quad x \notin u$,
- $\mathbf{apart}(x, u \cup v) \quad\Longleftrightarrow\quad \mathbf{apart}(x, u) \wedge \mathbf{apart}(x, v)$,
- $\mathbf{apart}(x, u) \wedge v \subseteq u \quad\Longrightarrow\quad \mathbf{apart}(x, v)$.

We lift the idea of the preceding concept so as to obtain a point-free version that avoids quantifiers.

Definition 8.4.1 Assume a set X and a relation $A : X \longrightarrow \mathbf{2}^X$—in addition to membership and singleton injection $\varepsilon, \sigma : X \longrightarrow \mathbf{2}^X$. This relation will then be called an **apartness**, provided

i)
$$\overline{\mathbb{I}} ; \sigma \subseteq A,$$
ii)
$$A \subseteq \overline{\varepsilon},$$
iii)
$$(A \circledless A) \subseteq A ; \mathfrak{J}_2^{\mathsf{T}},$$
iv)
$$A ; \Omega^{\mathsf{T}} \subseteq A.$$
□

One observation is immediate, namely that (iii) does not directly reflect the "$\Longleftrightarrow$" of the predicate-logic version when showing only "$\subseteq$". The reason is simply that the so conceived axioms are not independent. The following proposition shows that "$\supseteq$" is a consequence of (iv). We have therefore chosen not to mention this direction above in order to keep the definition clean.

Proposition 8.4.2 $A ; \Omega^{\mathsf{T}} \subseteq A$ *implies* $A ; \mathfrak{J}_2^{\mathsf{T}} \subseteq (A \circledless A)$.

Proof $A ; \mathfrak{J}_2^{\mathsf{T}} \subseteq (A \circledless A) = A ; \pi^{\mathsf{T}} \cap A ; \rho^{\mathsf{T}}$ may be reduced to $A ; \mathfrak{J}_2^{\mathsf{T}} \subseteq A ; \pi^{\mathsf{T}}$ and further shunted to $A ; \mathfrak{J}_2^{\mathsf{T}} ; \pi \subseteq A$. Now we recall that following Proposition 4.2.2.vii $\mathfrak{J}_2^{\mathsf{T}} ; \pi = \Omega^{\mathsf{T}}$, which completes the proof. □

We now provide examples of such apartnesses. First we establish the complement of the membership relation ε as a most trivial example.

Proposition 8.4.3 *Given any membership $\varepsilon : X \longrightarrow \mathbf{2}^X$, its complement $A := \overline{\varepsilon}$ is an apartness.*

Proof

i)

$$\overline{\mathbb{I}}{:}\sigma = \overline{\mathbb{I}}{:}\mathtt{syq}(\mathbb{I}, \varepsilon) = \overline{\mathbb{I}}{:}\mathtt{syq}(\overline{\mathbb{I}}, \overline{\varepsilon}) \subseteq \overline{\varepsilon} = A$$

ii) by definition

iii) $A{:}\mathfrak{J}_2^{\mathsf{T}} = \overline{\varepsilon}{:}\mathfrak{J}_2^{\mathsf{T}} = (\overline{\varepsilon} \circledless \overline{\varepsilon}) = (A \circledless A)$ using Proposition 4.3.2.i

iv) $A{:}\Omega^{\mathsf{T}} = \overline{\varepsilon}{:}\Omega^{\mathsf{T}} = \overline{\varepsilon{:}\overline{\varepsilon}^{\mathsf{T}}{:}\varepsilon} = \overline{\varepsilon} = A$ using membership deletion □

When looking at Definition 8.4.1 in more detail, it turns out that (i,iii,iv) offer a method of an expanding construction, however, somehow restricted by (ii). The start should obviously be positioned above $\overline{\mathbb{I}}{:}\sigma$. Then—resembling (iv)—the operation $f(A) : A \mapsto A{:}\Omega^{\mathsf{T}}$ should be applied. In addition—resembling (iii) in shunted form—$g(A) : A \mapsto A \cup (A \circledless A){:}\mathfrak{J}_2$ should take place until stability is reached; guaranteed at least in the finite case.

Two conclusions may be derived from this observation. The first should be seen in combination with the earlier Proposition 7.1.3 for Aumann contacts:

Remark 8.4.4 One should refrain from demanding (i). We underpin this suggestion starting the expanding iterations with the relation $A_0 := \overline{\mathbb{I}}{:}\sigma$.

	{}	{1}	{2}	{1,2}	{3}	{1,3}	{2,3}	{1,2,3}	{4}	{1,4}	{2,4}	{1,2,4}	{3,4}	{1,3,4}	{2,3,4}	{1,2,3,4}	{5}	{1,5}	{2,5}	{1,2,5}	{3,5}	{1,3,5}	{2,3,5}	{1,2,3,5}	{4,5}	{1,4,5}	{2,4,5}	{1,2,4,5}	{3,4,5}	{1,3,4,5}	{2,3,4,5}	{1,2,3,4,5}
1	0	0	1	0	1	0	0	0	1	0	0	0	0	0	0	0	1	0	0	0	0	0	0	0	0	0	0	0	0	0	0	0
2	0	1	0	0	1	0	0	0	1	0	0	0	0	0	0	0	1	0	0	0	0	0	0	0	0	0	0	0	0	0	0	0
3	0	1	1	0	0	0	0	0	1	0	0	0	0	0	0	0	1	0	0	0	0	0	0	0	0	0	0	0	0	0	0	0
4	0	1	1	0	1	0	0	0	0	0	0	0	0	0	0	0	1	0	0	0	0	0	0	0	0	0	0	0	0	0	0	0
5	0	1	1	0	1	0	0	0	1	0	0	0	0	0	0	0	0	0	0	0	0	0	0	0	0	0	0	0	0	0	0	0

The relation A_0 is now down-closed wrt. (iv) as $A_1 := f(A_0) = A_0{:}\Omega^{\mathsf{T}}$ obtaining

	{}	{1}	{2}	{1,2}	{3}	{1,3}	{2,3}	{1,2,3}	{4}	{1,4}	{2,4}	{1,2,4}	{3,4}	{1,3,4}	{2,3,4}	{1,2,3,4}	{5}	{1,5}	{2,5}	{1,2,5}	{3,5}	{1,3,5}	{2,3,5}	{1,2,3,5}	{4,5}	{1,4,5}	{2,4,5}	{1,2,4,5}	{3,4,5}	{1,3,4,5}	{2,3,4,5}	{1,2,3,4,5}
1	1	0	1	0	1	0	0	0	1	0	0	0	0	0	0	0	1	0	0	0	0	0	0	0	0	0	0	0	0	0	0	0
2	1	1	0	0	1	0	0	0	1	0	0	0	0	0	0	0	1	0	0	0	0	0	0	0	0	0	0	0	0	0	0	0
3	1	1	1	0	0	0	0	0	1	0	0	0	0	0	0	0	1	0	0	0	0	0	0	0	0	0	0	0	0	0	0	0
4	1	1	1	0	1	0	0	0	0	0	0	0	0	0	0	0	1	0	0	0	0	0	0	0	0	0	0	0	0	0	0	0
5	1	1	1	0	1	0	0	0	1	0	0	0	0	0	0	0	0	0	0	0	0	0	0	0	0	0	0	0	0	0	0	0

Finally the relation A_1 is also iteratively closed with respect to join-forming via $g(A) : A \mapsto A \cup (A \circledless A) ; \mathfrak{J}_2$. In this example the iteration is stable already after the second step obtaining simply $\overline{\varepsilon}$.

	{}	{1}	{2}	{1,2}	{3}	{1,3}	{2,3}	{1,2,3}	{4}	{1,4}	{2,4}	{1,2,4}	{3,4}	{1,3,4}	{2,3,4}	{1,2,3,4}	{5}	{1,5}	{2,5}	{1,2,5}	{3,5}	{1,3,5}	{2,3,5}	{1,2,3,5}	{4,5}	{1,4,5}	{2,4,5}	{1,2,4,5}	{3,4,5}	{1,3,4,5}	{2,3,4,5}	{1,2,3,4,5}
1	1	0	1	0	1	0	1	0	1	0	1	0	1	0	0	0	1	0	1	0	1	0	0	0	1	0	0	0	0	0	0	0
2	1	1	0	0	1	1	0	0	1	1	0	0	1	0	0	0	1	1	0	0	1	0	0	0	1	0	0	0	0	0	0	0
3	1	1	1	1	0	0	0	0	1	1	1	0	0	0	0	0	1	1	1	0	0	0	0	0	1	0	0	0	0	0	0	0
4	1	1	1	1	1	1	1	0	0	0	0	0	0	0	0	0	1	1	1	0	1	0	0	0	0	0	0	0	0	0	0	0
5	1	1	1	1	1	1	1	0	1	1	1	0	1	0	0	0	0	0	0	0	0	0	0	0	0	0	0	0	0	0	0	0

	{}	{1}	{2}	{1,2}	{3}	{1,3}	{2,3}	{1,2,3}	{4}	{1,4}	{2,4}	{1,2,4}	{3,4}	{1,3,4}	{2,3,4}	{1,2,3,4}	{5}	{1,5}	{2,5}	{1,2,5}	{3,5}	{1,3,5}	{2,3,5}	{1,2,3,5}	{4,5}	{1,4,5}	{2,4,5}	{1,2,4,5}	{3,4,5}	{1,3,4,5}	{2,3,4,5}	{1,2,3,4,5}
1	1	0	1	0	1	0	1	0	1	0	1	0	1	0	1	0	1	0	1	0	1	0	1	0	1	0	1	0	1	0	1	0
2	1	1	0	0	1	1	0	0	1	1	0	0	1	1	0	0	1	1	0	0	1	1	0	0	1	1	0	0	1	1	0	0
3	1	1	1	1	0	0	0	0	1	1	1	1	0	0	0	0	1	1	1	1	0	0	0	0	1	1	1	1	0	0	0	0
4	1	1	1	1	1	1	1	1	0	0	0	0	0	0	0	0	1	1	1	1	1	1	1	1	0	0	0	0	0	0	0	0
5	1	1	1	1	1	1	1	1	1	1	1	1	1	1	1	1	0	0	0	0	0	0	0	0	0	0	0	0	0	0	0	0

□

One will easily see that this must necessarily happen in the finite case. With $\overline{\mathbb{I}} ; \sigma$ all columns of the relation are indicated that belong to 1-element sets $\{x\}$ and consist of all elements $\neq x$. Then $f(A) : A \mapsto A ; \Omega^{\mathsf{T}}$ demands that all subsets of these are included. When closing this configuration with respect to forming binary joins $g(A) : A \mapsto A \cup (A \circledless A) ; \mathfrak{J}_2$ all of $\overline{\varepsilon}$ appears—at least in the finite case.

Demanding (i,ii,iii,iv) to hold—as we did—constitutes, therefore, not a reasonable definition and we are driven to drop axiom Definition 8.4.1.i in order to obtain more interesting models.

Earlier, we have seen that the complement of every membership ε is always an apartness. More generally, the complement of a topological Aumann contact shall now be tested as to which extent it is an apartness:

Proposition 8.4.5 *For any topological Aumann contact $C : X \longrightarrow \mathbf{2}^X$ the complement $A := \overline{C}$ satisfies properties (ii,iii,iv) of the definition Definition 8.4.1 of an apartness.*

Proof We follow the numbering of Definition 8.4.1. Then (ii) holds by definition of the contact, $\varepsilon \subseteq C = \overline{A}$.
iii) Using Proposition 7.1.2.iv, $A ; \mathfrak{J}_2^{\mathsf{T}} = \overline{C} ; \mathfrak{J}_2^{\mathsf{T}} = (\overline{C} \circledless \overline{C}) = (A \circledless A)$.
iv) Due to Proposition 7.1.2.ii $C ; \Omega \subseteq C$, so that $\overline{C} ; \Omega^{\mathsf{T}} \subseteq \overline{C}$. □

The modification of Proposition 7.1.3 would make it an apartness, but according to the preceding Remark 8.4.4, this will turn out to be just $\overline{\varepsilon}$.

The reverse statement will turn out to be only satisfied in specific situations, not in general:

Remark 8.4.6 Given an apartness $A : X \longrightarrow \mathbf{2}^X$ on a set X with at least two elements (algebraically: with $\overline{\mathbb{I}}{;}\mathbb{T} = \mathbb{T}$), its complement $C := \overline{A}$ need not form an Aumann contact relation. We would have to show $\varepsilon \subseteq C \subseteq \mathbb{T}{;}\varepsilon$ and $C^{\mathsf{T}}{;}\overline{C} \subseteq \varepsilon^{\mathsf{T}}{;}\overline{C}$ of which the first inclusion is trivial in view of Definition 8.4.1.ii.

For the second, we start with

$$\begin{array}{rll} \overline{C} \cup \mathbb{T}{;}\varepsilon &= A \cup \mathbb{T}{;}\varepsilon \supseteq A{;}\Omega^{\mathsf{T}} \cup \mathbb{T}{;}\varepsilon & \text{Definition 8.4.1.iv} \\ &\supseteq \overline{\mathbb{I}}{;}\sigma{;}\Omega^{\mathsf{T}} \cup \mathbb{T}{;}\varepsilon & \text{Definition 8.4.1.i} \\ &= \overline{\mathbb{I}}{;}(\sigma \cup \overline{\mathbb{T}{;}\varepsilon}) \cup \mathbb{T}{;}\varepsilon \supseteq \overline{\mathbb{I}}{;}\overline{\mathbb{T}{;}\varepsilon} \cup \mathbb{T}{;}\varepsilon & \text{Lemma 2.2.2.ii} \\ &= \overline{\mathbb{I}}{;}\overline{\mathbb{T}{;}\varepsilon} \cup \overline{\mathbb{I}}{;}\mathbb{T}{;}\varepsilon & \text{condition above} \\ &= \overline{\mathbb{I}}{;}(\overline{\mathbb{T}{;}\varepsilon} \cup \mathbb{T}{;}\varepsilon) & \\ &= \overline{\mathbb{I}}{;}\mathbb{T} = \mathbb{T} & \text{condition above} \end{array}$$

Concerning the third, we have only $A{;}\Omega^{\mathsf{T}} = A{;}\overline{\overline{\varepsilon}^{\mathsf{T}}{;}\varepsilon} \subseteq A$ from Definition 8.4.1.iv, which doesn't suffice to establish

$$\overline{A}^{\mathsf{T}}{;}A \subseteq \varepsilon^{\mathsf{T}}{;}A \quad\Longleftrightarrow\quad \overline{\varepsilon^{\mathsf{T}}{;}A}{;}A^{\mathsf{T}} \subseteq A^{\mathsf{T}} \quad\Longleftrightarrow\quad A{;}\overline{A^{\mathsf{T}}{;}\varepsilon} \subseteq A \qquad \square$$

The following example justifies this doubt.

Example 8.4.7 Using the iteration of Remark 8.4.4, we provide another example. The start is a simplistic relation $R \subseteq \overline{\varepsilon}$.

	{}	{1}	{2}	{1,2}	{3}	{1,3}	{2,3}	{1,2,3}	{4}	{1,4}	{2,4}	{1,2,4}	{3,4}	{1,3,4}	{2,3,4}	{1,2,3,4}	{5}	{1,5}	{2,5}	{1,2,5}	{3,5}	{1,3,5}	{2,3,5}	{1,2,3,5}	{4,5}	{1,4,5}	{2,4,5}	{1,2,4,5}	{3,4,5}	{1,3,4,5}	{2,3,4,5}	{1,2,3,4,5}
1	0	0	0	0	0	0	0	0	0	0	0	0	0	0	1	0	0	0	0	0	0	0	0	0	0	0	0	0	0	0	0	0
2	0	0	0	0	0	1	0	0	0	0	0	0	0	0	0	0	0	0	0	0	0	0	0	0	0	0	0	0	0	0	0	0
3	0	0	0	0	0	0	0	0	0	0	0	1	0	0	0	0	0	0	0	0	0	0	0	0	0	0	0	0	0	0	0	0
4	0	0	0	0	0	0	0	0	0	0	0	0	0	0	0	0	0	0	1	0	0	0	0	0	0	0	0	0	0	0	0	0
5	0	0	0	0	0	0	0	1	0	0	0	0	0	0	0	0	0	0	0	0	0	0	0	0	0	0	0	0	0	0	0	0

Iterating $X \mapsto f(X)$ and $X \mapsto g(X)$ until stability, we obtain the following apartness A without axiom (i). It turns out that its complement $\overline{A}$ fails to be an Aumann contact. We have, for instance, $(\{4\},\{1\}) \in \overline{A}^{\mathsf{T}}{;}A$ but $(\{4\},\{1\}) \notin \varepsilon^{\mathsf{T}}{;}A$.

	{}	{1}	{2}	{1,2}	{3}	{1,3}	{2,3}	{1,2,3}	{4}	{1,4}	{2,4}	{1,2,4}	{3,4}	{1,3,4}	{2,3,4}	{1,2,3,4}	{5}	{1,5}	{2,5}	{1,2,5}	{3,5}	{1,3,5}	{2,3,5}	{1,2,3,5}	{4,5}	{1,4,5}	{2,4,5}	{1,2,4,5}	{3,4,5}	{1,3,4,5}	{2,3,4,5}	{1,2,3,4,5}
1	1	0	1	0	1	0	1	0	1	0	1	0	1	0	1	0	0	0	0	0	0	0	0	0	0	0	0	0	0	0	0	0
2	1	1	0	0	1	1	0	0	0	0	0	0	0	0	0	0	0	0	0	0	0	0	0	0	0	0	0	0	0	0	0	0
3	1	1	1	1	0	0	0	0	1	1	1	1	0	0	0	0	0	0	0	0	0	0	0	0	0	0	0	0	0	0	0	0
4	1	0	1	0	0	0	0	0	0	0	0	0	0	0	0	0	1	0	1	0	0	0	0	0	0	0	0	0	0	0	0	0
5	1	1	1	1	1	1	1	1	0	0	0	0	0	0	0	0	0	0	0	0	0	0	0	0	0	0	0	0	0	0	0	0

$\square$

Connection Algebra Some other concepts have also been studied: Boolean contact algebras. Their definition in [GW14] is as follows:

Definition 8.4.8 A relation $C : \mathbf{2}^X \longrightarrow \mathbf{2}^X$, defined besides membership ε, powerset containment Ω and projections from pairs π, ρ, is called a **Boolean contact algebra** when

i) $\mathbb{T};\varepsilon \supseteq C$,
ii) $\mathbb{I} \cap \mathbb{T};\varepsilon \subseteq C$,
iii) $C^{\mathsf{T}} \subseteq C$,
iv) $C;\Omega \subseteq C$,
v) $(\overline{C} \otimes \overline{C});\mathfrak{J}_2 \subseteq \overline{C}$,
vi) $\mathtt{syq}(C, C) \subseteq \mathbb{I}$,
vii) $C;\mathcal{N};C \subseteq C$,
viii) $\mathcal{N} \cap \mathbb{T};\varepsilon \cap \mathbb{T};\overline{\varepsilon} \subseteq C$.

This obviously subsumes under our general theme, which we will, however, not elaborate on here. One should consider all these, nearness, proximity, versions of apartness, etc. as far as possible as cryptomorphic concepts, thus avoiding to study them in separate axiomatizations over and over again.

Chapter 9
Frames

There exists a scenario in computer science where intricate topological questions are discussed. The topic is best described considering a device we observe without any knowledge about its inner program or process structure. This means necessarily incomplete observations which are somehow ordered by precision. Handling such observations requires specific orderings and often entails employing topological concepts.

9.1 From a Topology to a Frame

To begin with, we recall a concept described, e.g. in [Vic89], starting with two sets S and X. For these a subset $\models\ \subseteq X \times S$ is considered, i.e., a relation denoted $x \models u$ when $(x,u) \in\ \models$. One then speaks of a **topological system**, when properties for $S, X, \models$ are satisfied as follows:

– If U is a finite subset of S, then

$$x \models \bigwedge U \iff x \models u \text{ for all } u \in U.$$

– If U is any subset of S, then

$$x \models \bigvee U \iff x \models u \text{ for some } u \in U.$$

Translating this setting to the notation developed in this text, we need for a given ordering E the univalent relation

$$\mathfrak{J} := \left[\mathtt{lub}_E(\varepsilon_1)\right]^{\mathsf{T}} = \mathtt{syq}(\mathtt{ubd}_E(\varepsilon_1), E^{\mathsf{T}})$$

G. Schmidt, M. Winter, *Relational Topology*, Lecture Notes
in Mathematics 2208, https://doi.org/10.1007/978-3-319-74451-3_9

according to Definition 4.2.1 delivering arbitrary joins that may exist or not. In a similar way, the univalent relation

$$\mathfrak{M}_2 := \mathtt{syq}((E \otimes E), E)$$

produces binary meets—if any. When $\mathfrak{J}$ is indeed a mapping, also $\mathfrak{M}_2$ will be a mapping. That the mapping property of $\mathfrak{J}$ implies that of meet $\mathfrak{M}_2$ is a lattice-theoretic standard result; see Proposition 4.2.4.i (Fig. 9.1).

Based on this context, the concept of a frame will be introduced, referring, e.g., to the text [Vic89] by Steven Vickers.

Definition 9.1.1 A relation $E : S \longrightarrow S$, accompanied by its binary meet $\mathfrak{M}_2$ and arbitrary join $\mathfrak{J}$, shall be called a **frame** when it satisfies the following:

i) E is an order,
ii) $\mathfrak{J}$ is a mapping,
iii) binary meets $\mathfrak{M}_2$ distribute over arbitrary joins $\mathfrak{J}$. □

Remember the definition $(\mathbb{I} \otimes \mathfrak{J}) ; \mathfrak{M}_2 = \mathtt{syq}(\mathfrak{M}_2^\mathsf{T} ; (\mathbb{I} \otimes \varepsilon_1), \varepsilon_1) ; \mathfrak{J}$ of such distributivity in Definition 4.3.9 (Fig. 9.2).

The idea behind this definition is directly derivable from a topology as the following proposition shows (Fig. 9.3).

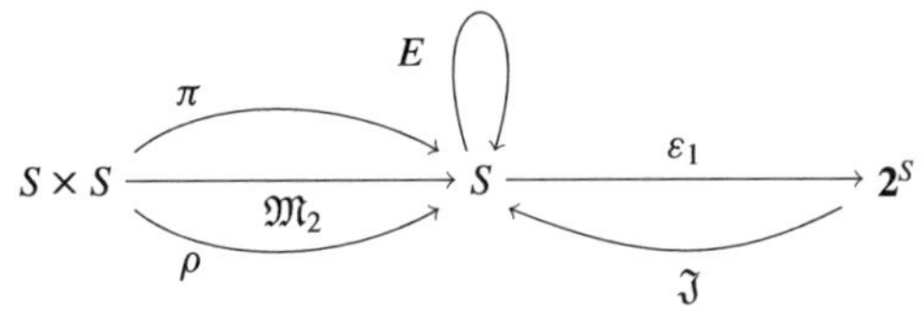

Fig. 9.1 Typing in case of the definition of a frame

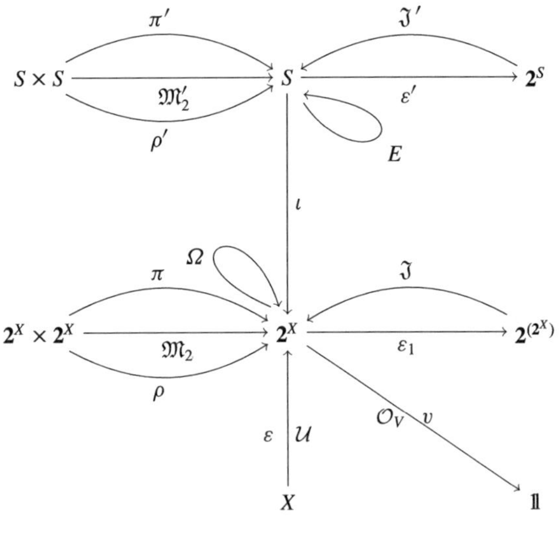

Fig. 9.2 Typing when a frame E is derived from a topology $\mathcal{U}$

$\mathcal{U} =$

	{}	{a}	{b}	{a,b}	{c}	{a,c}	{b,c}	{a,b,c}	{d}	{a,d}	{b,d}	{a,b,d}	{c,d}	{a,c,d}	{b,c,d}	{a,b,c,d}	{e}	{a,e}	{b,e}	{a,b,e}	{c,e}	{a,c,e}	{b,c,e}	{a,b,c,e}	{d,e}	{a,d,e}	{b,d,e}	{a,b,d,e}	{c,d,e}	{a,c,d,e}	{b,c,d,e}	{a,b,c,d,e}
a	0	0	0	0	0	0	0	0	0	0	0	0	0	0	0	0	0	0	0	1	0	0	0	1	0	0	0	1	0	0	0	1
b	0	0	1	1	0	0	1	1	0	0	1	1	0	0	1	1	0	0	1	1	0	0	1	1	0	0	1	1	0	0	1	1
c	0	0	0	0	1	1	1	1	0	0	0	0	1	1	1	1	0	0	0	0	1	1	1	1	0	0	0	0	1	1	1	1
d	0	0	0	0	0	0	0	0	0	0	0	0	0	0	0	0	0	0	0	0	0	0	0	0	0	0	0	1	0	0	0	1
e	0	0	0	0	0	0	0	0	0	0	0	0	0	0	0	0	1	1	1	1	1	1	1	1	1	1	1	1	1	1	1	1

$\mathcal{O}_V^{\mathsf{T}} =$

$$(1\ 0\ 1\ 0\ 1\ 0\ 1\ 0\ 0\ 0\ 0\ 0\ 0\ 0\ 0\ 0\ 1\ 0\ 1\ 1\ 1\ 0\ 1\ 1\ 0\ 0\ 0\ 1\ 0\ 0\ 0\ 1)$$

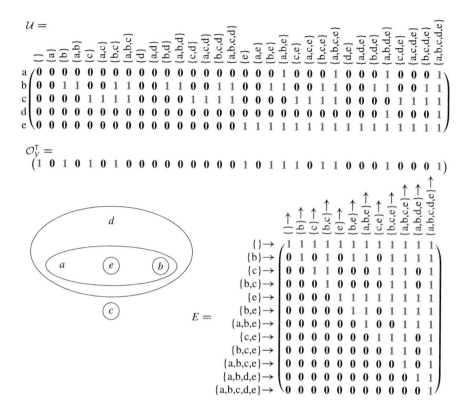

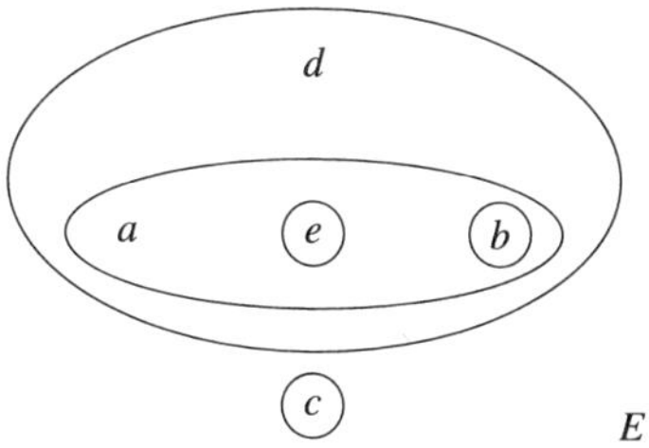

$E =$

	{}→	{b}→	{c}→	{b,c}→	{e}→	{b,e}→	{a,b,e}→	{c,e}→	{b,c,e}→	{a,b,c,e}→	{a,b,d,e}→	{a,b,c,d,e}→
{}→	1	1	1	1	1	1	1	1	1	1	1	1
{b}→	0	1	0	1	0	1	1	0	1	1	1	1
{c}→	0	0	1	1	0	0	0	1	1	1	0	1
{b,c}→	0	0	0	1	0	0	0	0	1	1	0	1
{e}→	0	0	0	0	1	1	1	1	1	1	1	1
{b,e}→	0	0	0	0	0	1	1	0	1	1	1	1
{a,b,e}→	0	0	0	0	0	0	1	0	0	1	1	1
{c,e}→	0	0	0	0	0	0	0	1	1	1	0	1
{b,c,e}→	0	0	0	0	0	0	0	0	1	1	0	1
{a,b,c,e}→	0	0	0	0	0	0	0	0	0	1	0	1
{a,b,d,e}→	0	0	0	0	0	0	0	0	0	0	1	1
{a,b,c,d,e}→	0	0	0	0	0	0	0	0	0	0	0	1

Fig. 9.3 Example for a frame E derived from a topology $\mathcal{U}$

Proposition 9.1.2 *Assume any neighborhood topology $\mathcal{U} : X \longrightarrow \mathbf{2}^X$ being given together with its vector $\mathcal{O}_V := \overline{\varepsilon^{\mathsf{T}} ; \overline{\mathcal{U}}} ; \mathbb{T}$ describing the open sets. Extruding $\mathcal{O}_V$ via the injection $\iota : S \longrightarrow \mathbf{2}^X$, one obtains*

$$E := \iota ; \Omega ; \iota^{\mathsf{T}},$$

which, together with arbitrary join and binary meet

$$\mathfrak{J}' := syq(ubd_E(\varepsilon'), E^{\mathsf{T}}), \qquad \mathfrak{M}_2' := syq((E \otimes E), E),$$

turns out to be a frame.

Proof

i) E is an order because it is reflexive, antisymmetric and transitive:

$$E = \iota ; \Omega ; \iota^{\mathsf{T}} \supseteq \iota ; \iota^{\mathsf{T}} = \mathbb{I}$$
$$E \cap E^{\mathsf{T}} = \iota ; \Omega ; \iota^{\mathsf{T}} \cap (\iota ; \Omega ; \iota^{\mathsf{T}})^{\mathsf{T}} = \iota ; (\Omega \cap \Omega^{\mathsf{T}}) ; \iota^{\mathsf{T}} = \iota ; \iota^{\mathsf{T}} = \mathbb{I}$$
$$E ; E = \iota ; \Omega ; \iota^{\mathsf{T}} ; \iota ; \Omega ; \iota^{\mathsf{T}} \subseteq \iota ; \Omega ; \Omega ; \iota^{\mathsf{T}} = \iota ; \Omega ; \iota^{\mathsf{T}} = E$$

ii) The obvious idea to prove that Definition 9.1.1.ii,iii hold is to show that $\mathfrak{M}'_2$, $\mathfrak{J}'$ obtained from E may be expressed via $\mathfrak{M}_2$ and $\mathfrak{J}$ obtained from the powerset ordering Ω as

$$\mathfrak{J}' = \vartheta_\iota ; \mathfrak{J} ; \iota^\mathsf{T} \qquad \mathfrak{M}'_2 = (\iota \otimes \iota) ; \mathfrak{M}_2 ; \iota^\mathsf{T}.$$

Join forming $\mathfrak{J}'$ for the order E is a univalent relation. The goal is to prove that the relation $\vartheta_\iota ; \mathfrak{J} ; \iota^\mathsf{T}$ is contained in $\mathfrak{J}'$ and is total, which will make $\mathfrak{J}'$ a mapping:

$\vartheta_\iota ; \mathfrak{J} ; \iota^\mathsf{T} = \vartheta_\iota ; \mathtt{syq}(\varepsilon ; \varepsilon_1, \varepsilon) ; \iota^\mathsf{T}$ by definition of $\mathfrak{J}$ in Definition 4.2.1
$= \mathtt{syq}(\varepsilon ; \varepsilon_1 ; \vartheta_\iota^\mathsf{T}, \varepsilon ; \iota^\mathsf{T})$ since ι and ϑ_ι are maps
$= \mathtt{syq}(\varepsilon ; \iota^\mathsf{T} ; \varepsilon', \varepsilon ; \iota^\mathsf{T})$ property Proposition 2.2.5.i of the existential image
$\subseteq \mathtt{syq}(\iota ; \overline{\varepsilon}^\mathsf{T} ; \varepsilon ; \iota^\mathsf{T} ; \varepsilon', \iota ; \overline{\varepsilon}^\mathsf{T} ; \varepsilon ; \iota^\mathsf{T})$ an additional common factor, Proposition 2.1.4.iv
$= \mathtt{syq}(\iota ; \overline{\Omega}^\mathsf{T} ; \iota^\mathsf{T} ; \varepsilon', \iota ; \overline{\Omega}^\mathsf{T} ; \iota^\mathsf{T}) = \mathtt{syq}(\overline{\iota ; \Omega^\mathsf{T} ; \iota^\mathsf{T}} ; \varepsilon', \overline{\iota ; \Omega^\mathsf{T} ; \iota^\mathsf{T}})$
$= \mathtt{syq}(\overline{E}^\mathsf{T} ; \varepsilon', \overline{E^\mathsf{T}}) = \mathtt{syq}(\mathrm{ubd}_E(\varepsilon'), \overline{E^\mathsf{T}}) = \mathfrak{J}'$

The relation contained in the univalent $\mathfrak{J}'$ now turns out to be total:

$\vartheta_\iota ; \mathfrak{J} ; \iota^\mathsf{T} ; \mathbb{T} = \vartheta_\iota ; \mathfrak{J} ; \mathcal{O}_V$ since ι extrudes $\mathcal{O}_V$
$\supseteq \vartheta_\iota ; \mathfrak{J} ; \overline{\mathfrak{J}^\mathsf{T} ; \overline{\varepsilon_1^\mathsf{T} ; \overline{\mathcal{O}_V}}}$ Proposition 5.3.5
$\supseteq \vartheta_\iota ; \overline{\varepsilon_1^\mathsf{T} ; \overline{\mathcal{O}_V}}$ join forming $\mathfrak{J}$ is total
$= \overline{\vartheta_\iota ; \varepsilon_1^\mathsf{T} ; \overline{\mathcal{O}_V}}$ ϑ_ι is a map
$= \overline{\varepsilon'^\mathsf{T} ; \iota ; \overline{\mathcal{O}_V}}$ property Proposition 2.2.5.i of the existential image
$= \overline{\varepsilon'^\mathsf{T} ; \mathbb{L}} = \mathbb{T}$ because ι extrudes $\mathcal{O}_V$, i.e. $\iota^\mathsf{T} ; \mathbb{T} = \mathcal{O}_V$

For the second claim again, equality $\mathfrak{M}'_2 = (\iota \otimes \iota) ; \mathfrak{M}_2 ; \iota^\mathsf{T}$ will hold, when $\supseteq$ and totality of the latter term can be proved, since $\mathfrak{M}'_2$ is a univalent relation by definition.

$\mathfrak{M}'_2 = \mathtt{syq}((E \otimes E), E)$
$= \mathtt{syq}((\iota ; \Omega ; \iota^\mathsf{T} \otimes \iota ; \Omega ; \iota^\mathsf{T}), \iota ; \Omega ; \iota^\mathsf{T})$
$= \mathtt{syq}((\iota ; \Omega \otimes \iota ; \Omega) ; (\iota \otimes \iota)^\mathsf{T}, \iota ; \Omega) ; \iota^\mathsf{T}$ since ι is a map
$= (\iota \otimes \iota)\, \mathtt{syq}((\iota ; \Omega \otimes \iota ; \Omega), \iota ; \Omega) ; \iota^\mathsf{T}$ again since ι is a map
$= (\iota \otimes \iota)\, \mathtt{syq}(\iota ; (\Omega \otimes \Omega), \iota ; \Omega) ; \iota^\mathsf{T}$ a third time since ι is a map
$\supseteq (\iota \otimes \iota)\, \mathtt{syq}((\Omega \otimes \Omega), \Omega) ; \iota^\mathsf{T}$ Proposition 2.1.4.iv
$= (\iota \otimes \iota) ; \mathfrak{M}_2 ; \iota^\mathsf{T}$ by definition of $\mathfrak{M}_2$

$(\iota \otimes \iota) ; \mathfrak{M}_2 ; \iota^\mathsf{T} ; \mathbb{T} = (\iota \otimes \iota) ; \mathfrak{M}_2 ; \mathcal{O}_V$ since ι extrudes $\mathcal{O}_V$
$\supseteq (\iota \otimes \iota)\, (\mathcal{O}_V \otimes \mathcal{O}_V)$ shunted version of Definition 5.3.2.iii
$= (\iota \otimes \iota)\, (\iota^\mathsf{T} ; \mathbb{T} \otimes \iota^\mathsf{T} ; \mathbb{T})$ again since ι extrudes $\mathcal{O}_V$
$= (\iota ; \iota^\mathsf{T} ; \mathbb{T} \otimes \iota ; \iota^\mathsf{T} ; \mathbb{T}) = (\mathbb{T} \otimes \mathbb{T}) = \mathbb{T}$ since ι is univalent and total

After this has been established, we will then also have

$$(\iota \otimes \iota) : \mathfrak{M}_2 : \iota^\mathsf{T} : \iota = (\iota \otimes \iota) : \mathfrak{M}_2 \qquad \vartheta_\iota : \mathfrak{J} : \iota^\mathsf{T} : \iota = \vartheta_\iota : \mathfrak{J},$$

mainly because ι as a map satisfies $\iota\iota^\mathsf{T}\iota = \iota$. Direction $\subseteq$ is obvious in both cases. For the reverse direction we have to invest that $\mathcal{O}_D = \iota^\mathsf{T} : \iota$ "moves through" $\mathfrak{M}_2$ resp. $\mathfrak{J}$:

$$\begin{aligned}
&(\iota \otimes \iota) : \mathfrak{M}_2 = (\iota : \iota^\mathsf{T} : \iota \otimes \iota : \iota^\mathsf{T} : \iota) : \mathfrak{M}_2 = (\iota : \mathcal{O}_D \otimes \iota : \mathcal{O}_D) : \mathfrak{M}_2 \\
&= (\iota \otimes \iota) : (\mathcal{O}_D \otimes \mathcal{O}_D) : \mathfrak{M}_2 \quad \text{since } \iota \text{ is univalent} \\
&\subseteq (\iota \otimes \iota) : \mathfrak{M}_2 : \mathcal{O}_D \quad \text{see Definition 5.3.3.iii} \\
&= (\iota \otimes \iota) : \mathfrak{M}_2 : \iota^\mathsf{T} : \iota \\
&\vartheta_\iota : \mathfrak{J} : \iota^\mathsf{T} : \iota = \vartheta_\iota : \mathfrak{J} : \mathcal{O}_D = \vartheta_\iota : \mathfrak{J} : (\mathcal{O}_V : \mathbb{T} \cap \mathbb{I}) \\
&= \vartheta_\iota : \mathfrak{J} : \mathcal{O}_V : \mathbb{T} \cap \vartheta_\iota : \mathfrak{J} = \mathbb{T} \cap \vartheta_\iota : \mathfrak{J} \quad \text{see above}
\end{aligned}$$

iii) Distributivity of $\mathfrak{M}_2'$ over $\mathfrak{J}'$ is a consequence of the following containment, since both sides are mappings:

$$\begin{aligned}
&\mathtt{syq}(\mathfrak{M}_2'^{\mathsf{T}} : (\mathbb{I} \otimes \varepsilon'), \varepsilon') : \mathfrak{J}' \\
&\subseteq \mathtt{syq}(\iota^\mathsf{T} : \mathfrak{M}_2'^{\mathsf{T}} : (\mathbb{I} \otimes \varepsilon'), \iota^\mathsf{T} : \varepsilon') : \mathfrak{J}' && \text{Proposition 2.1.4.iv} \\
&= \mathtt{syq}(\iota^\mathsf{T} : \mathfrak{M}_2'^{\mathsf{T}} : (\mathbb{I} \otimes \varepsilon'), \iota^\mathsf{T} : \varepsilon') : \vartheta_\iota : \mathfrak{J} : \iota^\mathsf{T} && \text{see above} \\
&= \mathtt{syq}(\iota^\mathsf{T} : \mathfrak{M}_2'^{\mathsf{T}} : (\mathbb{I} \otimes \varepsilon'), \iota^\mathsf{T} : \varepsilon') : \mathtt{syq}(\iota^\mathsf{T} : \varepsilon', \varepsilon_1) : \mathfrak{J} : \iota^\mathsf{T} && \text{expanding } \vartheta_\iota \\
&\subseteq \mathtt{syq}(\iota^\mathsf{T} : \mathfrak{M}_2'^{\mathsf{T}} : (\mathbb{I} \otimes \varepsilon'), \varepsilon_1) : \mathfrak{J} : \iota^\mathsf{T} && \text{Proposition 2.1.2} \\
&= \mathtt{syq}(\iota^\mathsf{T} : \iota : \mathfrak{M}_2^\mathsf{T} : (\iota^\mathsf{T} \otimes \iota^\mathsf{T}) : (\mathbb{I} \otimes \varepsilon'), \varepsilon_1) : \mathfrak{J} : \iota^\mathsf{T} && \text{expanding } \mathfrak{M}_2' \\
&= \mathtt{syq}(\mathfrak{M}_2^\mathsf{T} : (\iota^\mathsf{T} \otimes \iota^\mathsf{T}) : (\mathbb{I} \otimes \varepsilon'), \varepsilon_1) : \mathfrak{J} : \iota^\mathsf{T} && \text{see above} \\
&= \mathtt{syq}(\mathfrak{M}_2^\mathsf{T} : (\iota^\mathsf{T} \otimes \iota^\mathsf{T} : \varepsilon'), \varepsilon_1) : \mathfrak{J} : \iota^\mathsf{T} && \text{since } \iota \text{ is injective} \\
&= \mathtt{syq}(\mathfrak{M}_2^\mathsf{T} : (\iota^\mathsf{T} \otimes \varepsilon_1 : \vartheta_\iota^\mathsf{T}), \varepsilon_1) : \mathfrak{J} : \iota^\mathsf{T} && \text{Proposition 2.2.5.i} \\
&= \mathtt{syq}(\mathfrak{M}_2^\mathsf{T} : (\mathbb{I} \otimes \varepsilon_1) : (\iota^\mathsf{T} \otimes \vartheta_\iota^\mathsf{T}), \varepsilon_1) : \mathfrak{J} : \iota^\mathsf{T} && \iota, \vartheta_\iota \text{ are univalent} \\
&= (\iota \otimes \vartheta_\iota)\, \mathtt{syq}(\mathfrak{M}_2^\mathsf{T} : (\mathbb{I} \otimes \varepsilon_1), \varepsilon_1) : \mathfrak{J} : \iota^\mathsf{T} && \iota, \vartheta_\iota \text{ are maps} \\
&= (\iota \otimes \vartheta_\iota) : (\mathbb{I} \otimes \mathfrak{J}) : \mathfrak{M}_2 : \iota^\mathsf{T} && \text{Proposition 4.3.10} \\
&= (\iota \otimes \vartheta_\iota : \mathfrak{J}) : \mathfrak{M}_2 : \iota^\mathsf{T} && \iota, \vartheta_\iota \text{ are univalent} \\
&= (\iota \otimes \vartheta_\iota : \mathfrak{J} : \iota^\mathsf{T} : \iota) : \mathfrak{M}_2 : \iota^\mathsf{T} && \text{see above} \\
&= (\iota \otimes \mathfrak{J}' : \iota) : \mathfrak{M}_2 : \iota^\mathsf{T} \\
&= (\mathbb{I} \otimes \mathfrak{J}') : (\iota \otimes \iota) : \mathfrak{M}_2 : \iota^\mathsf{T} && \text{univalency of } \mathfrak{J}' \\
&= (\mathbb{I} \otimes \mathfrak{J}') : \mathfrak{M}_2'
\end{aligned}$$

□

This was the first direction, producing a frame out of a topology. The powerset ordering has been restricted so as to consider the order between the open sets only.

9.2 From a Frame to a Topology

The other direction is also possible. One may start with a frame E, together with its corresponding $\mathfrak{M}_2$, $\mathfrak{J}$ and—under certain circumstances—proceed to a closely related topology.

The next definition resembles this idea in quantifier-free form (We mention only in passing that even the extent relation—here just postulated to exist—might be constructively generated via prime filters.) (Fig. 9.4).

Definition 9.2.1 When a frame $E : S \longrightarrow S$ is given, we will call a relation $\varepsilon_F : X \longrightarrow S$ satisfying $E = \overline{\varepsilon_F^{\mathsf{T}} ; \overline{\varepsilon_F}}$ as well as $\varepsilon_F ; (E \circledless E) = (\varepsilon_F \circledless \varepsilon_F)$ and $\varepsilon_F ; \mathbb{T} = \mathbb{T}$ its **extent**. □

The extent relation enjoys an important property of a membership relation, because obviously

$$\mathtt{syq}(\varepsilon_F, \varepsilon_F) = E^{\mathsf{T}} \cap E = \mathbb{I}.$$

An example of a frame together with an extent and the resulting topology is provided with Fig. 9.5.

Requiring totality in Definition 9.2.1 is essential, since otherwise a **0**-row might be added to ε_F without changing E. Then, however, the $\mathcal{U}$ below would fail to be total. Figure 9.6 shows an example. From $\varepsilon_F ; (E \circledless E) = (\varepsilon_F \circledless \varepsilon_F)$ direction "$\subseteq$" is

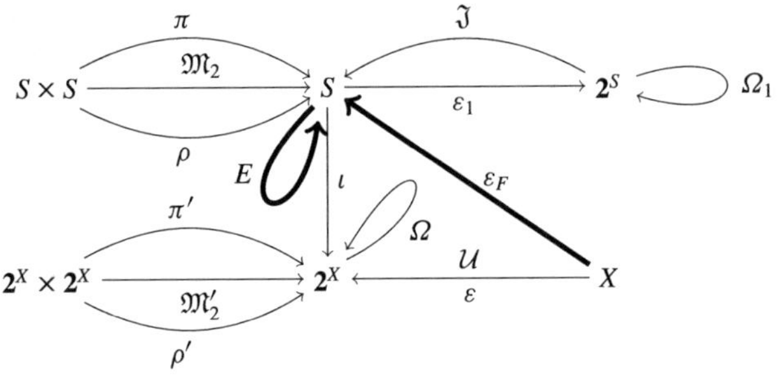

Fig. 9.4 Typing for frame E and extent ε_F

E:

	a	b	c	d	e
a	1	1	1	1	1
b	0	1	0	1	1
c	0	0	1	1	1
d	0	0	0	1	1
e	0	0	0	0	1

ε_F:

	a	b	c	d	e
1	0	1	0	1	1
2	0	0	0	0	1
3	0	1	0	1	1
4	0	0	1	1	1

$\mathcal{U}$:

	{}	{1}	{2}	{1,2}	{3}	{1,3}	{2,3}	{1,2,3}	{4}	{1,4}	{2,4}	{1,2,4}	{3,4}	{1,3,4}	{2,3,4}	{1,2,3,4}
1	0	0	0	0	0	1	0	1	0	0	0	0	0	1	0	1
2	0	0	0	0	0	0	0	0	0	0	0	0	0	0	0	1
3	0	0	0	0	0	1	0	1	0	0	0	0	0	1	0	1
4	0	0	0	0	0	0	0	0	1	1	1	1	1	1	1	1

Fig. 9.5 Example of obtaining a topology $\mathcal{U}$ from a frame E with extent ε_F

Fig. 9.6 A relation that also satisfies $E = \overline{\varepsilon_F^\mathsf{T} ; \overline{\varepsilon_F}}$ for the E in Fig. 9.5 but violates other requirements

$$\varepsilon_F = \begin{array}{c} \\ 1\\2\\3\\4\\5\\6\\7\\8 \end{array} \begin{array}{c} \begin{array}{ccccc} a & b & c & d & e \end{array} \\ \begin{pmatrix} 0 & 1 & 0 & 1 & 1 \\ 0 & 0 & 0 & 0 & 0 \\ 0 & 1 & 0 & 1 & 1 \\ 0 & 0 & 1 & 1 & 1 \\ 0 & 1 & 0 & 1 & 1 \\ 0 & 0 & 0 & 0 & 1 \\ 0 & 1 & 0 & 1 & 1 \\ 0 & 0 & 1 & 1 & 1 \end{pmatrix} \end{array}$$

trivial, while "$\supseteq$" seems to necessarily be postulated since it later requires some sort of a sharp factorization.

Proposition 9.2.2 *Let $E : S \longrightarrow S$ be a frame together with its binary meet $\mathfrak{M}_2$ and arbitrary join $\mathfrak{J}$ mappings according to Definition 9.1.1. Assume it is in addition endowed with an extent relation $\varepsilon_F : X \longrightarrow S$. Considering the generic membership with the corresponding powerset order*

$$\varepsilon : X \longrightarrow \mathbf{2}^X, \qquad \Omega := \overline{\varepsilon^\mathsf{T} ; \overline{\varepsilon}},$$

and defining—what is a mapping by construction—

$$\iota := \mathtt{syq}(\varepsilon_F, \varepsilon) : S \longrightarrow \mathbf{2}^X,$$

one will get the membership-in-open-sets topology

$$\varepsilon_{\mathcal{O}} := \varepsilon_F ; \iota : X \longrightarrow \mathbf{2}^X$$

and consequently the neighborhood topology

$$\mathcal{U} := \varepsilon_{\mathcal{O}} ; \Omega.$$

Proof We recall that in this setting

$$\begin{aligned} \varepsilon ; \iota^\mathsf{T} &= \varepsilon_F \\ \iota ; \iota^\mathsf{T} &= \mathtt{syq}(\varepsilon_F, \varepsilon_F) = E \cap E^\mathsf{T} = \mathbb{I} \quad \text{i.e., } \iota \text{ is an injective mapping} \\ \varepsilon_F ; \iota &= \varepsilon \cap \mathbb{T} ; \mathtt{syq}(\varepsilon_F, \varepsilon) = \varepsilon \cap \mathbb{T} ; \iota \quad \text{Proposition 2.1.1} \\ \iota ; \Omega ; \iota^\mathsf{T} &= \iota ; \overline{\varepsilon^\mathsf{T} ; \overline{\varepsilon ; \iota^\mathsf{T}}} = \overline{\varepsilon_F^\mathsf{T} ; \overline{\varepsilon_F}} = E \\ \varepsilon_F ; E &= \varepsilon_F ; \overline{\varepsilon_F^\mathsf{T} ; \overline{\varepsilon_F}} = \varepsilon_F \end{aligned}$$

In addition we have $\varepsilon_F ; (E \circledS E) = (\varepsilon_F \circledS \varepsilon_F)$, as explicitly postulated in Definition 9.2.1.

From these identities follows that meets $\mathfrak{M}_2$, $\mathfrak{M}_2'$ commute with ι as

$$(\iota \otimes \iota) ; \mathfrak{M}_2' = \mathfrak{M}_2 ; \iota.$$

Since both sides are mappings, it suffices to prove

$$\mathfrak{M}_2 \,;\, \iota \subseteq (\iota \otimes \iota) \,;\, \mathfrak{M}'_2 \quad \text{or the shunted version} \quad \mathfrak{M}_2 \subseteq (\iota \otimes \iota) \,;\, \mathfrak{M}'_2 \,;\, \iota^{\mathsf{T}}.$$

The right side is

$$\begin{aligned}
(\iota \otimes \iota) \,;\, \mathfrak{M}'_2 \,;\, \iota^{\mathsf{T}} &= (\iota \otimes \iota)\, \mathtt{syq}((\varepsilon \otimes \varepsilon), \varepsilon) \,;\, \iota^{\mathsf{T}} && \text{by definition} \\
&= \mathtt{syq}((\varepsilon \otimes \varepsilon) \,;\, (\iota \otimes \iota)^{\mathsf{T}}, \varepsilon \,;\, \iota^{\mathsf{T}}) \\
&= \mathtt{syq}((\varepsilon \,;\, \iota^{\mathsf{T}} \otimes \varepsilon \,;\, \iota^{\mathsf{T}}), \varepsilon \,;\, \iota^{\mathsf{T}}) \\
&= \mathtt{syq}((\varepsilon_F \otimes \varepsilon_F), \varepsilon_F) && (*) \\
&= \mathtt{syq}(\varepsilon_F \,;\, (E \otimes E), \varepsilon_F) && \text{Definition 9.2.1} \\
&= \mathtt{syq}(\varepsilon_F \,;\, E \,;\, \mathfrak{M}_2^{\mathsf{T}}, \varepsilon_F) && \text{Proposition 4.2.2.iii} \\
&= \mathfrak{M}_2 \,;\, \mathtt{syq}(\varepsilon_F \,;\, E, \varepsilon_F) \\
&= \mathfrak{M}_2 \,;\, \mathtt{syq}(\varepsilon_F, \varepsilon_F) = \mathfrak{M}_2 \,;\, \mathbb{I}
\end{aligned}$$

For the proof proper, we follow the numbering of Definition 5.2.1.

i) $\mathcal{U} \,;\, \mathbb{T} = \varepsilon_F \,;\, \iota \,;\, \Omega \,;\, \mathbb{T} = \varepsilon_F \,;\, \iota \,;\, \mathbb{T} = \varepsilon_F \,;\, \mathbb{T} = \mathbb{T}$ the latter is explicitly assumed

$$\mathcal{U} = \varepsilon_F \,;\, \iota \,;\, \Omega = \varepsilon_F \,;\, \mathtt{syq}(\varepsilon_F, \varepsilon) \,;\, \Omega \subseteq \varepsilon \,;\, \Omega = \varepsilon$$

ii)

$$\mathcal{U} \,;\, \Omega = \varepsilon_F \,;\, \iota \,;\, \Omega \,;\, \Omega = \varepsilon_F \,;\, \iota \,;\, \Omega = \mathcal{U}$$

iii)

$$\begin{aligned}
&(\mathcal{U} \otimes \mathcal{U}) \,;\, \mathfrak{M}'_2 = (\varepsilon_F \,;\, \iota \,;\, \Omega \otimes \varepsilon_F \,;\, \iota \,;\, \Omega) \,;\, \mathfrak{M}'_2 \\
&= (\varepsilon_F \,;\, \iota \otimes \varepsilon_F \,;\, \iota) \,;\, (\Omega \otimes \Omega) \,;\, \mathfrak{M}'_2 \quad \text{following Corollary 3.2.3} \\
&= (\varepsilon_F \,;\, \iota \otimes \varepsilon_F \,;\, \iota) \,;\, \mathfrak{M}'_2 \,;\, \Omega \quad \text{following Proposition 4.3.8.iv} \\
&= (\varepsilon_F \otimes \varepsilon_F) \,;\, (\iota \otimes \iota) \,;\, \mathfrak{M}'_2 \,;\, \Omega \\
&= (\varepsilon_F \otimes \varepsilon_F) \,;\, \mathfrak{M}_2 \,;\, \iota \,;\, \Omega \\
&= (\varepsilon_F \otimes \varepsilon_F) \,;\, \mathtt{syq}((\varepsilon_F \otimes \varepsilon_F), \varepsilon_F) \,;\, \iota \,;\, \Omega \quad \text{intermediate result } (*) \text{ above} \\
&\subseteq \varepsilon_F \,;\, \iota \,;\, \Omega \quad \text{cancellation; see Proposition 2.1.1.i} \\
&= \mathcal{U}
\end{aligned}$$

iv) We have to prove $\mathcal{U} \subseteq \mathcal{U} \,;\, \overline{\varepsilon^{\mathsf{T}} \,;\, \overline{\mathcal{U}}}$, i.e. $\varepsilon_F \,;\, \iota \,;\, \Omega \subseteq \varepsilon_F \,;\, \iota \,;\, \Omega \,;\, \overline{\varepsilon^{\mathsf{T}} \,;\, \overline{\varepsilon_F \,;\, \iota \,;\, \Omega}}$. Now trivially $\Omega \,;\, \overline{\varepsilon^{\mathsf{T}} \,;\, X} = \overline{\varepsilon^{\mathsf{T}} \,;\, X}$ since $\varepsilon \,;\, \Omega = \varepsilon$. This reduces the task to proving

$$\varepsilon_F \,;\, \iota \,;\, \Omega \subseteq \varepsilon_F \,;\, \iota \,;\, \overline{\varepsilon^{\mathsf{T}} \,;\, \overline{\varepsilon_F \,;\, \iota \,;\, \Omega}} = \varepsilon_F \,;\, \iota \,;\, \overline{\varepsilon^{\mathsf{T}} \,;\, \overline{\varepsilon_F \,;\, \iota \,;\, \Omega}} = \varepsilon_F \,;\, \overline{\varepsilon_F^{\mathsf{T}} \,;\, \overline{\varepsilon_F \,;\, \iota \,;\, \Omega}} = \varepsilon_F \,;\, \iota \,;\, \Omega,$$

where the rule $\mathtt{lbd}_R(\mathtt{ubd}_R(\mathtt{lbd}_R(Y))) = \mathtt{lbd}_R(Y)$ is used for $R := \overline{\varepsilon_F}$. □

$$E = \begin{array}{c} \\ a \\ b \\ c \\ d \\ e \\ f \\ g \\ h \\ i \\ j \end{array} \begin{array}{cccccccccc} a & b & c & d & e & f & g & h & i & j \\ 1 & 0 & 0 & 0 & 0 & 0 & 1 & 0 & 0 & 0 \\ 0 & 1 & 0 & 1 & 0 & 1 & 1 & 0 & 1 & 0 \\ 1 & 0 & 1 & 1 & 1 & 1 & 1 & 0 & 1 & 1 \\ 0 & 0 & 0 & 1 & 0 & 1 & 1 & 0 & 1 & 0 \\ 1 & 0 & 0 & 0 & 1 & 1 & 1 & 0 & 0 & 0 \\ 0 & 0 & 0 & 0 & 0 & 1 & 1 & 0 & 0 & 0 \\ 0 & 0 & 0 & 0 & 0 & 0 & 1 & 0 & 0 & 0 \\ 1 & 1 & 1 & 1 & 1 & 1 & 1 & 1 & 1 & 1 \\ 0 & 0 & 0 & 0 & 0 & 0 & 1 & 0 & 1 & 0 \\ 1 & 0 & 0 & 0 & 0 & 0 & 1 & 0 & 1 & 1 \end{array}$$

$$\varepsilon_F = \begin{array}{c} \\ 1 \\ 2 \\ 3 \\ 4 \end{array} \begin{array}{cccccccccc} a & b & c & d & e & f & g & h & i & j \\ 0 & 1 & 0 & 1 & 0 & 1 & 1 & 0 & 1 & 0 \\ 1 & 0 & 0 & 0 & 1 & 1 & 1 & 0 & 0 & 0 \\ 1 & 0 & 1 & 1 & 1 & 1 & 1 & 0 & 1 & 1 \\ 1 & 0 & 0 & 0 & 0 & 0 & 1 & 0 & 1 & 1 \end{array}$$

$$\mathcal{U} = \begin{array}{c} \\ 1 \\ 2 \\ 3 \\ 4 \end{array} \begin{array}{cccccccccccccccc} \{\} & \{1\} & \{2\} & \{1,2\} & \{3\} & \{1,3\} & \{2,3\} & \{1,2,3\} & \{4\} & \{1,4\} & \{2,4\} & \{1,2,4\} & \{3,4\} & \{1,3,4\} & \{2,3,4\} & \{1,2,3,4\} \\ 0 & 1 & 0 & 1 & 0 & 1 & 0 & 1 & 0 & 1 & 0 & 1 & 0 & 1 & 0 & 1 \\ 0 & 0 & 0 & 0 & 0 & 0 & 1 & 1 & 0 & 0 & 0 & 0 & 0 & 0 & 1 & 1 \\ 0 & 0 & 0 & 0 & 1 & 1 & 1 & 1 & 0 & 0 & 0 & 0 & 1 & 1 & 1 & 1 \\ 0 & 0 & 0 & 0 & 0 & 0 & 0 & 0 & 0 & 0 & 0 & 0 & 1 & 1 & 1 & 1 \end{array}$$

Fig. 9.7 Another frame E with extent ε_F and corresponding topology $\mathcal{U}$

A further example of a frame together with an extent is provided with Fig. 9.7. Also the resulting topology is shown.

The aspect studied here shows an important new development based on set-theoretic topological concepts. This made it particularly interesting to lift it to the level of equational reasoning without quantifiers.

Chapter 10
Simplicial Complexes

This section is intended to show how one might work relationally also for algebraic topology. We give a glimpse of simplicial complexes, usually subsumed under that topic.

Siegel writes in [Sie79] about his former Frankfurt colleague Max Dehn solving the Third Hilbert Problem: “We know that the areas of two given triangles can be proved equal by means of elementary geometry, i.e., without resorting to integral calculus or other limit processes. The question remained as to whether the same were possible for 3-dimensional figures; specifically, whether the volume of a tetrahedron could be rigorously defined without taking limits. This was one of the famous unsolved problems in mathematics posed by Hilbert at the international congress of mathematicians in Paris in 1900; Dehn was the first to have solved one of the Hilbert problems. The answer to the problem was in the negative, for Dehn showed that the theory of volume could not be developed on the basis of elementary geometry alone.” Dehn has simply constructed two equally voluminous polyhedra that he proved not to be *zerlegungsgleich* nor *ergänzungsgleich*, i.e. not equal by cutting it into pieces and recombining.

This remark has been inserted in order to prevent us from all too simplistic reasoning. Another hint in that direction are the astonishing titles of the four articles by Oskar Perron [Per40b, Per40c, Per40a, Per41]. They deal in slightly different versions with: How to exhaust the $\mathbb{R}^n$ with n-cubes—a task that one might consider being trivial.

10.1 Simplices

Several aspects of topology have been treated successfully using simplicial complexes. It seems that part of this can also be handled relationally. A non-oriented simplex is simply a finite set X with all subsets of it declared to be simplices. One

G. Schmidt, M. Winter, *Relational Topology*, Lecture Notes
in Mathematics 2208, https://doi.org/10.1007/978-3-319-74451-3_10

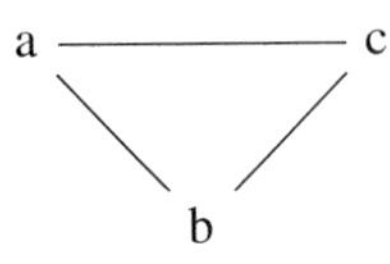

$$\begin{array}{r|cccccccc}
 & \{\} & \{a\} & \{b\} & \{a,b\} & \{c\} & \{a,c\} & \{b,c\} & \{a,b,c\} \\ \hline
\{\} & 1 & 1 & 1 & 1 & 1 & 1 & 1 & 1 \\
\{a\} & 0 & 1 & 0 & 1 & 0 & 1 & 0 & 1 \\
\{b\} & 0 & 0 & 1 & 1 & 0 & 0 & 1 & 1 \\
\{a,b\} & 0 & 0 & 0 & 1 & 0 & 0 & 0 & 1 \\
\{c\} & 0 & 0 & 0 & 0 & 1 & 1 & 1 & 1 \\
\{a,c\} & 0 & 0 & 0 & 0 & 0 & 1 & 0 & 1 \\
\{b,c\} & 0 & 0 & 0 & 0 & 0 & 0 & 1 & 1 \\
\{a,b,c\} & 0 & 0 & 0 & 0 & 0 & 0 & 0 & 1
\end{array}
\qquad
\begin{array}{cccccccc}
\{\} & \{a\} & \{b\} & \{a,b\} & \{c\} & \{a,c\} & \{b,c\} & \{a,b,c\} \\ \hline
0 & 1 & 1 & 0 & 1 & 0 & 0 & 0 \\
0 & 0 & 0 & 1 & 0 & 1 & 0 & 0 \\
0 & 0 & 0 & 1 & 0 & 0 & 1 & 0 \\
0 & 0 & 0 & 0 & 0 & 0 & 0 & 1 \\
0 & 0 & 0 & 0 & 0 & 1 & 1 & 0 \\
0 & 0 & 0 & 0 & 0 & 0 & 0 & 1 \\
0 & 0 & 0 & 0 & 0 & 0 & 0 & 1 \\
0 & 0 & 0 & 0 & 0 & 0 & 0 & 0
\end{array}$$

Fig. 10.1 Powerset ordering Ω and its Hasse relation H

then studies properties of the descent from a simplex of size n to all its subsets of size $n-1$. The Hasse relation

$$H = C \cap \overline{C;C} \quad \text{with} \quad C := \overline{\mathbb{I}} \cap \Omega,$$

of the powerset ordering Ω is obviously helpful. Its converse H^{T} leads from a subset precisely to subsets of one element less. An example is given in Fig. 10.1 with the set $X := \{a,b,c\}$ of which all subsets are considered as being simplices.

While Ω has a relatively obvious fractal generation, the corresponding fractal generation of H may seem less immediate:

$$\Omega_0 = (\mathbf{1}), \;\; \Omega_{n+1} = \begin{pmatrix} \Omega_n & \Omega_n \\ \mathbb{\perp\!\!\!\perp} & \Omega_n \end{pmatrix} \qquad H_0 = (\mathbf{0}), \;\; H_{n+1} = \begin{pmatrix} H_n & \mathbb{I} \\ \mathbb{\perp\!\!\!\perp} & H_n \end{pmatrix}.$$

Another example of a simplex is provided with Fig. 10.2.

10.2 Orientation

The next idea is to attach to all the as yet non-oriented simplices some orientation and to study how the descent mentioned behaves with regard to orientation.

Convention for the presentation of oriented simplices: For all the lower-dimensional simplices we demand that their tuples always be oriented according to the baseorder of the set X. An exception from this rule is made for the maximum-dimensional simplices: Since we usually give them as an input when studying some example, we accept for them the orientation as given in the input. □

Figure 10.3 illustrates this convention. We have typed $(2,1,3)$ providing an orientation indicated with the rotational arrow. In the cases of 1-dimensional arrows we always assume $(1,2)$, $(2,3)$, and $(1,3)$ etc. Obviously, (1,3) agrees with (2,1,3) in orientation, but (1,2) does not. By this we mean that the arrow from 1 to 3 is in

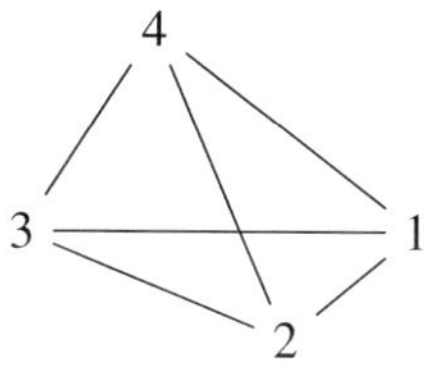

$H =$

	{}	{1}	{2}	{1,2}	{3}	{1,3}	{2,3}	{1,2,3}	{4}	{1,4}	{2,4}	{1,2,4}	{3,4}	{1,3,4}	{2,3,4}	{1,2,3,4}
{}	0	1	1	0	1	0	0	0	1	0	0	0	0	0	0	0
{1}	0	0	0	1	0	1	0	0	0	1	0	0	0	0	0	0
{2}	0	0	0	1	0	0	1	0	0	0	1	0	0	0	0	0
{1,2}	0	0	0	0	0	0	0	1	0	0	0	1	0	0	0	0
{3}	0	0	0	0	0	1	1	0	0	0	0	0	1	0	0	0
{1,3}	0	0	0	0	0	0	0	1	0	0	0	0	0	1	0	0
{2,3}	0	0	0	0	0	0	0	1	0	0	0	0	0	0	1	0
{1,2,3}	0	0	0	0	0	0	0	0	0	0	0	0	0	0	0	1
{4}	0	0	0	0	0	0	0	0	0	1	1	0	1	0	0	0
{1,4}	0	0	0	0	0	0	0	0	0	0	0	1	0	1	0	0
{2,4}	0	0	0	0	0	0	0	0	0	0	0	1	0	0	1	0
{1,2,4}	0	0	0	0	0	0	0	0	0	0	0	0	0	0	0	1
{3,4}	0	0	0	0	0	0	0	0	0	0	0	0	0	1	1	0
{1,3,4}	0	0	0	0	0	0	0	0	0	0	0	0	0	0	0	1
{2,3,4}	0	0	0	0	0	0	0	0	0	0	0	0	0	0	0	1
{1,2,3,4}	0	0	0	0	0	0	0	0	0	0	0	0	0	0	0	0

Fig. 10.2 Hasse relation H of the powerset order of a 3-dimensional simplex

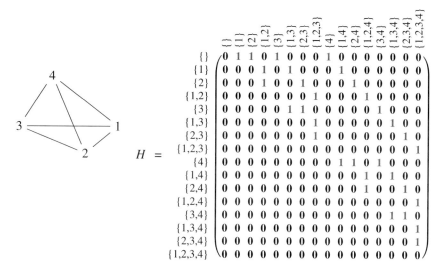

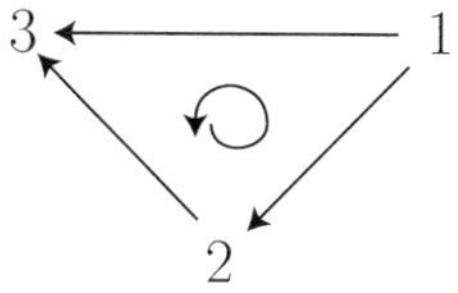

Fig. 10.3 Boundaries of a 2-dimensional oriented simplex

accordance with the rotational one, but the arrow from 1 to 2 is not, i.e., points in the opposite direction.

Based on this observation, transition to the boundary shall now be subdivided into two parts, which we call the positive/concordant as well as the negative/counter-rotating side.

Interlude An early attempt in this regard stems from classical homology of simplicial complexes: There, one is usually given an (additive) Abelian group G, and has to consider linear mappings sending the set of all oriented simplices into G. A mapping sending the n-dimensional oriented simplices into G is called an **n-chain** $\mathcal{C}_n$, provided $\mathcal{C}_n(-S) = -\mathcal{C}_n(S)$ for positively and negatively oriented versions of any simplex S. One is normally not interested in the values of these mappings beyond the combinatorial effect of applying a **boundary operator** ∂ to chains.

The boundary operator is a linear functional sending n-chains to $(n-1)$-chains. Since the boundary operator on chains is assumed to be linear, it needs only be defined for simplexes. If the n-simplex $(x_0, \ldots, x_n)$ gets by $\mathcal{C}$ assigned the value $g \in G$, we will for the moment denote this as $(x_0, \ldots, x_n)_g$. The definition of ∂ is then given showing to which lower-dimensional simplexes it contributes, written as a formal sum

$$(x_0, \ldots, x_n)_g \longmapsto \sum_i (-1)^i \, (x_0, \ldots, x_{i-1}, \; [\,\mathbf{x_i}, \; \textbf{deleted !}] \; x_{i+1}, \ldots, x_n)_g.$$

This suffices as a definition, since every chain may be decomposed down to the values it assigns to the single simplices. It means in particular that ∂ maps the value g assigned to (a, b, c) as

$$(a, b, c)_g \longmapsto (b, c)_g - (a, c)_g + (a, b)_g$$

and correspondingly $(a, b)_g \longmapsto (b)_g - (a)_g$. The main theorem then says that $\partial(\partial(x)) = 0$. To understand this result, we observe in this example how the contributions develop

$$\begin{aligned}(a, b, c)_g &\longmapsto (b, c)_g - (a, c)_g + (a, b)_g \\ &\longmapsto [(c)_g - (b)_g] - [(c)_g - (a)_g] + [(b)_g - (a)_g] = 0,\end{aligned}$$

regardless of how the chain C is actually defined, just following from the assumed linearity of ∂. □

We take our visualization from Fig. 10.14 and give a fairly "arbitrary" chain with group G equal to $\mathbb{Z}$ in Fig. 10.4. It shows the result of applying the boundary operation twice to a 2-chain getting a 0-chain assigning always $0 \in \mathbb{Z}$. What homology is intended to do using all this group theory is to keep track of the relative situations of the oriented simplices involved.

Working relationally, we are not in a position to *subtract* as above. We can, however, do some accounting or book-keeping of positive as well as of negative orientations and finally show that both sides result in the same.

Considering Fig. 10.4, we have in mind the subgroup of cycles, defined as having boundary 0 as well as the subgroup of boundaries, characterized as images of higher-dimensional chains. The quotient "cycles/boundaries" establishes the famous homology concept.

In a way corresponding to the boundary ∂, the converse $B := H^{\mathsf{T}}$ of H shall now be partitioned as in Fig. 10.5. This gives a boundary operation assigning to every simplex the set of all the oriented simplices that consist of precisely one element less and are oriented as described above: Positive boundaries of (a, b, c) are (b, c) and (a, b), while (a, c) is considered a negative one.

Linear Ordering of the Powerset When given an ordering $E : X \longrightarrow X$ on a baseset, one may wish to find an ordering $F : \mathbf{2}^X \longrightarrow \mathbf{2}^X$ on its powerset that respects E in some way. For comparison think of the pair of two ordered sets for which we are accustomed to work with the lexicographic ordering which is monotonic wrt. the first projection. An early approach to define an ordering on the powerset was made when studying semantics of nondeterminism and powerdomains. It brought forward the Egli-Milner orders; see [Win83]. However,

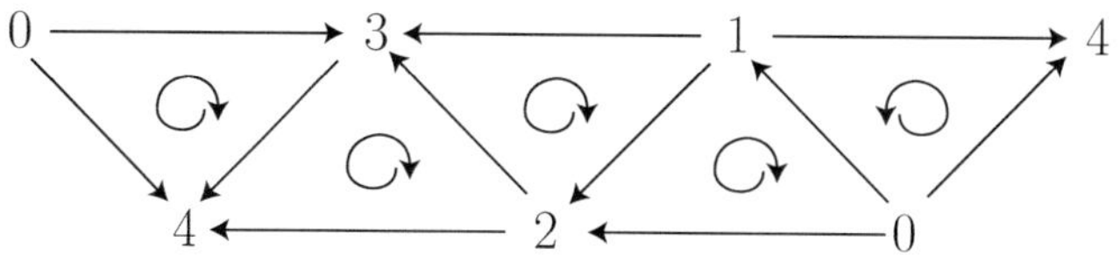

Given 2-chain C:

(4, 0, 3)	(1, 0, 4)	(4, 3, 2)	(2, 3, 1)	(2, 1, 0)
17	2	29	−3	5

its bounding 1-chain ∂C:

	(1, 4)	(0, 4)	(0, 3)	(3, 4)	(2, 4)	(2, 3)	(1, 3)	(1, 2)	(0, 2)	(0, 1)
	−2	−15	17	−12	29	−32	3	−8	5	−7
having obtained contributions from										
(4, 0, 3)		−17	17	17						
(1, 0, 4)	−2	2								−2
(4, 3, 2)				−29	29	−29				
(2, 3, 1)						−3	3	−3		
(2, 1, 0)								−5	5	−5

bounding 0-chain $\partial(\partial C)$ thereof:

	(0)	(1)	(2)	(3)	(4)
	0	0	0	0	0
having obtained contributions from					
(1, 4)		2			−2
(0, 4)	15				−15
(0, 3)	−17			17	
(3, 4)				12	−12
(2, 4)			−29		29
(2, 3)			32	−32	
(1, 3)		−3		3	
(1, 2)		8	−8		
(0, 2)	−5		5		
(0, 1)	7	−7			

Fig. 10.4 Applying the boundary operation ∂ twice to a 2-chain (cf. Fig. 10.14)

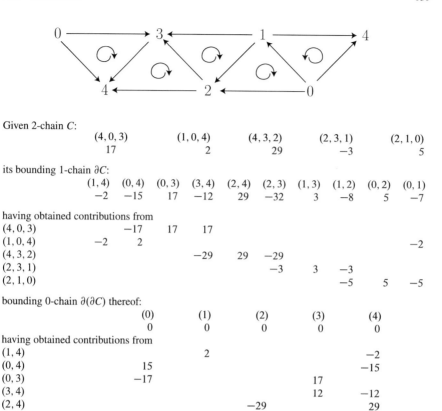

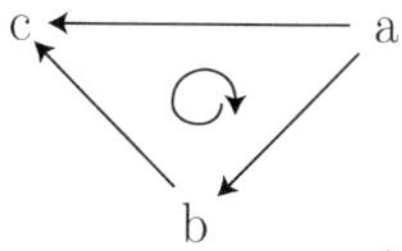

$B^{\Uparrow} =$

	{}	{a}	{b}	{a,b}	{c}	{a,c}	{b,c}	{a,b,c}
{}	0	0	0	0	0	0	0	0
{a}	1	0	0	0	0	0	0	0
{b}	1	0	0	0	0	0	0	0
{a,b}	0	0	1	0	0	0	0	0
{c}	1	0	0	0	0	0	0	0
{a,c}	0	0	0	0	1	0	0	0
{b,c}	0	0	0	0	1	0	0	0
{a,b,c}	0	0	0	1	0	0	1	0

$B^{\Downarrow\Uparrow} =$

{}	{a}	{b}	{a,b}	{c}	{a,c}	{b,c}	{a,b,c}
0	0	0	0	0	0	0	0
0	0	0	0	0	0	0	0
0	0	0	0	0	0	0	0
0	1	0	0	0	0	0	0
0	0	0	0	0	0	0	0
0	1	0	0	0	0	0	0
0	0	1	0	0	0	0	0
0	0	0	0	0	1	0	0

Fig. 10.5 Distinguishing positive and negative boundary: $H^{\top} = B^{\Uparrow} \cup B^{\Downarrow\Uparrow}$

these turned out to be just preorders even if E was a linear order. In addition, it was difficult to handle the empty set appropriately; see Chapt. 19 of [Sch11].

Example In Fig. 10.4, we study the transition to bounding chains for the Moebius strip.

□

In the following, we show how it is indeed possible to obtain a linear ordering on the powerset using relational means. We start with the linear baseorder

$$E = \begin{array}{c} \\ \mathrm{a} \\ \mathrm{b} \\ \mathrm{c} \end{array} \begin{array}{c} \begin{array}{ccc} \mathrm{a} & \mathrm{b} & \mathrm{c} \end{array} \\ \begin{pmatrix} 1 & 1 & 1 \\ 0 & 1 & 1 \\ 0 & 0 & 1 \end{pmatrix} \end{array} \qquad H_E = \begin{array}{c} \\ \mathrm{a} \\ \mathrm{b} \\ \mathrm{c} \end{array} \begin{array}{c} \begin{array}{ccc} \mathrm{a} & \mathrm{b} & \mathrm{c} \end{array} \\ \begin{pmatrix} 0 & 1 & 0 \\ 0 & 0 & 1 \\ 0 & 0 & 0 \end{pmatrix} \end{array}$$

consider its Hasse relation H_E and evaluate its decreasing sequence of points as

$$e_1 := \overline{\overline{E}^{\mathsf{T}};\mathbb{T}} = \begin{array}{c} \mathrm{a} \\ \mathrm{b} \\ \mathrm{c} \end{array} \begin{pmatrix} 0 \\ 0 \\ 1 \end{pmatrix} \quad e_2 := H_E;e_1 = \begin{array}{c} \mathrm{a} \\ \mathrm{b} \\ \mathrm{c} \end{array} \begin{pmatrix} 0 \\ 1 \\ 0 \end{pmatrix} \quad e_3 := H_E;e_2 = \begin{array}{c} \mathrm{a} \\ \mathrm{b} \\ \mathrm{c} \end{array} \begin{pmatrix} 1 \\ 0 \\ 0 \end{pmatrix}$$

Herefrom, we get sets "above" as

$$v_1 := \varepsilon^{\mathsf{T}};e_1 = \begin{array}{c} \{\} \\ \{\mathrm{a}\} \\ \{\mathrm{b}\} \\ \{\mathrm{a,b}\} \\ \{\mathrm{c}\} \\ \{\mathrm{a,c}\} \\ \{\mathrm{b,c}\} \\ \{\mathrm{a,b,c}\} \end{array} \begin{pmatrix} 0 \\ 0 \\ 0 \\ 0 \\ 1 \\ 1 \\ 1 \\ 1 \end{pmatrix} \quad v_2 := \varepsilon^{\mathsf{T}};e_2 = \begin{pmatrix} 0 \\ 0 \\ 1 \\ 1 \\ 0 \\ 0 \\ 1 \\ 1 \end{pmatrix} \quad v_3 := \varepsilon^{\mathsf{T}};e_3 = \begin{pmatrix} 0 \\ 1 \\ 0 \\ 1 \\ 0 \\ 1 \\ 0 \\ 1 \end{pmatrix}$$

This allows us to form

$$q_1 := \overline{v_1};v_1^{\mathsf{T}} \qquad q_2 := \overline{v_2};v_2^{\mathsf{T}} \cap \overline{q_1}^{\mathsf{T}} \qquad q_3 := \overline{v_3};v_3^{\mathsf{T}} \cap \overline{q_1}^{\mathsf{T}} \cap \overline{q_2}^{\mathsf{T}}$$

$$\begin{array}{c} \\ \{\} \\ \{\mathrm{a}\} \\ \{\mathrm{b}\} \\ \{\mathrm{a,b}\} \\ \{\mathrm{c}\} \\ \{\mathrm{a,c}\} \\ \{\mathrm{b,c}\} \\ \{\mathrm{a,b,c}\} \end{array} \begin{array}{c} \scriptstyle \{\}\ \{a\}\ \{b\}\ \{a,b\}\ \{c\}\ \{a,c\}\ \{b,c\}\ \{a,b,c\} \\ \begin{pmatrix} 0&0&0&0&1&1&1&1 \\ 0&0&0&0&1&1&1&1 \\ 0&0&0&0&1&1&1&1 \\ 0&0&0&0&1&1&1&1 \\ 0&0&0&0&0&0&0&0 \\ 0&0&0&0&0&0&0&0 \\ 0&0&0&0&0&0&0&0 \\ 0&0&0&0&0&0&0&0 \end{pmatrix} \end{array} \begin{array}{c} \scriptstyle \{\}\ \{a\}\ \{b\}\ \{a,b\}\ \{c\}\ \{a,c\}\ \{b,c\}\ \{a,b,c\} \\ \begin{pmatrix} 0&0&1&1&0&0&1&1 \\ 0&0&1&1&0&0&1&1 \\ 0&0&0&0&0&0&0&0 \\ 0&0&0&0&0&0&0&0 \\ 0&0&0&0&0&0&1&1 \\ 0&0&0&0&0&0&1&1 \\ 0&0&0&0&0&0&0&0 \\ 0&0&0&0&0&0&0&0 \end{pmatrix} \end{array} \begin{array}{c} \scriptstyle \{\}\ \{a\}\ \{b\}\ \{a,b\}\ \{c\}\ \{a,c\}\ \{b,c\}\ \{a,b,c\} \\ \begin{pmatrix} 0&1&0&1&0&1&0&1 \\ 0&0&0&0&0&0&0&0 \\ 0&0&0&1&0&1&0&1 \\ 0&0&0&0&0&0&0&0 \\ 0&0&0&0&0&1&0&1 \\ 0&0&0&0&0&0&0&0 \\ 0&0&0&0&0&0&0&1 \\ 0&0&0&0&0&0&0&0 \end{pmatrix} \end{array}$$

from which we finally obtain

$$F = \begin{array}{r} \\ \{\} \\ \{a\} \\ \{b\} \\ \{a,b\} \\ \{c\} \\ \{a,c\} \\ \{b,c\} \\ \{a,b,c\} \end{array} \begin{array}{c} \begin{array}{cccccccc} \{\} & \{a\} & \{b\} & \{a,b\} & \{c\} & \{a,c\} & \{b,c\} & \{a,b,c\} \end{array} \\ \begin{pmatrix} 1 & 1 & 1 & 1 & 1 & 1 & 1 & 1 \\ \mathbf{0} & 1 & 1 & 1 & 1 & 1 & 1 & 1 \\ \mathbf{0} & \mathbf{0} & 1 & 1 & 1 & 1 & 1 & 1 \\ \mathbf{0} & \mathbf{0} & \mathbf{0} & 1 & 1 & 1 & 1 & 1 \\ \mathbf{0} & \mathbf{0} & \mathbf{0} & \mathbf{0} & 1 & 1 & 1 & 1 \\ \mathbf{0} & \mathbf{0} & \mathbf{0} & \mathbf{0} & \mathbf{0} & 1 & 1 & 1 \\ \mathbf{0} & \mathbf{0} & \mathbf{0} & \mathbf{0} & \mathbf{0} & \mathbf{0} & 1 & 1 \\ \mathbf{0} & \mathbf{0} & \mathbf{0} & \mathbf{0} & \mathbf{0} & \mathbf{0} & \mathbf{0} & 1 \end{pmatrix} \end{array} = \mathbb{I} \cup q_1 \cup q_2 \cup q_3$$

This is—as of yet—an ugly iteration that should be simplified; but it shows that we have $F = \mathbb{I} \cup q_1 \cup q_2 \cup q_3$ finally evaluated by a relational construction from the linear order E in a way comparable with a lexicographic ordering for a product of linear orders.

Quite obviously, E and F satisfy $E ; \varepsilon = \varepsilon ; F$. Furthermore, F may recursively be generated as

$$F_0 := (1) \qquad F_{n+1} := \begin{pmatrix} F_n & \top \\ \bot & F_n \end{pmatrix},$$

which may be proved using this recursion:

$$E_1 = (1) \quad \varepsilon_1 = (\mathbf{0}\ 1) \quad F_1 = \begin{pmatrix} 1 & 1 \\ \mathbf{0} & 1 \end{pmatrix} \qquad E_{n+1} = \begin{pmatrix} E_n & \top \\ \bot & 1 \end{pmatrix} \quad \varepsilon_{n+1} = \begin{pmatrix} \varepsilon_n & \varepsilon_n \\ \bot & \top \end{pmatrix}$$

$$E_{n+1} ; \varepsilon_{n+1} = \begin{pmatrix} E_n ; \varepsilon_n & \top \\ \bot & \top \end{pmatrix} = \begin{pmatrix} \varepsilon_n ; F_n & \top \\ \bot & \top \end{pmatrix} = \varepsilon_{n+1} ; F_{n+1}.$$

Relational Evaluation of Boundaries Using the F thus obtained, it becomes possible to evaluate boundary relations such as $B^{\Uparrow}$, $B^{\Updownarrow}$ based on H and the order E of the set X, following the idea of homology theory. A first contribution to positive boundaries is given by taking row-wise the greatest elements of H^{T} according to F:

$$B^{\Uparrow}{}_1 = \mathtt{greR}_F(H^{\mathsf{T}}) =$$

	{}	{a}	{b}	{a,b}	{c}	{a,c}	{b,c}	{a,b,c}
{}	0	0	0	0	0	0	0	0
{a}	1	0	0	0	0	0	0	0
{b}	1	0	0	0	0	0	0	0
{a,b}	0	0	1	0	0	0	0	0
{c}	1	0	0	0	0	0	0	0
{a,c}	0	0	0	0	1	0	0	0
{b,c}	0	0	0	0	1	0	0	0
{a,b,c}	0	0	0	0	0	0	1	0

$$B^{\Updownarrow}{}_1 = \mathtt{greR}_F(H^{\mathsf{T}} \cap \overline{B^{\Uparrow}{}_1}) =$$

	{}	{a}	{b}	{a,b}	{c}	{a,c}	{b,c}	{a,b,c}
{}	0	0	0	0	0	0	0	0
{a}	0	0	0	0	0	0	0	0
{b}	0	0	0	0	0	0	0	0
{a,b}	0	1	0	0	0	0	0	0
{c}	0	0	0	0	0	0	0	0
{a,c}	0	1	0	0	0	0	0	0
{b,c}	0	0	1	0	0	0	0	0
{a,b,c}	0	0	0	0	0	1	0	0

$$B^{\Uparrow}{}_2 = B^{\Uparrow}{}_1 \cup \mathtt{greR}_F(H^{\mathsf{T}} \cap \overline{B^{\Uparrow}{}_1 \cup B^{\Updownarrow}{}_1}) =$$

	{}	{a}	{b}	{a,b}	{c}	{a,c}	{b,c}	{a,b,c}
{}	0	0	0	0	0	0	0	0
{a}	1	0	0	0	0	0	0	0
{b}	1	0	0	0	0	0	0	0
{a,b}	0	0	1	0	0	0	0	0
{c}	1	0	0	0	0	0	0	0
{a,c}	0	0	0	0	1	0	0	0
{b,c}	0	0	0	0	1	0	0	0
{a,b,c}	0	0	0	1	0	0	1	0

etc.

The idea how to proceed is evident. Already at this early point we have stability with $H^{\mathsf{T}} = B^{\Uparrow} \cup B^{\Updownarrow}$, where $B^{\Uparrow} = B^{\Uparrow}{}_2$ and $B^{\Updownarrow} = B^{\Updownarrow}{}_1$.

10.3 Simplicial Complexes

A simplicial complex in topology is usually defined on a set X of which subsets are declared to be simplices. Whenever a simplex is identified, all its subsets have to be simplices again.[1] One then studies in particular the descent from one simplex of size

[1] Should X be non-finite, one usually requires that every element be contained in only a finite number of subsets: locally finite.

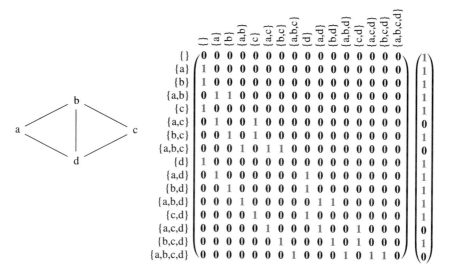

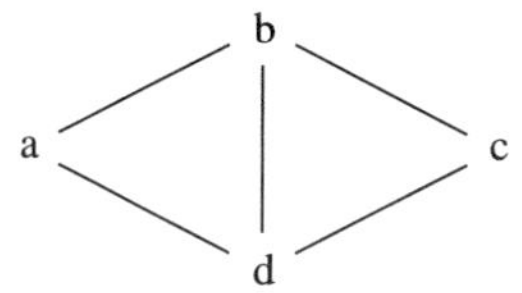

	{}	{a}	{b}	{a,b}	{c}	{a,c}	{b,c}	{a,b,c}	{d}	{a,d}	{b,d}	{a,b,d}	{c,d}	{a,c,d}	{b,c,d}	{a,b,c,d}	s
{}	0	0	0	0	0	0	0	0	0	0	0	0	0	0	0	0	1
{a}	1	0	0	0	0	0	0	0	0	0	0	0	0	0	0	0	1
{b}	1	0	0	0	0	0	0	0	0	0	0	0	0	0	0	0	1
{a,b}	0	1	1	0	0	0	0	0	0	0	0	0	0	0	0	0	1
{c}	1	0	0	0	0	0	0	0	0	0	0	0	0	0	0	0	1
{a,c}	0	1	0	0	1	0	0	0	0	0	0	0	0	0	0	0	0
{b,c}	0	0	1	0	1	0	0	0	0	0	0	0	0	0	0	0	1
{a,b,c}	0	0	0	1	0	1	1	0	0	0	0	0	0	0	0	0	0
{d}	1	0	0	0	0	0	0	0	0	0	0	0	0	0	0	0	1
{a,d}	0	1	0	0	0	0	0	0	1	0	0	0	0	0	0	0	1
{b,d}	0	0	1	0	0	0	0	0	1	0	0	0	0	0	0	0	1
{a,b,d}	0	0	0	1	0	0	0	0	0	1	1	0	0	0	0	0	1
{c,d}	0	0	0	0	1	0	0	0	1	0	0	0	0	0	0	0	1
{a,c,d}	0	0	0	0	0	1	0	0	0	1	0	0	1	0	0	0	0
{b,c,d}	0	0	0	0	0	0	1	0	0	0	1	0	1	0	0	0	1
{a,b,c,d}	0	0	0	0	0	0	0	1	0	0	0	1	0	1	1	0	0

Fig. 10.6 Concept of a simplicial complex with boundary operator H^{T} and s

n to all its subsimplices of size $n-1$. This leads us to conceive a simplicial complex on a set X as some vector s along its powerset $\mathbf{2}^X$. It must be down-closed, i.e., must satisfy $\Omega : s \subseteq s$ (with Ω the powerset ordering).

An example to which an orientation has not yet been attached is provided with Fig. 10.6. Later, we will forget the vector s and restrict the boundary relation correspondingly omitting rows and columns; see e.g. Fig. 10.7.[2]

Next, we intend to give orientation not just to a single simplex, but to a whole simplicial complex and we start with the most trivial example of Fig. 10.7. Every oriented simplex imposes an orientation on its bounding simplices; for instance running (c,b,d) means running along (c,b), (b,d), and (d,c). We have, however, agreed upon orienting the lower-dimensional simplices according to the baseorder, so that the first and the last, (b,c) and (c,d), contradict the rotational orientation, but the middle one, (b,d), agrees. Orientations may thus agree or may disagree, so that we have chosen to define the disjoint partition $B = H^{\mathsf{T}} = B^{\Uparrow} \cup B^{\Updownarrow}$ indicating agreement resp. non-agreement.

The two oriented triangles will further be said to have the same orientation because the vertical arrow gets a counter-running orientation from the orientations of the two triangles. This resembles the idea that then the vertical arrow might be removed, leaving us with a common circuit orientation (c,b,a,d).

[2]The rectangles in the figure indicate in which way the relation will later for reasons of size be reduced to the descent from highest to next lower dimension.

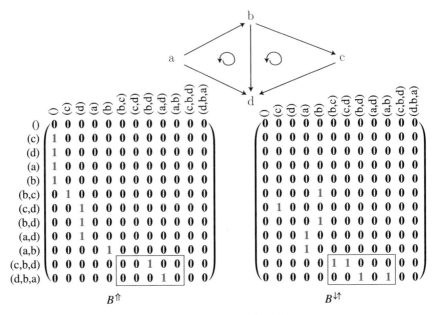

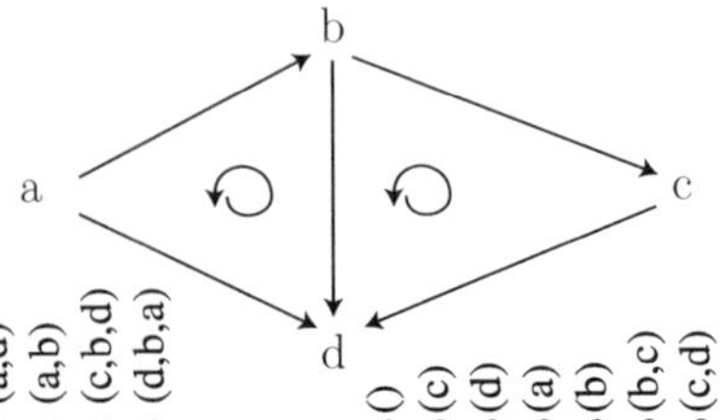

	()	(c)	(d)	(a)	(b)	(b,c)	(c,d)	(b,d)	(a,d)	(a,b)	(c,b,d)	(d,b,a)
()	0	0	0	0	0	0	0	0	0	0	0	0
(c)	1	0	0	0	0	0	0	0	0	0	0	0
(d)	1	0	0	0	0	0	0	0	0	0	0	0
(a)	1	0	0	0	0	0	0	0	0	0	0	0
(b)	1	0	0	0	0	0	0	0	0	0	0	0
(b,c)	0	1	0	0	0	0	0	0	0	0	0	0
(c,d)	0	0	1	0	0	0	0	0	0	0	0	0
(b,d)	0	0	1	0	0	0	0	0	0	0	0	0
(a,d)	0	0	1	0	0	0	0	0	0	0	0	0
(a,b)	0	0	0	0	1	0	0	0	0	0	0	0
(c,b,d)	0	0	0	0	0	0	0	1	0	0	0	0
(d,b,a)	0	0	0	0	0	0	0	0	1	0	0	0

$B^{\Uparrow}$

	()	(c)	(d)	(a)	(b)	(b,c)	(c,d)	(b,d)	(a,d)	(a,b)	(c,b,d)	(d,b,a)
()	0	0	0	0	0	0	0	0	0	0	0	0
(c)	0	0	0	0	0	0	0	0	0	0	0	0
(d)	0	0	0	0	0	0	0	0	0	0	0	0
(a)	0	0	0	0	0	0	0	0	0	0	0	0
(b)	0	0	0	0	0	0	0	0	0	0	0	0
(b,c)	0	0	0	0	1	0	0	0	0	0	0	0
(c,d)	0	1	0	0	0	0	0	0	0	0	0	0
(b,d)	0	0	0	0	1	0	0	0	0	0	0	0
(a,d)	0	0	0	1	0	0	0	0	0	0	0	0
(a,b)	0	0	0	1	0	0	0	0	0	0	0	0
(c,b,d)	0	0	0	0	0	1	1	0	0	0	0	0
(d,b,a)	0	0	0	0	0	0	0	1	0	1	0	0

$B^{\Downarrow\Uparrow}$

Fig. 10.7 Oriented complex with boundary operators $B^{\Uparrow}$, $B^{\Updownarrow}$

Definition 10.3.1 (Oriented Boundary Operators) For any finite set X consider the powerset ordering $\Omega : \mathbf{2}^X \longrightarrow \mathbf{2}^X$ and the converse B of its Hasse relation H. When a disjoint partition $B = B^{\Uparrow} \cup B^{\Updownarrow}$ is given that satisfies

$$B^{\Uparrow} ; B^{\Uparrow} \cup B^{\Updownarrow} ; B^{\Updownarrow} = B^{\Uparrow} ; B^{\Updownarrow} \cup B^{\Updownarrow} ; B^{\Uparrow} ,$$

we decide to call

- $B^{\Uparrow}$ the **positively oriented boundary operator** and
- $B^{\Updownarrow}$ the **negatively oriented boundary operator**. □

Unfortunately, we have to pay attention also to whether the original maximum-dimensional simplex is positively oriented or not. To cope with orientation in a general fashion, one will consider all the simplices in a two-fold form, namely as positively as well as negatively oriented.

The proper relational tool for such a consideration is the extrusion of the *full* subset obtaining its injection $\theta := \theta_{\mathbb{T}} : \Theta \longrightarrow \mathbf{2}^X$. We thus have the set Θ of negatively oriented versions for all the simplices, thereby generating for the simplex (c,b,d), e.g., its negatively oriented version (c,b,d)-, so that we get symbolically $\theta(\text{(c,b,d)-}) = \text{(c,b,d)}$. We recall that as an injection the extrusion mapping θ satisfies $\theta^{\mathsf{T}} ; \theta = \mathbb{I}_{\mathbf{2}^X}, \quad \theta ; \theta^{\mathsf{T}} = \mathbb{I}_{\Theta}$.

So, we form the direct sum of these two copies introducing the injections

$$\iota : \mathbf{2}^X \longrightarrow \mathbf{2}^X + \Theta \quad \text{and} \quad \kappa : \Theta \longrightarrow \mathbf{2}^X + \Theta,$$

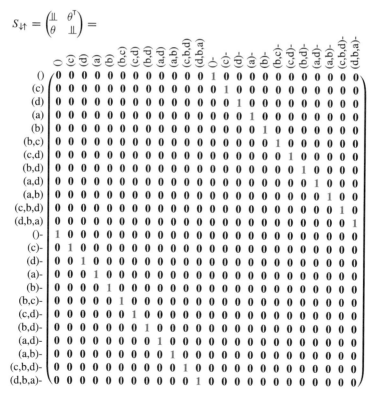

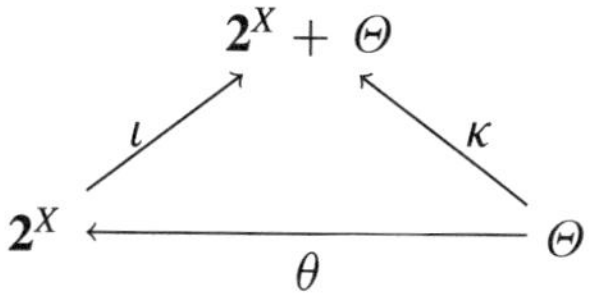

and thus having all positively as well as all negatively oriented simplices in one set. The **orientation flip relation**

$$S_{\Updownarrow} := \iota^{\mathsf{T}} ; \theta^{\mathsf{T}} ; \kappa \cup \kappa^{\mathsf{T}} ; \theta ; \iota, \qquad \text{satisfying} \qquad S_{\Updownarrow}{}^2 = \mathbb{I}_{2^X + \Theta},$$

obviously governs the transition to the differently oriented counterpart; see, e.g., Fig. 10.8.

Using this basic configuration, we will now define matrices of relations to express, e.g., in submatrix position (1,2) that we go with $B^{\Updownarrow} ; \theta^{\mathsf{T}}$ from a positively

$$S_{\downarrow\uparrow} = \begin{pmatrix} \mathbb{I} & \theta^{\mathsf{T}} \\ \theta & \mathbb{I} \end{pmatrix} =$$

	()	(c)	(d)	(a)	(b)	(b,c)	(c,d)	(b,d)	(a,d)	(a,b)	(c,b,d)	(d,b,a)	()-	(c)-	(d)-	(a)-	(b)-	(b,c)-	(c,d)-	(b,d)-	(a,d)-	(a,b)-	(c,b,d)-	(d,b,a)-
()	0	0	0	0	0	0	0	0	0	0	0	0	1	0	0	0	0	0	0	0	0	0	0	0
(c)	0	0	0	0	0	0	0	0	0	0	0	0	0	1	0	0	0	0	0	0	0	0	0	0
(d)	0	0	0	0	0	0	0	0	0	0	0	0	0	0	1	0	0	0	0	0	0	0	0	0
(a)	0	0	0	0	0	0	0	0	0	0	0	0	0	0	0	1	0	0	0	0	0	0	0	0
(b)	0	0	0	0	0	0	0	0	0	0	0	0	0	0	0	0	1	0	0	0	0	0	0	0
(b,c)	0	0	0	0	0	0	0	0	0	0	0	0	0	0	0	0	0	1	0	0	0	0	0	0
(c,d)	0	0	0	0	0	0	0	0	0	0	0	0	0	0	0	0	0	0	1	0	0	0	0	0
(b,d)	0	0	0	0	0	0	0	0	0	0	0	0	0	0	0	0	0	0	0	1	0	0	0	0
(a,d)	0	0	0	0	0	0	0	0	0	0	0	0	0	0	0	0	0	0	0	0	1	0	0	0
(a,b)	0	0	0	0	0	0	0	0	0	0	0	0	0	0	0	0	0	0	0	0	0	1	0	0
(c,b,d)	0	0	0	0	0	0	0	0	0	0	0	0	0	0	0	0	0	0	0	0	0	0	1	0
(d,b,a)	0	0	0	0	0	0	0	0	0	0	0	0	0	0	0	0	0	0	0	0	0	0	0	1
()-	1	0	0	0	0	0	0	0	0	0	0	0	0	0	0	0	0	0	0	0	0	0	0	0
(c)-	0	1	0	0	0	0	0	0	0	0	0	0	0	0	0	0	0	0	0	0	0	0	0	0
(d)-	0	0	1	0	0	0	0	0	0	0	0	0	0	0	0	0	0	0	0	0	0	0	0	0
(a)-	0	0	0	1	0	0	0	0	0	0	0	0	0	0	0	0	0	0	0	0	0	0	0	0
(b)-	0	0	0	0	1	0	0	0	0	0	0	0	0	0	0	0	0	0	0	0	0	0	0	0
(b,c)-	0	0	0	0	0	1	0	0	0	0	0	0	0	0	0	0	0	0	0	0	0	0	0	0
(c,d)-	0	0	0	0	0	0	1	0	0	0	0	0	0	0	0	0	0	0	0	0	0	0	0	0
(b,d)-	0	0	0	0	0	0	0	1	0	0	0	0	0	0	0	0	0	0	0	0	0	0	0	0
(a,d)-	0	0	0	0	0	0	0	0	1	0	0	0	0	0	0	0	0	0	0	0	0	0	0	0
(a,b)-	0	0	0	0	0	0	0	0	0	1	0	0	0	0	0	0	0	0	0	0	0	0	0	0
(c,b,d)-	0	0	0	0	0	0	0	0	0	0	1	0	0	0	0	0	0	0	0	0	0	0	0	0
(d,b,a)-	0	0	0	0	0	0	0	0	0	0	0	1	0	0	0	0	0	0	0	0	0	0	0	0

Fig. 10.8 The orientation flip operator $S_{\Updownarrow}$ for Fig. 10.7

oriented simplex to its negative boundary. This together with the two boundary operators gives four relations, positive/negative versus positive/negative. However, instead of using the former relations $B^{\Uparrow}$, $B^{\Updownarrow}$ of Fig. 10.7, we embed them in the new configuration as

	positive	negative
positive	$B^{\Uparrow}$	$B^{\Updownarrow};\theta^{\mathsf{T}}$
negative	$\theta;B^{\Updownarrow}$	$\theta;B^{\Uparrow};\theta^{\mathsf{T}}$.

The orientation flip operator obviously subsumes to this general scheme.

Led by this idea, we define the relation

$$\mathfrak{B} := \iota^{\mathsf{T}};B^{\Uparrow};\iota \cup \iota^{\mathsf{T}};B^{\Updownarrow};\theta^{\mathsf{T}};\kappa \cup \kappa^{\mathsf{T}};\theta;B^{\Updownarrow};\iota \cup \kappa^{\mathsf{T}};\theta;B^{\Uparrow};\theta^{\mathsf{T}};\kappa,$$

or more pictorially as a matrix

$$\mathfrak{B} = \begin{pmatrix} B^{\Uparrow} & B^{\Updownarrow};\theta^{\mathsf{T}} \\ \theta;B^{\Updownarrow} & \theta;B^{\Uparrow};\theta^{\mathsf{T}} \end{pmatrix},$$

to be conceived as a **joint boundary relation**. In Fig. 10.9, it can be seen how the negatively oriented $(2,1,3)$–, for instance, has the positive boundary $(1,2)$ via $\theta;B^{\Updownarrow}$. One consequence follows directly from this definition:

$$\mathfrak{B};S_{\Updownarrow} = S_{\Updownarrow};\mathfrak{B}$$

We will further recognize that when computing the square $\mathfrak{B}^2$ the two sub-matrices in the diagonal as well as the two outside are built in the same way—up to indications via $\theta, \theta^{\mathsf{T}}$. Even more: Due to Definition 10.3.1, all sub-matrices turn out to be the same.

$$\begin{aligned}\mathfrak{B}^2 &= \begin{pmatrix} B^{\Uparrow};B^{\Uparrow} \cup B^{\Updownarrow};\theta^{\mathsf{T}};\theta;B^{\Updownarrow} & B^{\Uparrow};B^{\Updownarrow};\theta^{\mathsf{T}} \cup B^{\Updownarrow};\theta^{\mathsf{T}};\theta;B^{\Uparrow};\theta^{\mathsf{T}} \\ \theta;B^{\Updownarrow};B^{\Uparrow} \cup \theta;B^{\Uparrow};\theta^{\mathsf{T}};\theta;B^{\Updownarrow} & \theta;B^{\Updownarrow};B^{\Updownarrow};\theta^{\mathsf{T}} \cup \theta;B^{\Uparrow};\theta^{\mathsf{T}};\theta;B^{\Uparrow};\theta^{\mathsf{T}} \end{pmatrix} \\ &= \begin{pmatrix} B^{\Uparrow};B^{\Uparrow} \cup B^{\Updownarrow};B^{\Updownarrow} & (B^{\Uparrow};B^{\Updownarrow} \cup B^{\Updownarrow};B^{\Uparrow});\theta^{\mathsf{T}} \\ \theta;(B^{\Updownarrow};B^{\Uparrow} \cup B^{\Uparrow};B^{\Updownarrow}) & \theta;(B^{\Updownarrow};B^{\Updownarrow} \cup B^{\Uparrow};B^{\Uparrow});\theta^{\mathsf{T}} \end{pmatrix}\end{aligned}$$

In algebraic topology, one would use the chains with values in a group. Their values annihilate one another after double application $s \mapsto \partial(\partial(s))$ of the boundary operator bringing the result 0. This effect is here reflected by providing the same result in two different ways; see Figs. 10.9 and 10.10:

$$B^{\Uparrow};B^{\Uparrow} \cup B^{\Updownarrow};B^{\Updownarrow} = B^{\Uparrow};B^{\Updownarrow} \cup B^{\Updownarrow};B^{\Uparrow}$$

	()	(3)	(0)	(1)	(2)	(2,3)	(0,3)	(1,3)	(0,1)	(0,2)	(1,2)	(2,1,3)	(2,3,0)	(0,3,1)	(0,1,2)	()-	(3)-	(0)-	(1)-	(2)-	(2,3)-	(0,3)-	(1,3)-	(0,1)-	(0,2)-	(1,2)-	(2,1,3)-	(2,3,0)-	(0,3,1)-	(0,1,2)-
()	0	0	0	0	0	0	0	0	0	0	0	0	0	0	0	0	0	0	0	0	0	0	0	0	0	0	0	0	0	0
(3)	1	0	0	0	0	0	0	0	0	0	0	0	0	0	0	0	0	0	0	0	0	0	0	0	0	0	0	0	0	0
(0)	1	0	0	0	0	0	0	0	0	0	0	0	0	0	0	0	0	0	0	0	0	0	0	0	0	0	0	0	0	0
(1)	1	0	0	0	0	0	0	0	0	0	0	0	0	0	0	0	0	0	0	0	0	0	0	0	0	0	0	0	0	0
(2)	1	0	0	0	0	0	0	0	0	0	0	0	0	0	0	0	0	0	0	0	0	0	0	0	0	0	0	0	0	0
(2,3)	0	1	0	0	0	0	0	0	0	0	0	0	0	0	0	0	0	0	0	1	0	0	0	0	0	0	0	0	0	0
(0,3)	0	1	0	0	0	0	0	0	0	0	0	0	0	0	0	0	0	1	0	0	0	0	0	0	0	0	0	0	0	0
(1,3)	0	1	0	0	0	0	0	0	0	0	0	0	0	0	0	0	0	0	1	0	0	0	0	0	0	0	0	0	0	0
(0,1)	0	0	0	1	0	0	0	0	0	0	0	0	0	0	0	0	0	1	0	0	0	0	0	0	0	0	0	0	0	0
(0,2)	0	0	0	0	1	0	0	0	0	0	0	0	0	0	0	0	0	1	0	0	0	0	0	0	0	0	0	0	0	0
(1,2)	0	0	0	0	1	0	0	0	0	0	0	0	0	0	0	0	0	0	1	0	0	0	0	0	0	0	0	0	0	0
(2,1,3)	0	0	0	0	0	0	0	1	0	0	0	0	0	0	0	0	0	0	0	0	1	0	0	0	0	1	0	0	0	0
(2,3,0)	0	0	0	0	0	1	0	0	0	1	0	0	0	0	0	0	0	0	0	0	0	1	0	0	0	0	0	0	0	0
(0,3,1)	0	0	0	0	0	0	1	0	0	0	0	0	0	0	0	0	0	0	0	0	0	0	1	1	0	0	0	0	0	0
(0,1,2)	0	0	0	0	0	0	0	0	1	0	1	0	0	0	0	0	0	0	0	0	0	0	0	0	1	0	0	0	0	0
()-	0	0	0	0	0	0	0	0	0	0	0	0	0	0	0	0	0	0	0	0	0	0	0	0	0	0	0	0	0	0
(3)-	0	0	0	0	0	0	0	0	0	0	0	0	0	0	0	1	0	0	0	0	0	0	0	0	0	0	0	0	0	0
(0)-	0	0	0	0	0	0	0	0	0	0	0	0	0	0	0	1	0	0	0	0	0	0	0	0	0	0	0	0	0	0
(1)-	0	0	0	0	0	0	0	0	0	0	0	0	0	0	0	1	0	0	0	0	0	0	0	0	0	0	0	0	0	0
(2)-	0	0	0	0	0	0	0	0	0	0	0	0	0	0	0	1	0	0	0	0	0	0	0	0	0	0	0	0	0	0
(2,3)-	0	0	0	0	1	0	0	0	0	0	0	0	0	0	0	0	1	0	0	0	0	0	0	0	0	0	0	0	0	0
(0,3)-	0	0	1	0	0	0	0	0	0	0	0	0	0	0	0	0	1	0	0	0	0	0	0	0	0	0	0	0	0	0
(1,3)-	0	0	0	1	0	0	0	0	0	0	0	0	0	0	0	0	1	0	0	0	0	0	0	0	0	0	0	0	0	0
(0,1)-	0	0	1	0	0	0	0	0	0	0	0	0	0	0	0	0	0	0	1	0	0	0	0	0	0	0	0	0	0	0
(0,2)-	0	0	1	0	0	0	0	0	0	0	0	0	0	0	0	0	0	0	0	1	0	0	0	0	0	0	0	0	0	0
(1,2)-	0	0	0	1	0	0	0	0	0	0	0	0	0	0	0	0	0	0	0	1	0	0	0	0	0	0	0	0	0	0
(2,1,3)-	0	0	0	0	0	1	0	0	0	0	1	0	0	0	0	0	0	0	0	0	0	0	1	0	0	0	0	0	0	0
(2,3,0)-	0	0	0	0	0	0	1	0	0	0	0	0	0	0	0	0	0	0	0	0	1	0	0	0	1	0	0	0	0	0
(0,3,1)-	0	0	0	0	0	0	0	1	1	0	0	0	0	0	0	0	0	0	0	0	0	1	0	0	0	0	0	0	0	0
(0,1,2)-	0	0	0	0	0	0	0	0	0	1	0	0	0	0	0	0	0	0	0	0	0	0	0	1	0	1	0	0	0	0

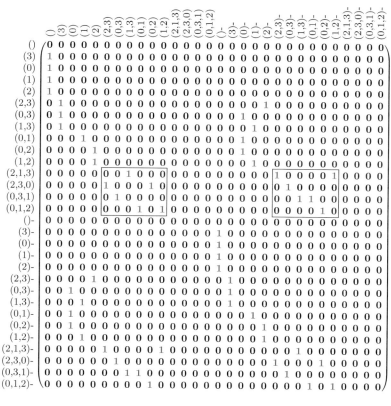

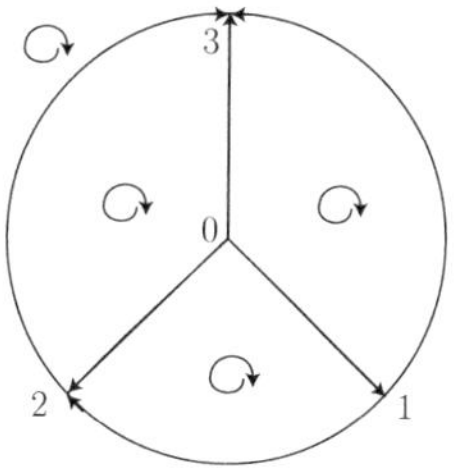

$$\mathfrak{B} = \begin{pmatrix} B^{\Uparrow} & B^{\downarrow\uparrow};\theta^{\top} \\ \theta; B^{\downarrow\uparrow} & \theta; B^{\Uparrow};\theta^{\top} \end{pmatrix}$$

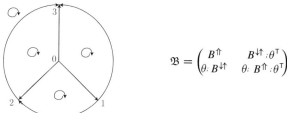

Fig. 10.9 Boundary relation $\mathfrak{B}$ evaluated for all directions positive/negative

10.4 Orientability of a Simplicial Complex

The concept of orientability has already shown up together with Fig. 10.7. We had considered the two oriented triangles as having a *common* orientation: They were adjacent, meaning that they had sub-simplex (b,d) in common, and this common sub-simplex (b,d) agreed in orientation with one of them and disagreed with the other. We are about to reformulate this relationally considering $\mathfrak{B}; S_{\Updownarrow}; \mathfrak{B}^{\top}$, i.e., considering two simplices as having the same orientation when their coinciding boundary obtains opposite orientations.

	()	(3)	(0)	(1)	(2)	(2,3)	(0,3)	(1,3)	(0,1)	(0,2)	(1,2)	(2,1,3)	(2,3,0)	(0,3,1)	(0,1,2)	()-	(3)-	(0)-	(1)-	(2)-	(2,3)-	(0,3)-	(1,3)-	(0,1)-	(0,2)-	(1,2)-	(2,1,3)-	(2,3,0)-	(0,3,1)-	(0,1,2)-
()	0	0	0	0	0	0	0	0	0	0	0	0	0	0	0	0	0	0	0	0	0	0	0	0	0	0	0	0	0	0
(3)	0	0	0	0	0	0	0	0	0	0	0	0	0	0	0	0	0	0	0	0	0	0	0	0	0	0	0	0	0	0
(0)	0	0	0	0	0	0	0	0	0	0	0	0	0	0	0	0	0	0	0	0	0	0	0	0	0	0	0	0	0	0
(1)	0	0	0	0	0	0	0	0	0	0	0	0	0	0	0	0	0	0	0	0	0	0	0	0	0	0	0	0	0	0
(2)	0	0	0	0	0	0	0	0	0	0	0	0	0	0	0	0	0	0	0	0	0	0	0	0	0	0	0	0	0	0
(2,3)	1	0	0	0	0	0	0	0	0	0	0	0	0	0	0	1	0	0	0	0	0	0	0	0	0	0	0	0	0	0
(0,3)	1	0	0	0	0	0	0	0	0	0	0	0	0	0	0	1	0	0	0	0	0	0	0	0	0	0	0	0	0	0
(1,3)	1	0	0	0	0	0	0	0	0	0	0	0	0	0	0	1	0	0	0	0	0	0	0	0	0	0	0	0	0	0
(0,1)	1	0	0	0	0	0	0	0	0	0	0	0	0	0	0	1	0	0	0	0	0	0	0	0	0	0	0	0	0	0
(0,2)	1	0	0	0	0	0	0	0	0	0	0	0	0	0	0	1	0	0	0	0	0	0	0	0	0	0	0	0	0	0
(1,2)	1	0	0	0	0	0	0	0	0	0	0	0	0	0	0	1	0	0	0	0	0	0	0	0	0	0	0	0	0	0
(2,1,3)	0	1	0	1	1	0	0	0	0	0	0	0	0	0	0	0	1	0	1	1	0	0	0	0	0	0	0	0	0	0
(2,3,0)	0	1	1	0	1	0	0	0	0	0	0	0	0	0	0	0	1	1	0	1	0	0	0	0	0	0	0	0	0	0
(0,3,1)	0	1	1	1	0	0	0	0	0	0	0	0	0	0	0	0	1	1	1	0	0	0	0	0	0	0	0	0	0	0
(0,1,2)	0	0	1	1	1	0	0	0	0	0	0	0	0	0	0	0	0	1	1	1	0	0	0	0	0	0	0	0	0	0
()-	0	0	0	0	0	0	0	0	0	0	0	0	0	0	0	0	0	0	0	0	0	0	0	0	0	0	0	0	0	0
(3)-	0	0	0	0	0	0	0	0	0	0	0	0	0	0	0	0	0	0	0	0	0	0	0	0	0	0	0	0	0	0
(0)-	0	0	0	0	0	0	0	0	0	0	0	0	0	0	0	0	0	0	0	0	0	0	0	0	0	0	0	0	0	0
(1)-	0	0	0	0	0	0	0	0	0	0	0	0	0	0	0	0	0	0	0	0	0	0	0	0	0	0	0	0	0	0
(2)-	0	0	0	0	0	0	0	0	0	0	0	0	0	0	0	0	0	0	0	0	0	0	0	0	0	0	0	0	0	0
(2,3)-	1	0	0	0	0	0	0	0	0	0	0	0	0	0	0	1	0	0	0	0	0	0	0	0	0	0	0	0	0	0
(0,3)-	1	0	0	0	0	0	0	0	0	0	0	0	0	0	0	1	0	0	0	0	0	0	0	0	0	0	0	0	0	0
(1,3)-	1	0	0	0	0	0	0	0	0	0	0	0	0	0	0	1	0	0	0	0	0	0	0	0	0	0	0	0	0	0
(0,1)-	1	0	0	0	0	0	0	0	0	0	0	0	0	0	0	1	0	0	0	0	0	0	0	0	0	0	0	0	0	0
(0,2)-	1	0	0	0	0	0	0	0	0	0	0	0	0	0	0	1	0	0	0	0	0	0	0	0	0	0	0	0	0	0
(1,2)-	1	0	0	0	0	0	0	0	0	0	0	0	0	0	0	1	0	0	0	0	0	0	0	0	0	0	0	0	0	0
(2,1,3)-	0	1	0	1	1	0	0	0	0	0	0	0	0	0	0	0	1	0	1	1	0	0	0	0	0	0	0	0	0	0
(2,3,0)-	0	1	1	0	1	0	0	0	0	0	0	0	0	0	0	0	1	1	0	1	0	0	0	0	0	0	0	0	0	0
(0,3,1)-	0	1	1	1	0	0	0	0	0	0	0	0	0	0	0	0	1	1	1	0	0	0	0	0	0	0	0	0	0	0
(0,1,2)-	0	0	1	1	1	0	0	0	0	0	0	0	0	0	0	0	0	1	1	1	0	0	0	0	0	0	0	0	0	0

Fig. 10.10 Boundary relation $\mathfrak{B}$ of Fig. 10.9 applied twice: $\mathfrak{B}^2$

There is, however, a minor obstacle when looking at $(0, 3, 1)$ and $(0, 3, 1)$⊢, which we should consider as being differently oriented; but

$(0, 3, 1)$ has among others via $B^{\Uparrow}$ the boundary $(0, 3)$ and

$(0, 3, 1)$⊢ has via $\theta ; B^{\Uparrow} ; \theta^{\mathsf{T}}$ among others the boundary $(0, 3)$⊢.

Thus, $(0, 3, 1)$ goes via upper left $B^{\Uparrow}$, flip θ^{T}, and lower right $\theta ; B^{\Uparrow\,\mathsf{T}} ; \theta^{\mathsf{T}}$, i.e.,

$$B^{\Uparrow} ; \theta^{\mathsf{T}} ; \theta ; B^{\Uparrow\,\mathsf{T}} ; \theta^{\mathsf{T}} = B^{\Uparrow} ; B^{\Uparrow\,\mathsf{T}} ; \theta^{\mathsf{T}}$$

to its inverse $(0, 3, 1)$⊢. Such an immediate change of orientation is considered uninteresting. With the following idea we try to get rid of it.

Definition 10.4.1 Given an oriented simplicial complex with its joint boundary operator $\mathfrak{B}$, the construct

$$\varGamma := \mathfrak{B} ; S_{\Updownarrow} ; \mathfrak{B}^{\mathsf{T}} \cap \overline{S_{\Updownarrow}}$$

will be called its corresponding relation of **orientation adjacency**. □

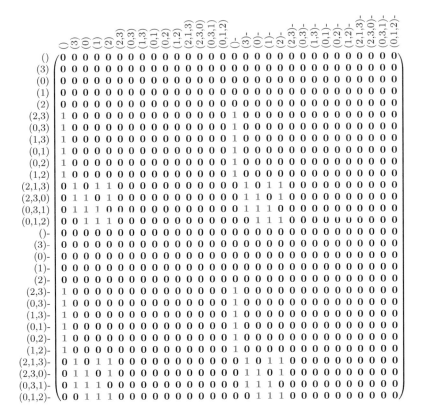

	()	(3)	(0)	(1)	(2)	(2,3)	(0,3)	(1,3)	(0,1)	(0,2)	(1,2)	(2,1,3)	(2,3,0)	(0,3,1)	(0,1,2)	()-	(3)-	(0)-	(1)-	(2)-	(2,3)-	(0,3)-	(1,3)-	(0,1)-	(0,2)-	(1,2)-	(2,1,3)-	(2,3,0)-	(0,3,1)-	(0,1,2)-
()	0	0	0	0	0	0	0	0	0	0	0	0	0	0	0	0	0	0	0	0	0	0	0	0	0	0	0	0	0	0
(3)	0	0	0	0	0	0	0	0	0	0	0	0	0	0	0	0	0	1	1	1	0	0	0	0	0	0	0	0	0	0
(0)	0	0	0	0	0	0	0	0	0	0	0	0	0	0	0	0	1	0	1	1	0	0	0	0	0	0	0	0	0	0
(1)	0	0	0	0	0	0	0	0	0	0	0	0	0	0	0	0	1	1	0	1	0	0	0	0	0	0	0	0	0	0
(2)	0	0	0	0	0	0	0	0	0	0	0	0	0	0	0	0	1	1	1	0	0	0	0	0	0	0	0	0	0	0
(2,3)	0	0	0	0	0	0	0	0	0	1	1	0	0	0	0	0	0	0	0	0	0	1	1	0	0	0	0	0	0	0
(0,3)	0	0	0	0	0	0	0	0	0	0	0	0	0	0	0	0	0	0	0	0	1	0	1	1	1	0	0	0	0	0
(1,3)	0	0	0	0	0	0	0	0	1	0	0	0	0	0	0	0	0	0	0	0	1	1	0	0	0	1	0	0	0	0
(0,1)	0	0	0	0	0	0	0	1	0	0	1	0	0	0	0	0	0	0	0	0	0	1	0	0	1	0	0	0	0	0
(0,2)	0	0	0	0	0	1	0	0	0	0	0	0	0	0	0	0	0	0	0	0	0	1	0	1	0	1	0	0	0	0
(1,2)	0	0	0	0	0	1	0	0	1	0	0	0	0	0	0	0	0	0	0	0	0	0	1	0	1	0	0	0	0	0
(2,1,3)	0	0	0	0	0	0	0	0	0	0	0	0	1	1	1	0	0	0	0	0	0	0	0	0	0	0	0	0	0	0
(2,3,0)	0	0	0	0	0	0	0	0	0	0	0	1	0	1	1	0	0	0	0	0	0	0	0	0	0	0	0	0	0	0
(0,3,1)	0	0	0	0	0	0	0	0	0	0	0	1	1	0	1	0	0	0	0	0	0	0	0	0	0	0	0	0	0	0
(0,1,2)	0	0	0	0	0	0	0	0	0	0	0	1	1	1	0	0	0	0	0	0	0	0	0	0	0	0	0	0	0	0
()-	0	0	0	0	0	0	0	0	0	0	0	0	0	0	0	0	0	0	0	0	0	0	0	0	0	0	0	0	0	0
(3)-	0	0	1	1	1	0	0	0	0	0	0	0	0	0	0	0	0	0	0	0	0	0	0	0	0	0	0	0	0	0
(0)-	0	1	0	1	1	0	0	0	0	0	0	0	0	0	0	0	0	0	0	0	0	0	0	0	0	0	0	0	0	0
(1)-	0	1	1	0	1	0	0	0	0	0	0	0	0	0	0	0	0	0	0	0	0	0	0	0	0	0	0	0	0	0
(2)-	0	1	1	1	0	0	0	0	0	0	0	0	0	0	0	0	0	0	0	0	0	0	0	0	0	0	0	0	0	0
(2,3)-	0	0	0	0	0	0	1	1	0	0	0	0	0	0	0	0	0	0	0	0	0	0	0	0	1	1	0	0	0	0
(0,3)-	0	0	0	0	0	1	0	1	1	1	0	0	0	0	0	0	0	0	0	0	0	0	0	0	0	0	0	0	0	0
(1,3)-	0	0	0	0	0	1	1	0	0	0	1	0	0	0	0	0	0	0	0	0	0	0	0	1	0	0	0	0	0	0
(0,1)-	0	0	0	0	0	0	1	0	0	1	0	0	0	0	0	0	0	0	0	0	0	0	1	0	0	1	0	0	0	0
(0,2)-	0	0	0	0	0	0	1	0	1	0	1	0	0	0	0	0	0	0	0	0	1	0	0	0	0	0	0	0	0	0
(1,2)-	0	0	0	0	0	0	0	1	0	1	0	0	0	0	0	0	0	0	0	0	1	0	0	1	0	0	0	0	0	0
(2,1,3)-	0	0	0	0	0	0	0	0	0	0	0	0	0	0	0	0	0	0	0	0	0	0	0	0	0	0	0	1	1	1
(2,3,0)-	0	0	0	0	0	0	0	0	0	0	0	0	0	0	0	0	0	0	0	0	0	0	0	0	0	0	1	0	1	1
(0,3,1)-	0	0	0	0	0	0	0	0	0	0	0	0	0	0	0	0	0	0	0	0	0	0	0	0	0	0	1	1	0	1
(0,1,2)-	0	0	0	0	0	0	0	0	0	0	0	0	0	0	0	0	0	0	0	0	0	0	0	0	0	0	1	1	1	0

Fig. 10.11 Orientation adjacency Γ for Fig. 10.9

Disregarding their orientation, they shall in addition be different, so that one cannot proceed from the first to the second via the orientation flip operator $S_{\Updownarrow}$. Figure 10.11 shows an example.

While it is possible to proceed from, e.g., $(2, 3)$ via Γ^* to its counter-oriented version $(2, 3)$-, the corresponding is—in the present example—impossible for highest-dimensional simplices. In contrast, it is possible for all lower-dimensional ones.

Size Restriction Examples will soon get large, and we are interested in a less spacious representation. Therefore, we refer back to Fig. 10.7 and mainly focus on the largest dimension and the one below for $\mathfrak{B}$. Since the sub-matrices $\mathfrak{B}_{1,1}$ and $\mathfrak{B}_{2,2}$ as well as $\mathfrak{B}_{1,2}$ and $\mathfrak{B}_{2,1}$ are equal, just two framed excisions are necessary, which we denote as $B^{\Uparrow}{}_{\square}$ and $B^{\Updownarrow}{}_{\square}$.

Correspondingly, we provide reductions for Γ, Γ^*, where we focus on the largest dimension only; see the squares in Figs. 10.11 and 10.12. Since Γ is necessarily symmetric, we need just the framed parts, which we are going to denote as $\Gamma_{\boxtimes}$

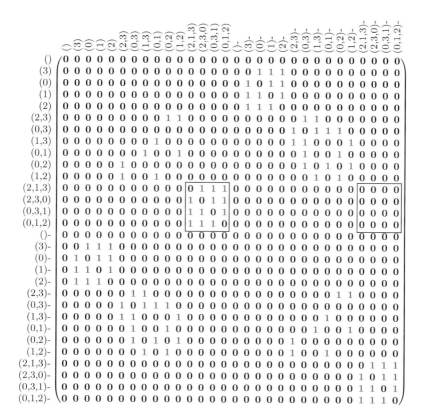

	()	(3)	(0)	(1)	(2)	(2,3)	(0,3)	(1,3)	(0,1)	(0,2)	(1,2)	(2,1,3)	(2,3,0)	(0,3,1)	(0,1,2)	()-	(3)-	(0)-	(1)-	(2)-	(2,3)-	(0,3)-	(1,3)-	(0,1)-	(0,2)-	(1,2)-	(2,1,3)-	(2,3,0)-	(0,3,1)-	(0,1,2)-
()	1	0	0	0	0	0	0	0	0	0	0	0	0	0	0	0	0	0	0	0	0	0	0	0	0	0	0	0	0	0
(3)	0	1	1	1	1	0	0	0	0	0	0	0	0	0	0	0	1	1	1	1	0	0	0	0	0	0	0	0	0	0
(0)	0	1	1	1	1	0	0	0	0	0	0	0	0	0	0	0	1	1	1	1	0	0	0	0	0	0	0	0	0	0
(1)	0	1	1	1	1	0	0	0	0	0	0	0	0	0	0	0	1	1	1	1	0	0	0	0	0	0	0	0	0	0
(2)	0	1	1	1	1	0	0	0	0	0	0	0	0	0	0	0	1	1	1	1	0	0	0	0	0	0	0	0	0	0
(2,3)	0	0	0	0	0	1	1	1	1	1	1	0	0	0	0	0	0	0	0	0	1	1	1	1	1	1	0	0	0	0
(0,3)	0	0	0	0	0	1	1	1	1	1	1	0	0	0	0	0	0	0	0	0	1	1	1	1	1	1	0	0	0	0
(1,3)	0	0	0	0	0	1	1	1	1	1	1	0	0	0	0	0	0	0	0	0	1	1	1	1	1	1	0	0	0	0
(0,1)	0	0	0	0	0	1	1	1	1	1	1	0	0	0	0	0	0	0	0	0	1	1	1	1	1	1	0	0	0	0
(0,2)	0	0	0	0	0	1	1	1	1	1	1	0	0	0	0	0	0	0	0	0	1	1	1	1	1	1	0	0	0	0
(1,2)	0	0	0	0	0	1	1	1	1	1	1	0	0	0	0	0	0	0	0	0	1	1	1	1	1	1	0	0	0	0
(2,1,3)	0	0	0	0	0	0	0	0	0	0	0	1	1	1	1	0	0	0	0	0	0	0	0	0	0	0	0	0	0	0
(2,3,0)	0	0	0	0	0	0	0	0	0	0	0	1	1	1	1	0	0	0	0	0	0	0	0	0	0	0	0	0	0	0
(0,3,1)	0	0	0	0	0	0	0	0	0	0	0	1	1	1	1	0	0	0	0	0	0	0	0	0	0	0	0	0	0	0
(0,1,2)	0	0	0	0	0	0	0	0	0	0	0	1	1	1	1	0	0	0	0	0	0	0	0	0	0	0	0	0	0	0
()-	0	0	0	0	0	0	0	0	0	0	0	0	0	0	0	1	0	0	0	0	0	0	0	0	0	0	0	0	0	0
(3)-	0	1	1	1	1	0	0	0	0	0	0	0	0	0	0	0	1	1	1	1	0	0	0	0	0	0	0	0	0	0
(0)-	0	1	1	1	1	0	0	0	0	0	0	0	0	0	0	0	1	1	1	1	0	0	0	0	0	0	0	0	0	0
(1)-	0	1	1	1	1	0	0	0	0	0	0	0	0	0	0	0	1	1	1	1	0	0	0	0	0	0	0	0	0	0
(2)-	0	1	1	1	1	0	0	0	0	0	0	0	0	0	0	0	1	1	1	1	0	0	0	0	0	0	0	0	0	0
(2,3)-	0	0	0	0	0	1	1	1	1	1	1	0	0	0	0	0	0	0	0	0	1	1	1	1	1	1	0	0	0	0
(0,3)-	0	0	0	0	0	1	1	1	1	1	1	0	0	0	0	0	0	0	0	0	1	1	1	1	1	1	0	0	0	0
(1,3)-	0	0	0	0	0	1	1	1	1	1	1	0	0	0	0	0	0	0	0	0	1	1	1	1	1	1	0	0	0	0
(0,1)-	0	0	0	0	0	1	1	1	1	1	1	0	0	0	0	0	0	0	0	0	1	1	1	1	1	1	0	0	0	0
(0,2)-	0	0	0	0	0	1	1	1	1	1	1	0	0	0	0	0	0	0	0	0	1	1	1	1	1	1	0	0	0	0
(1,2)-	0	0	0	0	0	1	1	1	1	1	1	0	0	0	0	0	0	0	0	0	1	1	1	1	1	1	0	0	0	0
(2,1,3)-	0	0	0	0	0	0	0	0	0	0	0	0	0	0	0	0	0	0	0	0	0	0	0	0	0	0	1	1	1	1
(2,3,0)-	0	0	0	0	0	0	0	0	0	0	0	0	0	0	0	0	0	0	0	0	0	0	0	0	0	0	1	1	1	1
(0,3,1)-	0	0	0	0	0	0	0	0	0	0	0	0	0	0	0	0	0	0	0	0	0	0	0	0	0	0	1	1	1	1
(0,1,2)-	0	0	0	0	0	0	0	0	0	0	0	0	0	0	0	0	0	0	0	0	0	0	0	0	0	0	1	1	1	1

Fig. 10.12 Reflexive-transitive closure Γ^* for Γ of Fig. 10.11

to indicate that it belongs to the diagonal and $\Gamma_{\square}$. Analogously, $\Gamma^*_{\boxtimes}$ and $\Gamma^*_{\square}$ (Fig. 10.13). □

We will observe several identities:

$$\mathfrak{B}^\top = \begin{pmatrix} B^{\Uparrow\top} & B^{\Updownarrow\top};\theta^\top \\ \theta;B^{\Updownarrow\top} & \theta;B^{\Uparrow\top};\theta^\top \end{pmatrix} \quad S_{\Updownarrow};\mathfrak{B}^\top = \begin{pmatrix} B^{\Updownarrow\top} & B^{\Uparrow\top};\theta^\top \\ \theta;B^{\Uparrow\top} & \theta;B^{\Updownarrow\top};\theta^\top \end{pmatrix}$$

$$\Gamma = \mathfrak{B};S_{\Updownarrow};\mathfrak{B}^\top \cap \overline{S_{\Updownarrow}}$$

$$= \begin{pmatrix} B^{\Uparrow};B^{\Updownarrow\top} \cup B^{\Updownarrow};B^{\Uparrow\top} & \left[\overline{\mathbb{I}} \cap (B^{\Uparrow};B^{\Uparrow\top} \cup B^{\Updownarrow};B^{\Updownarrow\top})\right];\theta^\top \\ \theta;\left[\overline{\mathbb{I}} \cap (B^{\Updownarrow};B^{\Updownarrow\top} \cup B^{\Uparrow};B^{\Uparrow\top})\right] & \theta;(B^{\Updownarrow};B^{\Uparrow\top} \cup B^{\Uparrow};B^{\Updownarrow\top});\theta^\top \end{pmatrix}$$

This leads us to the concept of orientability of a simplicial complex, which concerns the highest dimension only.

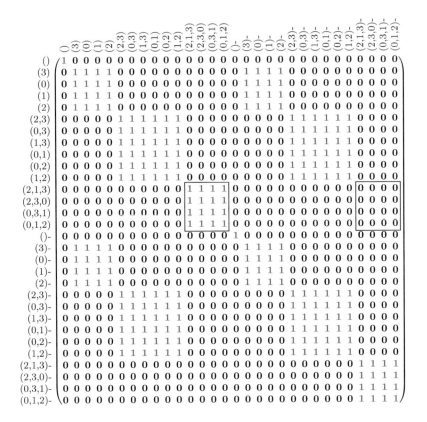

	(2,3)	(0,3)	(1,3)	(0,1)	(0,2)	(1,2)
(2,1,3)	0	0	1	0	0	0
(2,3,0)	1	0	0	0	1	0
(0,3,1)	0	1	0	0	0	0
(0,1,2)	0	0	0	1	0	1

$B^{\Uparrow}{}_{\square}$

	(2,3)-	(0,3)-	(1,3)-	(0,1)-	(0,2)-	(1,2)-
(2,1,3)	1	0	0	0	0	1
(2,3,0)	0	1	0	0	0	0
(0,3,1)	0	0	1	1	0	0
(0,1,2)	0	0	0	0	1	0

$B^{\downarrow\uparrow}{}_{\square}$

	(2,1,3)	(2,3,0)	(0,3,1)	(0,1,2)	(2,1,3)-	(2,3,0)-	(0,3,1)-	(0,1,2)-
(2,1,3)	0	1	1	1	0	0	0	0
(2,3,0)	1	0	1	1	0	0	0	0
(0,3,1)	1	1	0	1	0	0	0	0
(0,1,2)	1	1	1	0	0	0	0	0

$\varGamma_{\boxtimes}$, $\varGamma_{\square}$

	(2,1,3)	(2,3,0)	(0,3,1)	(0,1,2)	(2,1,3)-	(2,3,0)-	(0,3,1)-	(0,1,2)-
(2,1,3)	1	1	1	1	0	0	0	0
(2,3,0)	1	1	1	1	0	0	0	0
(0,3,1)	1	1	1	1	0	0	0	0
(0,1,2)	1	1	1	1	0	0	0	0

$\varGamma^*_{\boxtimes}$, $\varGamma^*_{\square}$

Fig. 10.13 Size-restricted parts of Figs. 10.9, 10.11, 10.12

Definition 10.4.2 Let be given a simplicial complex together with the vector v along $\mathbf{2}^X + \Theta$ that characterizes the highest-dimensional entries, we will call it **orientable** if

$$\varGamma^* \cap v\,{:}\,v^{\mathsf{T}} \subseteq \overline{S_{\Updownarrow}},$$

or, speaking in terms of the size-restricted matrices, when

$$\varGamma^*_{\square} \subseteq \overline{\mathbb{I}};$$

i.e., if orientation adjacency iterated never switches orientation of a highest-dimensional simplex. □

In several examples in dimension 2 as well as 3, we will now evaluate this criterion. Some of these examples are well-known; the last one is probably not.

To present a complex graphically, one often draws vertices multiply in the graphics for clarity. They have then to be identified; see, e.g., the necessary gluing of the two copies of 0 as well as those of 4 to obtain the Moebius strip below.

Moebius Strip A first example of non-orientability is the Moebius strip of Fig. 10.14–that need not be introduced in more detail. A last time we provide the corresponding relation $\mathfrak{B}$ together with the lower dimensional simplices.

The zones $B^{\Uparrow}{}_{\square}$ and $B^{\Updownarrow}{}_{\square}$ are then extracted for the first line of Fig. 10.15. The second line shows $\varGamma_{\boxtimes}$, $\varGamma_{\square}$ and $\varGamma^*_{\boxtimes}$, $\varGamma^*_{\square}$. The Moebius strip is—as could be expected—not orientable.

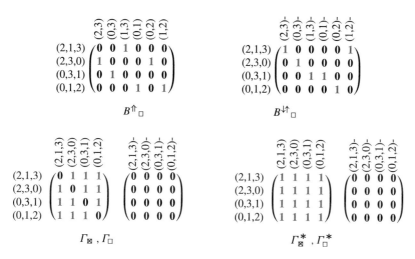

	()	(4)	(3)	(2)	(0)	(1)	(1,4)	(0,4)	(0,3)	(3,4)	(2,4)	(2,3)	(1,3)	(1,2)	(0,2)	(0,1)	(1,0,4)	(4,0,3)	(4,3,2)	(2,3,1)	(2,1,0)	()-	(4)-	(3)-	(2)-	(0)-	(1)-	(1,4)-	(0,4)-	(0,3)-	(3,4)-	(2,4)-	(2,3)-	(1,3)-	(1,2)-	(0,2)-	(0,1)-	(1,0,4)-	(4,0,3)-	(4,3,2)-	(2,3,1)-	(2,1,0)-
()	0	0	0	0	0	0	0	0	0	0	0	0	0	0	0	0	0	0	0	0	0	0	0	0	0	0	0	0	0	0	0	0	0	0	0	0	0	0	0	0	0	0
(4)	1	0	0	0	0	0	0	0	0	0	0	0	0	0	0	0	0	0	0	0	0	0	0	0	0	0	0	0	0	0	0	0	0	0	0	0	0	0	0	0	0	0
(3)	1	0	0	0	0	0	0	0	0	0	0	0	0	0	0	0	0	0	0	0	0	0	0	0	0	0	0	0	0	0	0	0	0	0	0	0	0	0	0	0	0	0
(2)	1	0	0	0	0	0	0	0	0	0	0	0	0	0	0	0	0	0	0	0	0	0	0	0	0	0	0	0	0	0	0	0	0	0	0	0	0	0	0	0	0	0
(0)	1	0	0	0	0	0	0	0	0	0	0	0	0	0	0	0	0	0	0	0	0	0	0	0	0	0	0	0	0	0	0	0	0	0	0	0	0	0	0	0	0	0
(1)	1	0	0	0	0	0	0	0	0	0	0	0	0	0	0	0	0	0	0	0	0	0	0	0	0	0	0	0	0	0	0	0	0	0	0	0	0	0	0	0	0	0
(1,4)	0	1	0	0	0	0	0	0	0	0	0	0	0	0	0	0	0	0	0	0	0	0	0	0	0	0	1	0	0	0	0	0	0	0	0	0	0	0	0	0	0	0
(0,4)	0	1	0	0	0	0	0	0	0	0	0	0	0	0	0	0	0	0	0	0	0	0	0	0	0	1	0	0	0	0	0	0	0	0	0	0	0	0	0	0	0	0
(0,3)	0	0	1	0	0	0	0	0	0	0	0	0	0	0	0	0	0	0	0	0	0	0	0	0	0	1	0	0	0	0	0	0	0	0	0	0	0	0	0	0	0	0
(3,4)	0	1	0	0	0	0	0	0	0	0	0	0	0	0	0	0	0	0	0	0	0	0	0	1	0	0	0	0	0	0	0	0	0	0	0	0	0	0	0	0	0	0
(2,4)	0	1	0	0	0	0	0	0	0	0	0	0	0	0	0	0	0	0	0	0	0	0	0	0	1	0	0	0	0	0	0	0	0	0	0	0	0	0	0	0	0	0
(2,3)	0	0	1	0	0	0	0	0	0	0	0	0	0	0	0	0	0	0	0	0	0	0	0	0	1	0	0	0	0	0	0	0	0	0	0	0	0	0	0	0	0	0
(1,3)	0	0	1	0	0	0	0	0	0	0	0	0	0	0	0	0	0	0	0	0	0	0	0	0	0	0	1	0	0	0	0	0	0	0	0	0	0	0	0	0	0	0
(1,2)	0	0	0	1	0	0	0	0	0	0	0	0	0	0	0	0	0	0	0	0	0	0	0	0	0	0	1	0	0	0	0	0	0	0	0	0	0	0	0	0	0	0
(0,2)	0	0	0	1	0	0	0	0	0	0	0	0	0	0	0	0	0	0	0	0	0	0	0	0	0	1	0	0	0	0	0	0	0	0	0	0	0	0	0	0	0	0
(0,1)	0	0	0	0	0	1	0	0	0	0	0	0	0	0	0	0	0	0	0	0	0	0	0	0	0	1	0	0	0	0	0	0	0	0	0	0	0	0	0	0	0	0
(1,0,4)	0	0	0	0	0	0	0	1	0	0	0	0	0	0	0	0	0	0	0	0	0	0	0	0	0	0	0	1	0	0	0	0	0	0	0	0	1	0	0	0	0	0
(4,0,3)	0	0	0	0	0	0	0	0	1	1	0	0	0	0	0	0	0	0	0	0	0	0	0	0	0	0	0	0	1	0	0	0	0	0	0	0	0	0	0	0	0	0
(4,3,2)	0	0	0	0	0	0	0	0	0	0	1	0	0	0	0	0	0	0	0	0	0	0	0	0	0	0	0	0	0	0	1	0	1	0	0	0	0	0	0	0	0	0
(2,3,1)	0	0	0	0	0	0	0	0	0	0	0	1	0	1	0	0	0	0	0	0	0	0	0	0	0	0	0	0	0	0	0	0	0	1	0	0	0	0	0	0	0	0
(2,1,0)	0	0	0	0	0	0	0	0	0	0	0	0	0	0	1	0	0	0	0	0	0	0	0	0	0	0	0	0	0	0	0	0	0	0	1	0	1	0	0	0	0	0
()-	0	0	0	0	0	0	0	0	0	0	0	0	0	0	0	0	0	0	0	0	0	0	0	0	0	0	0	0	0	0	0	0	0	0	0	0	0	0	0	0	0	0
(4)-	0	0	0	0	0	0	0	0	0	0	0	0	0	0	0	0	0	0	0	0	0	1	0	0	0	0	0	0	0	0	0	0	0	0	0	0	0	0	0	0	0	0
(3)-	0	0	0	0	0	0	0	0	0	0	0	0	0	0	0	0	0	0	0	0	0	1	0	0	0	0	0	0	0	0	0	0	0	0	0	0	0	0	0	0	0	0
(2)-	0	0	0	0	0	0	0	0	0	0	0	0	0	0	0	0	0	0	0	0	0	1	0	0	0	0	0	0	0	0	0	0	0	0	0	0	0	0	0	0	0	0
(0)-	0	0	0	0	0	0	0	0	0	0	0	0	0	0	0	0	0	0	0	0	0	1	0	0	0	0	0	0	0	0	0	0	0	0	0	0	0	0	0	0	0	0
(1)-	0	0	0	0	0	0	0	0	0	0	0	0	0	0	0	0	0	0	0	0	0	1	0	0	0	0	0	0	0	0	0	0	0	0	0	0	0	0	0	0	0	0
(1,4)-	0	0	0	0	0	1	0	0	0	0	0	0	0	0	0	0	0	0	0	0	0	0	1	0	0	0	0	0	0	0	0	0	0	0	0	0	0	0	0	0	0	0
(0,4)-	0	0	0	0	1	0	0	0	0	0	0	0	0	0	0	0	0	0	0	0	0	0	1	0	0	0	0	0	0	0	0	0	0	0	0	0	0	0	0	0	0	0
(0,3)-	0	0	0	0	1	0	0	0	0	0	0	0	0	0	0	0	0	0	0	0	0	0	0	1	0	0	0	0	0	0	0	0	0	0	0	0	0	0	0	0	0	0
(3,4)-	0	0	1	0	0	0	0	0	0	0	0	0	0	0	0	0	0	0	0	0	0	0	1	0	0	0	0	0	0	0	0	0	0	0	0	0	0	0	0	0	0	0
(2,4)-	0	0	0	1	0	0	0	0	0	0	0	0	0	0	0	0	0	0	0	0	0	0	1	0	0	0	0	0	0	0	0	0	0	0	0	0	0	0	0	0	0	0
(2,3)-	0	0	0	1	0	0	0	0	0	0	0	0	0	0	0	0	0	0	0	0	0	0	0	1	0	0	0	0	0	0	0	0	0	0	0	0	0	0	0	0	0	0
(1,3)-	0	0	0	0	0	1	0	0	0	0	0	0	0	0	0	0	0	0	0	0	0	0	0	1	0	0	0	0	0	0	0	0	0	0	0	0	0	0	0	0	0	0
(1,2)-	0	0	0	0	0	1	0	0	0	0	0	0	0	0	0	0	0	0	0	0	0	0	0	0	1	0	0	0	0	0	0	0	0	0	0	0	0	0	0	0	0	0
(0,2)-	0	0	0	0	1	0	0	0	0	0	0	0	0	0	0	0	0	0	0	0	0	0	0	0	1	0	0	0	0	0	0	0	0	0	0	0	0	0	0	0	0	0
(0,1)-	0	0	0	0	1	0	0	0	0	0	0	0	0	0	0	0	0	0	0	0	0	0	0	0	0	0	1	0	0	0	0	0	0	0	0	0	0	0	0	0	0	0
(1,0,4)-	0	0	0	0	0	0	1	0	0	0	0	0	0	0	0	1	0	0	0	0	0	0	0	0	0	0	0	0	1	0	0	0	0	0	0	0	0	0	0	0	0	0
(4,0,3)-	0	0	0	0	0	0	0	1	0	0	0	0	0	0	0	0	0	0	0	0	0	0	0	0	0	0	0	0	0	1	1	0	0	0	0	0	0	0	0	0	0	0
(4,3,2)-	0	0	0	0	0	0	0	0	0	1	0	1	0	0	0	0	0	0	0	0	0	0	0	0	0	0	0	0	0	0	0	1	0	0	0	0	0	0	0	0	0	0
(2,3,1)-	0	0	0	0	0	0	0	0	0	0	0	0	1	0	0	0	0	0	0	0	0	0	0	0	0	0	0	0	0	0	0	0	1	0	1	0	0	0	0	0	0	0
(2,1,0)-	0	0	0	0	0	0	0	0	0	0	0	0	0	1	0	1	0	0	0	0	0	0	0	0	0	0	0	0	0	0	0	0	0	0	0	1	0	0	0	0	0	0

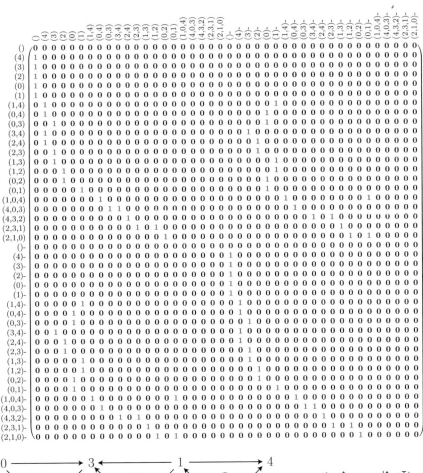

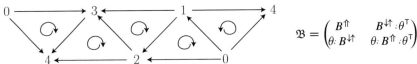

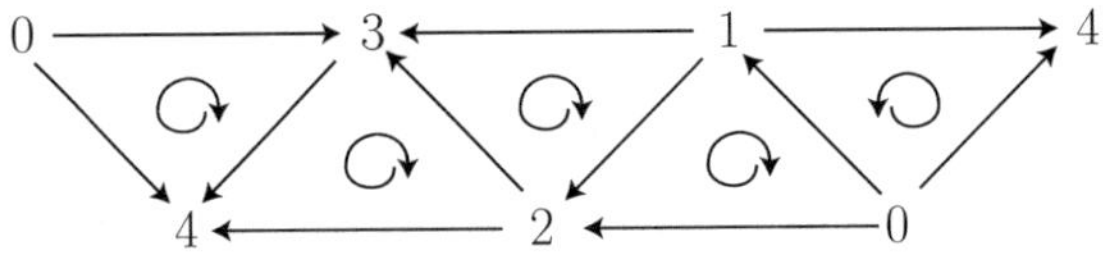

$$\mathfrak{B} = \begin{pmatrix} B^{\Uparrow} & B^{\downarrow\uparrow};\theta^{\mathsf{T}} \\ \theta;B^{\downarrow\uparrow} & \theta;B^{\Uparrow};\theta^{\mathsf{T}} \end{pmatrix}$$

Fig. 10.14 Joint boundary relation evaluated for a Moebius strip

Torus Our next example will be the torus as shown in Fig. 10.16. When restricting size as indicated, the boundary processes for the torus are mainly directed by the two relations of Fig. 10.18 (Fig. 10.17).

Having computed the orientation adjacency of Fig. 10.18, we will obtain $\Gamma^*_{\boxtimes} = \mathbb{T}$ and $\Gamma^*_{\square} = \perp\!\!\!\perp$, which need not be presented.

	(1,4)	(0,4)	(0,3)	(3,4)	(2,4)	(2,3)	(1,3)	(1,2)	(0,2)	(0,1)
(1,0,4)	0	1	0	0	0	0	0	0	0	0
(4,0,3)	0	0	1	1	0	0	0	0	0	0
(4,3,2)	0	0	0	0	1	0	0	0	0	0
(2,3,1)	0	0	0	0	0	1	0	1	0	0
(2,1,0)	0	0	0	0	0	0	0	0	1	0

$B^{\Uparrow}{}_{\Box}$

(1,4)-	(0,4)-	(0,3)-	(3,4)-	(2,4)-	(2,3)-	(1,3)-	(1,2)-	(0,2)-	(0,1)-
1	0	0	0	0	0	0	0	0	1
0	1	0	0	0	0	0	0	0	0
0	0	0	1	0	1	0	0	0	0
0	0	0	0	0	0	1	0	0	0
0	0	0	0	0	0	0	1	0	1

$B^{\Downarrow\Uparrow}{}_{\Box}$

	(1,0,4)	(4,0,3)	(4,3,2)	(2,3,1)	(2,1,0)
(1,0,4)	0	1	0	0	0
(4,0,3)	1	0	1	0	0
(4,3,2)	0	1	0	1	0
(2,3,1)	0	0	1	0	1
(2,1,0)	0	0	0	1	0

(1,0,4)-	(4,0,3)-	(4,3,2)-	(2,3,1)-	(2,1,0)-
0	0	0	0	1
0	0	0	0	0
0	0	0	0	0
0	0	0	0	0
1	0	0	0	0

$\Gamma_{\boxtimes}\ , \Gamma_{\Box}$

(1,0,4)	(4,0,3)	(4,3,2)	(2,3,1)	(2,1,0)
1	1	1	1	1
1	1	1	1	1
1	1	1	1	1
1	1	1	1	1
1	1	1	1	1

(1,0,4)-	(4,0,3)-	(4,3,2)-	(2,3,1)-	(2,1,0)-
1	1	1	1	1
1	1	1	1	1
1	1	1	1	1
1	1	1	1	1
1	1	1	1	1

$\Gamma^*_{\boxtimes}\ , \Gamma^*_{\Box}$

Fig. 10.15 Size-restricted results for the Moebius strip

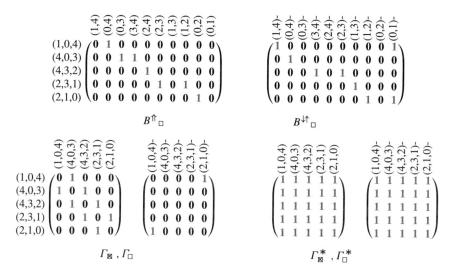

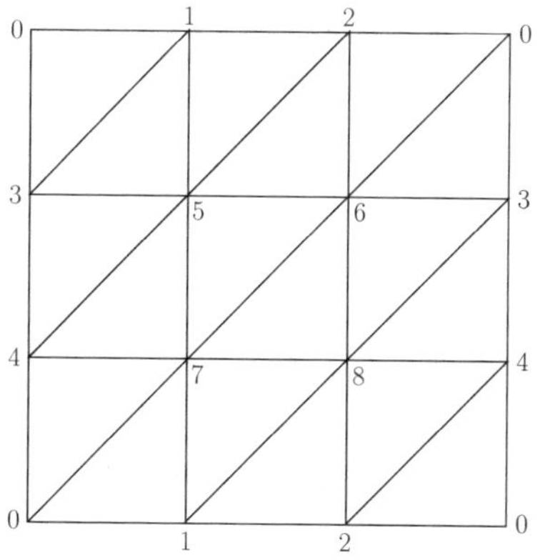

Fig. 10.16 A torus—after identifications left/right and top/down

Projective Plane Also the projective plane is usually depicted using duplicate vertices. In this case, the outermost circuit resembles the unit circle that is taken to represent directions of straight lines through the origin. The opposite endpoints of such lines have, thus, to be identified. From Fig. 10.19, we directly proceed to $\Gamma_{\boxtimes}$ and $\Gamma_{\Box}{}^{\mathsf{T}}$ of Fig. 10.20.

Then we get by calculation that $\Gamma^*_{\boxtimes} = \mathbb{T}$ and also $\Gamma^*_{\Box} = \mathbb{T}$, which is not shown. The projective plain is, thus, not orientable.

	(2,4)	(2,8)	(1,8)	(1,7)	(0,7)	(0,4)	(4,8)	(3,8)	(7,8)	(6,8)	(6,7)	(4,7)	(5,7)	(4,5)	(3,4)	(3,6)	(0,6)	(0,2)	(5,6)	(2,6)	(2,5)	(1,2)	(3,5)	(1,5)	(1,3)	(0,3)	(0,1)
(2,0,4)	0	0	0	0	0	1	0	0	0	0	0	0	0	0	0	0	0	0	0	0	0	0	0	0	0	0	0
(2,4,8)	1	0	0	0	0	0	1	0	0	0	0	0	0	0	0	0	0	0	0	0	0	0	0	0	0	0	0
(1,2,8)	0	1	0	0	0	0	0	0	0	0	0	0	0	0	0	0	0	0	0	0	0	1	0	0	0	0	0
(1,8,7)	0	0	1	0	0	0	0	0	0	0	0	0	0	0	0	0	0	0	0	0	0	0	0	0	0	0	0
(0,1,7)	0	0	0	1	0	0	0	0	0	0	0	0	0	0	0	0	0	0	0	0	0	0	0	0	0	0	1
(0,7,4)	0	0	0	0	1	0	0	0	0	0	0	0	0	0	0	0	0	0	0	0	0	0	0	0	0	0	0
(8,4,3)	0	0	0	0	0	0	0	1	0	0	0	0	0	0	0	0	0	0	0	0	0	0	0	0	0	0	0
(8,3,6)	0	0	0	0	0	0	0	0	0	1	0	0	0	0	0	1	0	0	0	0	0	0	0	0	0	0	0
(7,8,6)	0	0	0	0	0	0	0	0	1	0	1	0	0	0	0	0	0	0	0	0	0	0	0	0	0	0	0
(7,6,5)	0	0	0	0	0	0	0	0	0	0	0	0	1	0	0	0	0	0	0	0	0	0	0	0	0	0	0
(4,7,5)	0	0	0	0	0	0	0	0	0	0	0	1	0	0	0	0	0	0	0	0	0	0	0	0	0	0	0
(4,5,3)	0	0	0	0	0	0	0	0	0	0	0	0	0	1	1	0	0	0	0	0	0	0	0	0	0	0	0
(6,3,0)	0	0	0	0	0	0	0	0	0	0	0	0	0	0	0	0	1	0	0	0	0	0	0	0	0	0	0
(6,0,2)	0	0	0	0	0	0	0	0	0	0	0	0	0	0	0	0	0	1	0	1	0	0	0	0	0	0	0
(5,6,2)	0	0	0	0	0	0	0	0	0	0	0	0	0	0	0	0	0	0	1	0	1	0	0	0	0	0	0
(5,2,1)	0	0	0	0	0	0	0	0	0	0	0	0	0	0	0	0	0	0	0	0	0	0	0	1	0	0	0
(3,5,1)	0	0	0	0	0	0	0	0	0	0	0	0	0	0	0	0	0	0	0	0	0	0	1	0	1	0	0
(3,1,0)	0	0	0	0	0	0	0	0	0	0	0	0	0	0	0	0	0	0	0	0	0	0	0	0	0	1	0

	(2,4)-	(2,8)-	(1,8)-	(1,7)-	(0,7)-	(0,4)-	(4,8)-	(3,8)-	(7,8)-	(6,8)-	(6,7)-	(4,7)-	(5,7)-	(4,5)-	(3,4)-	(3,6)-	(0,6)-	(0,2)-	(5,6)-	(2,6)-	(2,5)-	(1,2)-	(3,5)-	(1,5)-	(1,3)-	(0,3)-	(0,1)-
(2,0,4)	1	0	0	0	0	0	0	0	0	0	0	0	0	0	0	0	0	1	0	0	0	0	0	0	0	0	0
(2,4,8)	0	1	0	0	0	0	0	0	0	0	0	0	0	0	0	0	0	0	0	0	0	0	0	0	0	0	0
(1,2,8)	0	0	1	0	0	0	0	0	0	0	0	0	0	0	0	0	0	0	0	0	0	0	0	0	0	0	0
(1,8,7)	0	0	0	1	0	0	0	0	1	0	0	0	0	0	0	0	0	0	0	0	0	0	0	0	0	0	0
(0,1,7)	0	0	0	0	1	0	0	0	0	0	0	0	0	0	0	0	0	0	0	0	0	0	0	0	0	0	0
(0,7,4)	0	0	0	0	0	1	0	0	0	0	0	1	0	0	0	0	0	0	0	0	0	0	0	0	0	0	0
(8,4,3)	0	0	0	0	0	0	1	0	0	0	0	0	0	0	1	0	0	0	0	0	0	0	0	0	0	0	0
(8,3,6)	0	0	0	0	0	0	0	1	0	0	0	0	0	0	0	0	0	0	0	0	0	0	0	0	0	0	0
(7,8,6)	0	0	0	0	0	0	0	0	0	1	0	0	0	0	0	0	0	0	0	0	0	0	0	0	0	0	0
(7,6,5)	0	0	0	0	0	0	0	0	0	0	1	0	0	0	0	0	0	0	1	0	0	0	0	0	0	0	0
(4,7,5)	0	0	0	0	0	0	0	0	0	0	0	0	1	1	0	0	0	0	0	0	0	0	0	0	0	0	0
(4,5,3)	0	0	0	0	0	0	0	0	0	0	0	0	0	0	0	0	0	0	0	0	0	0	1	0	0	0	0
(6,3,0)	0	0	0	0	0	0	0	0	0	0	0	0	0	0	0	1	0	0	0	0	0	0	0	0	0	1	0
(6,0,2)	0	0	0	0	0	0	0	0	0	0	0	0	0	0	0	0	1	0	0	0	0	0	0	0	0	0	0
(5,6,2)	0	0	0	0	0	0	0	0	0	0	0	0	0	0	0	0	0	0	0	1	0	0	0	0	0	0	0
(5,2,1)	0	0	0	0	0	0	0	0	0	0	0	0	0	0	0	0	0	0	0	0	1	1	0	0	0	0	0
(3,5,1)	0	0	0	0	0	0	0	0	0	0	0	0	0	0	0	0	0	0	0	0	0	0	0	1	0	0	0
(3,1,0)	0	0	0	0	0	0	0	0	0	0	0	0	0	0	0	0	0	0	0	0	0	0	0	0	1	0	1

$B^{\Uparrow}{}_{\square}$ above and $B^{\downarrow\uparrow}{}_{\square}$ below

Fig. 10.17 Size-restricted boundary relations for the torus

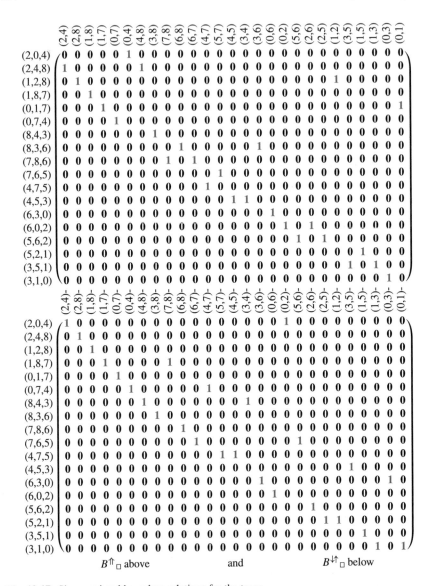

	(2,0,4)	(2,4,8)	(1,2,8)	(1,8,7)	(0,1,7)	(0,7,4)	(8,4,3)	(8,3,6)	(7,8,6)	(7,6,5)	(4,7,5)	(4,5,3)	(6,3,0)	(6,0,2)	(5,6,2)	(5,2,1)	(3,5,1)	(3,1,0)
(2,0,4)	0	1	0	0	0	1	0	0	0	0	0	0	0	1	0	0	0	0
(2,4,8)	1	0	1	0	0	0	1	0	0	0	0	0	0	0	0	0	0	0
(1,2,8)	0	1	0	1	0	0	0	0	0	0	0	0	0	0	0	1	0	0
(1,8,7)	0	0	1	0	1	0	0	0	1	0	0	0	0	0	0	0	0	0
(0,1,7)	0	0	0	1	0	1	0	0	0	0	0	0	0	0	0	0	0	1
(0,7,4)	1	0	0	0	1	0	0	0	0	0	1	0	0	0	0	0	0	0
(8,4,3)	0	1	0	0	0	0	0	1	0	0	0	1	0	0	0	0	0	0
(8,3,6)	0	0	0	0	0	0	1	0	1	0	0	0	1	0	0	0	0	0
(7,8,6)	0	0	0	1	0	0	0	1	0	1	0	0	0	0	0	0	0	0
(7,6,5)	0	0	0	0	0	0	0	0	1	0	1	0	0	0	1	0	0	0
(4,7,5)	0	0	0	0	0	1	0	0	0	1	0	1	0	0	0	0	0	0
(4,5,3)	0	0	0	0	0	0	1	0	0	0	1	0	0	0	0	0	1	0
(6,3,0)	0	0	0	0	0	0	0	1	0	0	0	0	0	1	0	0	0	1
(6,0,2)	1	0	0	0	0	0	0	0	0	0	0	0	1	0	1	0	0	0
(5,6,2)	0	0	0	0	0	0	0	0	0	1	0	0	0	1	0	1	0	0
(5,2,1)	0	0	1	0	0	0	0	0	0	0	0	0	0	0	1	0	1	0
(3,5,1)	0	0	0	0	0	0	0	0	0	0	0	1	0	0	0	1	0	1
(3,1,0)	0	0	0	0	1	0	0	0	0	0	0	1	0	0	0	0	1	0

$\Gamma_{\boxtimes}$

(2,0,4)-	(2,4,8)-	(1,2,8)-	(1,8,7)-	(0,1,7)-	(0,7,4)-	(8,4,3)-	(8,3,6)-	(7,8,6)-	(7,6,5)-	(4,7,5)-	(4,5,3)-	(6,3,0)-	(6,0,2)-	(5,6,2)-	(5,2,1)-	(3,5,1)-	(3,1,0)-
0	0	0	0	0	0	0	0	0	0	0	0	0	0	0	0	0	0
0	0	0	0	0	0	0	0	0	0	0	0	0	0	0	0	0	0
0	0	0	0	0	0	0	0	0	0	0	0	0	0	0	0	0	0
0	0	0	0	0	0	0	0	0	0	0	0	0	0	0	0	0	0
0	0	0	0	0	0	0	0	0	0	0	0	0	0	0	0	0	0
0	0	0	0	0	0	0	0	0	0	0	0	0	0	0	0	0	0
0	0	0	0	0	0	0	0	0	0	0	0	0	0	0	0	0	0
0	0	0	0	0	0	0	0	0	0	0	0	0	0	0	0	0	0
0	0	0	0	0	0	0	0	0	0	0	0	0	0	0	0	0	0
0	0	0	0	0	0	0	0	0	0	0	0	0	0	0	0	0	0
0	0	0	0	0	0	0	0	0	0	0	0	0	0	0	0	0	0
0	0	0	0	0	0	0	0	0	0	0	0	0	0	0	0	0	0
0	0	0	0	0	0	0	0	0	0	0	0	0	0	0	0	0	0
0	0	0	0	0	0	0	0	0	0	0	0	0	0	0	0	0	0
0	0	0	0	0	0	0	0	0	0	0	0	0	0	0	0	0	0
0	0	0	0	0	0	0	0	0	0	0	0	0	0	0	0	0	0
0	0	0	0	0	0	0	0	0	0	0	0	0	0	0	0	0	0
0	0	0	0	0	0	0	0	0	0	0	0	0	0	0	0	0	0

$\Gamma_{\square}$

Fig. 10.18 Size-restricted orientation adjacency of the torus

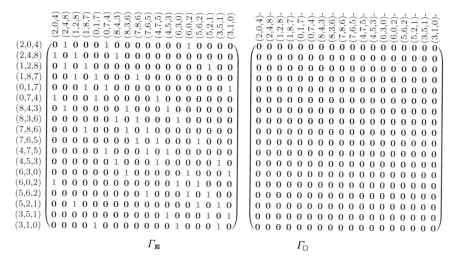

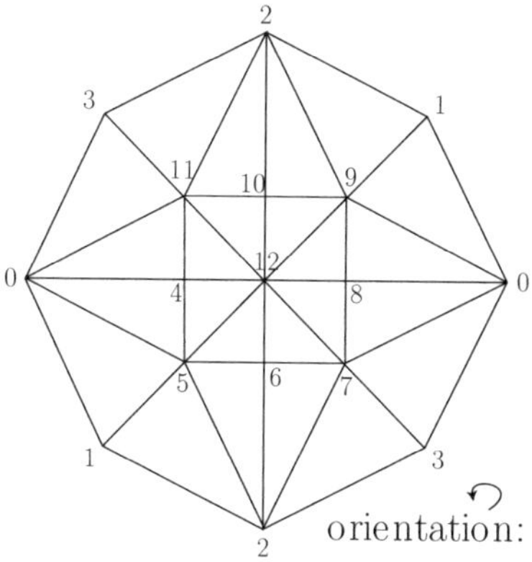

Fig. 10.19 Projective plane—equally named vertices to be identified

2-Pretzel The following example shows the well-known 2-pretzel with its triangulation. As before, equally named vertices have to be identified.

After a first folding of Figs. 10.21 and 10.22, further identifications are necessary. When one identifies the two 0's of the horizontal middle ellipse, two tangent holes will appear, into which the so far open left and right ends of the "pipe" may be glued, ending with the 2-pretzel announced. It turns out that this pretzel is indeed orientable.

We only present $\Gamma_{\boxtimes}$ and omit $\Gamma_{\square}$, since $\Gamma_{\square} = \perp\!\!\!\perp$. It turns out—as was to be expected—that $\Gamma_{\boxtimes}^* = \mathbb{T}$ and $\Gamma_{\square}^* = \perp\!\!\!\perp$ (Fig. 10.23).

	(3,0,7)	(2,3,7)	(2,7,6)	(2,6,5)	(1,2,5)	(0,1,5)	(7,0,8)	(7,8,12)	(6,7,12)	(5,6,12)	(5,12,4)	(0,5,4)	(8,0,9)	(12,8,9)	(12,9,10)	(12,10,11)	(4,12,11)	(0,4,11)	(9,0,1)	(9,1,2)	(10,9,2)	(11,10,2)	(11,2,3)	(0,11,3)
(3,0,7)	0	1	0	0	0	0	1	0	0	0	0	0	0	0	0	0	0	0	0	0	0	0	0	0
(2,3,7)	1	0	1	0	0	0	0	0	0	0	0	0	0	0	0	0	0	0	0	0	0	0	0	0
(2,7,6)	0	1	0	1	0	0	0	0	1	0	0	0	0	0	0	0	0	0	0	0	0	0	0	0
(2,6,5)	0	0	1	0	1	0	0	0	0	1	0	0	0	0	0	0	0	0	0	0	0	0	0	0
(1,2,5)	0	0	0	1	0	1	0	0	0	0	0	0	0	0	0	0	0	0	0	0	0	0	0	0
(0,1,5)	0	0	0	0	1	0	0	0	0	0	0	1	0	0	0	0	0	0	0	0	0	0	0	0
(7,0,8)	1	0	0	0	0	0	0	1	0	0	0	0	1	0	0	0	0	0	0	0	0	0	0	0
(7,8,12)	0	0	0	0	0	0	1	0	1	0	0	0	0	1	0	0	0	0	0	0	0	0	0	0
(6,7,12)	0	0	1	0	0	0	0	1	0	1	0	0	0	0	0	0	0	0	0	0	0	0	0	0
(5,6,12)	0	0	0	1	0	0	0	0	1	0	1	0	0	0	0	0	0	0	0	0	0	0	0	0
(5,12,4)	0	0	0	0	0	0	0	0	0	1	0	1	0	0	0	0	1	0	0	0	0	0	0	0
(0,5,4)	0	0	0	0	0	1	0	0	0	0	1	0	0	0	0	0	0	1	0	0	0	0	0	0
(8,0,9)	0	0	0	0	0	0	1	0	0	0	0	0	0	1	0	0	0	0	1	0	0	0	0	0
(12,8,9)	0	0	0	0	0	0	0	1	0	0	0	0	1	0	1	0	0	0	0	0	0	0	0	0
(12,9,10)	0	0	0	0	0	0	0	0	0	0	0	0	0	1	0	1	0	0	0	0	1	0	0	0
(12,10,11)	0	0	0	0	0	0	0	0	0	0	0	0	0	0	1	0	1	0	0	0	0	1	0	0
(4,12,11)	0	0	0	0	0	0	0	0	0	0	1	0	0	0	0	1	0	1	0	0	0	0	0	0
(0,4,11)	0	0	0	0	0	0	0	0	0	0	0	1	0	0	0	0	1	0	0	0	0	0	0	1
(9,0,1)	0	0	0	0	0	0	0	0	0	0	0	0	1	0	0	0	0	0	0	1	0	0	0	0
(9,1,2)	0	0	0	0	0	0	0	0	0	0	0	0	0	0	0	0	0	0	1	0	1	0	0	0
(10,9,2)	0	0	0	0	0	0	0	0	0	0	0	0	0	0	1	0	0	0	0	1	0	1	0	0
(11,10,2)	0	0	0	0	0	0	0	0	0	0	0	0	0	0	0	1	0	0	0	0	1	0	1	0
(11,2,3)	0	0	0	0	0	0	0	0	0	0	0	0	0	0	0	0	0	0	0	0	0	1	0	1
(0,11,3)	0	0	0	0	0	0	0	0	0	0	0	0	0	0	0	0	0	1	0	0	0	0	1	0
(3,0,7)-	0	0	0	0	0	0	0	0	0	0	0	0	0	0	0	0	0	0	0	0	0	0	0	1
(2,3,7)-	0	0	0	0	0	0	0	0	0	0	0	0	0	0	0	0	0	0	0	0	0	0	1	0
(2,7,6)-	0	0	0	0	0	0	0	0	0	0	0	0	0	0	0	0	0	0	0	0	0	0	0	0
(2,6,5)-	0	0	0	0	0	0	0	0	0	0	0	0	0	0	0	0	0	0	0	0	0	0	0	0
(1,2,5)-	0	0	0	0	0	0	0	0	0	0	0	0	0	0	0	0	0	0	0	1	0	0	0	0
(0,1,5)-	0	0	0	0	0	0	0	0	0	0	0	0	0	0	0	0	0	0	1	0	0	0	0	0
(7,0,8)-	0	0	0	0	0	0	0	0	0	0	0	0	0	0	0	0	0	0	0	0	0	0	0	0
(7,8,12)-	0	0	0	0	0	0	0	0	0	0	0	0	0	0	0	0	0	0	0	0	0	0	0	0
(6,7,12)-	0	0	0	0	0	0	0	0	0	0	0	0	0	0	0	0	0	0	0	0	0	0	0	0
(5,6,12)-	0	0	0	0	0	0	0	0	0	0	0	0	0	0	0	0	0	0	0	0	0	0	0	0
(5,12,4)-	0	0	0	0	0	0	0	0	0	0	0	0	0	0	0	0	0	0	0	0	0	0	0	0
(0,5,4)-	0	0	0	0	0	0	0	0	0	0	0	0	0	0	0	0	0	0	0	0	0	0	0	0
(8,0,9)-	0	0	0	0	0	0	0	0	0	0	0	0	0	0	0	0	0	0	0	0	0	0	0	0
(12,8,9)-	0	0	0	0	0	0	0	0	0	0	0	0	0	0	0	0	0	0	0	0	0	0	0	0
(12,9,10)-	0	0	0	0	0	0	0	0	0	0	0	0	0	0	0	0	0	0	0	0	0	0	0	0
(12,10,11)-	0	0	0	0	0	0	0	0	0	0	0	0	0	0	0	0	0	0	0	0	0	0	0	0
(4,12,11)-	0	0	0	0	0	0	0	0	0	0	0	0	0	0	0	0	0	0	0	0	0	0	0	0
(0,4,11)-	0	0	0	0	0	0	0	0	0	0	0	0	0	0	0	0	0	0	0	0	0	0	0	0
(9,0,1)-	0	0	0	0	0	1	0	0	0	0	0	0	0	0	0	0	0	0	0	0	0	0	0	0
(9,1,2)-	0	0	0	0	1	0	0	0	0	0	0	0	0	0	0	0	0	0	0	0	0	0	0	0
(10,9,2)-	0	0	0	0	0	0	0	0	0	0	0	0	0	0	0	0	0	0	0	0	0	0	0	0
(11,10,2)-	0	0	0	0	0	0	0	0	0	0	0	0	0	0	0	0	0	0	0	0	0	0	0	0
(11,2,3)-	0	1	0	0	0	0	0	0	0	0	0	0	0	0	0	0	0	0	0	0	0	0	0	0
(0,11,3)-	1	0	0	0	0	0	0	0	0	0	0	0	0	0	0	0	0	0	0	0	0	0	0	0

Fig. 10.20 $\Gamma_{\boxtimes}$ (above) and $\Gamma_{\square}{}^{\top}$ (below) of the projective plane of Fig. 10.19

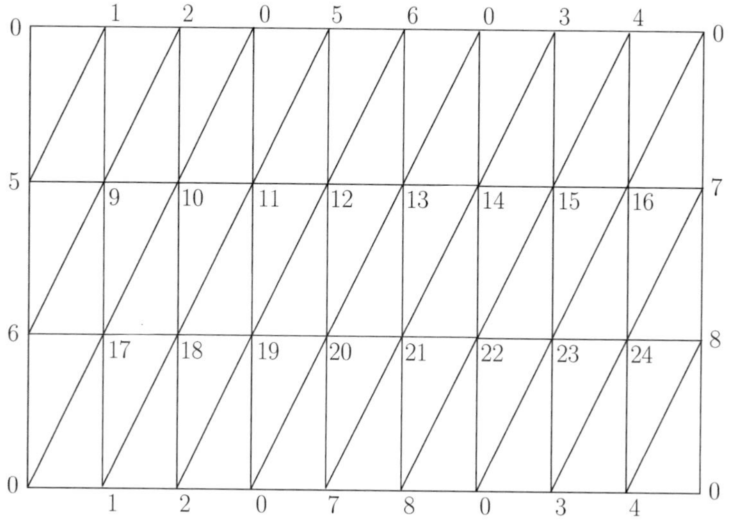

Fig. 10.21 A triangulation of the 2-hole-pretzel

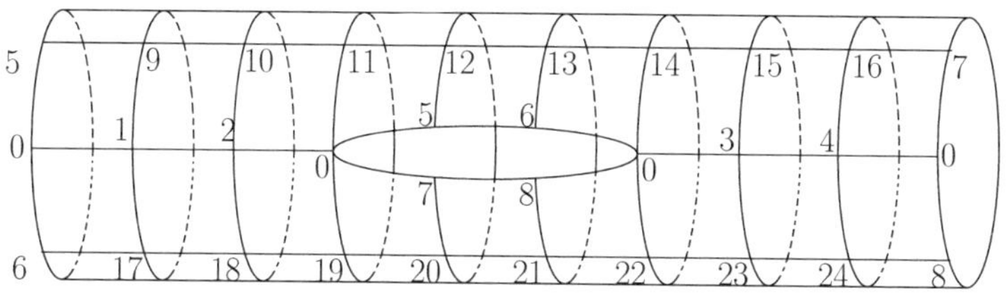

Fig. 10.22 Partial folding to obtain the 2-hole-pretzel

Cube One will observe that the triangulation given for the 3-dimensional cube in Fig. 10.24 is fully determined by the dashed space diagonal and the three square diagonals emanating from each of its endpoints, and is, thus, far from symmetric in the elementary sense.

For the first time the maximum dimension will be 3; the highest simplices are, thus, tetrahedra.

Clearly, the number of simplices required to represent an n-cube increases as a factorial!

We don't get $\Gamma^*_{\boxtimes} = \mathbb{T}$. Look at the tetrahedra (1,2,4,5) and (1,3,4,5): "Screw-shifting" (1,2,4) as well as (1,3,4) towards 5 does *not* lead to an orientation-adjacency between the two via the triangle (1,4,5).

	(4,0,8)	(4,8,24)	(3,4,24)	(3,24,23)	(0,3,23)	(0,23,22)	(8,0,22)	(8,22,21)	(7,8,21)	(7,21,20)	(0,7,20)	(0,20,19)	(2,0,19)	(2,19,18)	(1,2,18)	(1,18,17)	(0,1,17)	(0,17,6)	(24,8,7)	(24,7,16)	(23,24,16)	(23,16,15)	(22,23,15)	(22,15,14)	(21,22,14)	(21,14,13)	(20,21,13)	(20,13,12)	(19,20,12)	(19,12,11)	(18,19,11)	(18,11,10)	(17,18,10)	(17,10,9)	(6,17,9)	(6,9,5)	(16,7,0)	(16,0,4)	(15,16,4)	(15,4,3)	(14,15,3)	(14,3,0)	(13,14,0)	(13,0,6)	(12,13,6)	(12,6,5)	(11,12,5)	(11,5,0)	(10,11,0)	(10,0,2)	(9,10,2)	(9,2,1)	(5,9,1)	(5,1,0)
(4,0,8)	0	1	0	0	0	0	1	0	0	0	0	0	0	0	0	0	0	0	0	0	0	0	0	0	0	0	0	0	0	0	0	0	0	0	0	0	0	1	0	0	0	0	0	0	0	0	0	0	0	0	0	0	0	0
(4,8,24)	1	0	1	0	0	0	0	0	0	0	0	0	0	0	0	0	0	0	1	0	0	0	0	0	0	0	0	0	0	0	0	0	0	0	0	0	0	0	0	0	0	0	0	0	0	0	0	0	0	0	0	0	0	0
(3,4,24)	0	1	0	1	0	0	0	0	0	0	0	0	0	0	0	0	0	0	0	0	0	0	0	0	0	0	0	0	0	0	0	0	0	0	0	0	0	0	0	1	0	0	0	0	0	0	0	0	0	0	0	0	0	0
(3,24,23)	0	0	1	0	1	0	0	0	0	0	0	0	0	0	0	0	0	0	0	0	1	0	0	0	0	0	0	0	0	0	0	0	0	0	0	0	0	0	0	0	0	0	0	0	0	0	0	0	0	0	0	0	0	0
(0,3,23)	0	0	0	1	0	1	0	0	0	0	0	0	0	0	0	0	0	0	0	0	0	0	0	0	0	0	0	0	0	0	0	0	0	0	0	0	0	0	0	0	0	1	0	0	0	0	0	0	0	0	0	0	0	0
(0,23,22)	0	0	0	0	1	0	1	0	0	0	0	0	0	0	0	0	0	0	0	0	0	0	1	0	0	0	0	0	0	0	0	0	0	0	0	0	0	0	0	0	0	0	0	0	0	0	0	0	0	0	0	0	0	0
(8,0,22)	1	0	0	0	0	1	0	1	0	0	0	0	0	0	0	0	0	0	0	0	0	0	0	0	0	0	0	0	0	0	0	0	0	0	0	0	0	0	0	0	0	0	0	0	0	0	0	0	0	0	0	0	0	0
(8,22,21)	0	0	0	0	0	0	1	0	1	0	0	0	0	0	0	0	0	0	0	0	0	0	0	0	1	0	0	0	0	0	0	0	0	0	0	0	0	0	0	0	0	0	0	0	0	0	0	0	0	0	0	0	0	0
(7,8,21)	0	0	0	0	0	0	0	1	0	1	0	0	0	0	0	0	0	0	1	0	0	0	0	0	0	0	0	0	0	0	0	0	0	0	0	0	0	0	0	0	0	0	0	0	0	0	0	0	0	0	0	0	0	0
(7,21,20)	0	0	0	0	0	0	0	0	1	0	1	0	0	0	0	0	0	0	0	0	0	0	0	0	0	0	1	0	0	0	0	0	0	0	0	0	0	0	0	0	0	0	0	0	0	0	0	0	0	0	0	0	0	0
(0,7,20)	0	0	0	0	0	0	0	0	0	1	0	1	0	0	0	0	0	0	0	0	0	0	0	0	0	0	0	0	0	0	0	0	0	0	0	0	1	0	0	0	0	0	0	0	0	0	0	0	0	0	0	0	0	0
(0,20,19)	0	0	0	0	0	0	0	0	0	0	1	0	1	0	0	0	0	0	0	0	0	0	0	0	0	0	0	0	1	0	0	0	0	0	0	0	0	0	0	0	0	0	0	0	0	0	0	0	0	0	0	0	0	0
(2,0,19)	0	0	0	0	0	0	0	0	0	0	0	1	0	1	0	0	0	0	0	0	0	0	0	0	0	0	0	0	0	0	0	0	0	0	0	0	0	0	0	0	0	0	0	0	0	0	0	0	0	1	0	0	0	0
(2,19,18)	0	0	0	0	0	0	0	0	0	0	0	0	1	0	1	0	0	0	0	0	0	0	0	0	0	0	0	0	0	0	1	0	0	0	0	0	0	0	0	0	0	0	0	0	0	0	0	0	0	0	0	0	0	0
(1,2,18)	0	0	0	0	0	0	0	0	0	0	0	0	0	1	0	1	0	0	0	0	0	0	0	0	0	0	0	0	0	0	0	0	0	0	0	0	0	0	0	0	0	0	0	0	0	0	0	0	0	0	0	1	0	0
(1,18,17)	0	0	0	0	0	0	0	0	0	0	0	0	0	0	1	0	1	0	0	0	0	0	0	0	0	0	0	0	0	0	0	0	1	0	0	0	0	0	0	0	0	0	0	0	0	0	0	0	0	0	0	0	0	0
(0,1,17)	0	0	0	0	0	0	0	0	0	0	0	0	0	0	0	1	0	1	0	0	0	0	0	0	0	0	0	0	0	0	0	0	0	0	0	0	0	0	0	0	0	0	0	0	0	0	0	0	0	0	0	0	0	1
(0,17,6)	0	0	0	0	0	0	0	0	0	0	0	0	0	0	0	0	1	0	0	0	0	0	0	0	0	0	0	0	0	0	0	0	0	0	1	0	0	0	0	0	0	0	0	1	0	0	0	0	0	0	0	0	0	0
(24,8,7)	0	1	0	0	0	0	0	0	1	0	0	0	0	0	0	0	0	0	0	1	0	0	0	0	0	0	0	0	0	0	0	0	0	0	0	0	0	0	0	0	0	0	0	0	0	0	0	0	0	0	0	0	0	0
(24,7,16)	0	0	0	0	0	0	0	0	0	0	0	0	0	0	0	0	0	0	1	0	1	0	0	0	0	0	0	0	0	0	0	0	0	0	0	0	1	0	0	0	0	0	0	0	0	0	0	0	0	0	0	0	0	0
(23,24,16)	0	0	0	1	0	0	0	0	0	0	0	0	0	0	0	0	0	0	0	1	0	1	0	0	0	0	0	0	0	0	0	0	0	0	0	0	0	0	0	0	0	0	0	0	0	0	0	0	0	0	0	0	0	0
(23,16,15)	0	0	0	0	0	0	0	0	0	0	0	0	0	0	0	0	0	0	0	0	1	0	1	0	0	0	0	0	0	0	0	0	0	0	0	0	0	0	1	0	0	0	0	0	0	0	0	0	0	0	0	0	0	0
(22,23,15)	0	0	0	0	0	1	0	0	0	0	0	0	0	0	0	0	0	0	0	0	0	1	0	1	0	0	0	0	0	0	0	0	0	0	0	0	0	0	0	0	0	0	0	0	0	0	0	0	0	0	0	0	0	0
(22,15,14)	0	0	0	0	0	0	0	0	0	0	0	0	0	0	0	0	0	0	0	0	0	0	1	0	1	0	0	0	0	0	0	0	0	0	0	0	0	0	0	0	1	0	0	0	0	0	0	0	0	0	0	0	0	0
(21,22,14)	0	0	0	0	0	0	0	1	0	0	0	0	0	0	0	0	0	0	0	0	0	0	0	1	0	1	0	0	0	0	0	0	0	0	0	0	0	0	0	0	0	0	0	0	0	0	0	0	0	0	0	0	0	0
(21,14,13)	0	0	0	0	0	0	0	0	0	0	0	0	0	0	0	0	0	0	0	0	0	0	0	0	1	0	1	0	0	0	0	0	0	0	0	0	0	0	0	0	0	0	1	0	0	0	0	0	0	0	0	0	0	0
(20,21,13)	0	0	0	0	0	0	0	0	0	1	0	0	0	0	0	0	0	0	0	0	0	0	0	0	0	1	0	1	0	0	0	0	0	0	0	0	0	0	0	0	0	0	0	0	0	0	0	0	0	0	0	0	0	0
(20,13,12)	0	0	0	0	0	0	0	0	0	0	0	0	0	0	0	0	0	0	0	0	0	0	0	0	0	0	1	0	1	0	0	0	0	0	0	0	0	0	0	0	0	0	0	0	1	0	0	0	0	0	0	0	0	0
(19,20,12)	0	0	0	0	0	0	0	0	0	0	0	1	0	0	0	0	0	0	0	0	0	0	0	0	0	0	0	1	0	1	0	0	0	0	0	0	0	0	0	0	0	0	0	0	0	0	0	0	0	0	0	0	0	0
(19,12,11)	0	0	0	0	0	0	0	0	0	0	0	0	0	0	0	0	0	0	0	0	0	0	0	0	0	0	0	0	1	0	1	0	0	0	0	0	0	0	0	0	0	0	0	0	0	0	1	0	0	0	0	0	0	0
(18,19,11)	0	0	0	0	0	0	0	0	0	0	0	0	0	1	0	0	0	0	0	0	0	0	0	0	0	0	0	0	0	1	0	1	0	0	0	0	0	0	0	0	0	0	0	0	0	0	0	0	0	0	0	0	0	0
(18,11,10)	0	0	0	0	0	0	0	0	0	0	0	0	0	0	0	0	0	0	0	0	0	0	0	0	0	0	0	0	0	0	1	0	1	0	0	0	0	0	0	0	0	0	0	0	0	0	0	0	1	0	0	0	0	0
(17,18,10)	0	0	0	0	0	0	0	0	0	0	0	0	0	0	0	1	0	0	0	0	0	0	0	0	0	0	0	0	0	0	0	1	0	1	0	0	0	0	0	0	0	0	0	0	0	0	0	0	0	0	0	0	0	0
(17,10,9)	0	0	0	0	0	0	0	0	0	0	0	0	0	0	0	0	0	0	0	0	0	0	0	0	0	0	0	0	0	0	0	0	1	0	1	0	0	0	0	0	0	0	0	0	0	0	0	0	0	0	1	0	0	0
(6,17,9)	0	0	0	0	0	0	0	0	0	0	0	0	0	0	0	0	0	1	0	0	0	0	0	0	0	0	0	0	0	0	0	0	0	1	0	1	0	0	0	0	0	0	0	0	0	0	0	0	0	0	0	0	0	0
(6,9,5)	0	0	0	0	0	0	0	0	0	0	0	0	0	0	0	0	0	0	0	0	0	0	0	0	0	0	0	0	0	0	0	0	0	0	1	0	0	0	0	0	0	0	0	0	0	1	0	0	0	0	0	0	1	0
(16,7,0)	0	0	0	0	0	0	0	0	0	0	1	0	0	0	0	0	0	0	0	1	0	0	0	0	0	0	0	0	0	0	0	0	0	0	0	0	0	1	0	0	0	0	0	0	0	0	0	0	0	0	0	0	0	0
(16,0,4)	1	0	0	0	0	0	0	0	0	0	0	0	0	0	0	0	0	0	0	0	0	0	0	0	0	0	0	0	0	0	0	0	0	0	0	0	1	0	1	0	0	0	0	0	0	0	0	0	0	0	0	0	0	0
(15,16,4)	0	0	0	0	0	0	0	0	0	0	0	0	0	0	0	0	0	0	0	0	0	1	0	0	0	0	0	0	0	0	0	0	0	0	0	0	0	1	0	1	0	0	0	0	0	0	0	0	0	0	0	0	0	0
(15,4,3)	0	0	1	0	0	0	0	0	0	0	0	0	0	0	0	0	0	0	0	0	0	0	0	0	0	0	0	0	0	0	0	0	0	0	0	0	0	0	1	0	1	0	0	0	0	0	0	0	0	0	0	0	0	0
(14,15,3)	0	0	0	0	0	0	0	0	0	0	0	0	0	0	0	0	0	0	0	0	0	0	0	1	0	0	0	0	0	0	0	0	0	0	0	0	0	0	0	1	0	1	0	0	0	0	0	0	0	0	0	0	0	0
(14,3,0)	0	0	0	0	1	0	0	0	0	0	0	0	0	0	0	0	0	0	0	0	0	0	0	0	0	0	0	0	0	0	0	0	0	0	0	0	0	0	0	0	1	0	1	0	0	0	0	0	0	0	0	0	0	0
(13,14,0)	0	0	0	0	0	0	0	0	0	0	0	0	0	0	0	0	0	0	0	0	0	0	0	0	0	1	0	0	0	0	0	0	0	0	0	0	0	0	0	0	0	1	0	1	0	0	0	0	0	0	0	0	0	0
(13,0,6)	0	0	0	0	0	0	0	0	0	0	0	0	0	0	0	0	0	1	0	0	0	0	0	0	0	0	0	0	0	0	0	0	0	0	0	0	0	0	0	0	0	0	1	0	1	0	0	0	0	0	0	0	0	0
(12,13,6)	0	0	0	0	0	0	0	0	0	0	0	0	0	0	0	0	0	0	0	0	0	0	0	0	0	0	0	1	0	0	0	0	0	0	0	0	0	0	0	0	0	0	0	1	0	1	0	0	0	0	0	0	0	0
(12,6,5)	0	0	0	0	0	0	0	0	0	0	0	0	0	0	0	0	0	0	0	0	0	0	0	0	0	0	0	0	0	0	0	0	0	0	0	1	0	0	0	0	0	0	0	0	1	0	1	0	0	0	0	0	0	0
(11,12,5)	0	0	0	0	0	0	0	0	0	0	0	0	0	0	0	0	0	0	0	0	0	0	0	0	0	0	0	0	0	1	0	0	0	0	0	0	0	0	0	0	0	0	0	0	0	1	0	1	0	0	0	0	0	0
(11,5,0)	0	0	0	0	0	0	0	0	0	0	0	0	0	0	0	0	0	0	0	0	0	0	0	0	0	0	0	0	0	0	0	0	0	0	0	0	0	0	0	0	0	0	0	0	0	0	1	0	1	0	0	0	0	1
(10,11,0)	0	0	0	0	0	0	0	0	0	0	0	0	0	0	0	0	0	0	0	0	0	0	0	0	0	0	0	0	0	0	0	1	0	0	0	0	0	0	0	0	0	0	0	0	0	0	0	1	0	1	0	0	0	0
(10,0,2)	0	0	0	0	0	0	0	0	0	0	0	0	1	0	0	0	0	0	0	0	0	0	0	0	0	0	0	0	0	0	0	0	0	0	0	0	0	0	0	0	0	0	0	0	0	0	0	0	1	0	1	0	0	0
(9,10,2)	0	0	0	0	0	0	0	0	0	0	0	0	0	0	0	0	0	0	0	0	0	0	0	0	0	0	0	0	0	0	0	0	0	1	0	0	0	0	0	0	0	0	0	0	0	0	0	0	0	1	0	1	0	0
(9,2,1)	0	0	0	0	0	0	0	0	0	0	0	0	0	0	1	0	0	0	0	0	0	0	0	0	0	0	0	0	0	0	0	0	0	0	0	0	0	0	0	0	0	0	0	0	0	0	0	0	0	0	1	0	1	0
(5,9,1)	0	0	0	0	0	0	0	0	0	0	0	0	0	0	0	0	0	0	0	0	0	0	0	0	0	0	0	0	0	0	0	0	0	0	0	1	0	0	0	0	0	0	0	0	0	0	0	0	0	0	0	1	0	1
(5,1,0)	0	0	0	0	0	0	0	0	0	0	0	0	0	0	0	0	1	0	0	0	0	0	0	0	0	0	0	0	0	0	0	0	0	0	0	0	0	0	0	0	0	0	0	0	0	0	0	1	0	0	0	0	1	0

Fig. 10.23 $\Gamma_{\boxtimes}$ of the 2-pretzel

This may bring the possibility for work in knot theory. The idea is to tesselate a part of the 3-dimensional space to the extent that a given knot may be properly represented in it. When considering a knot as a closed file or wire, i.e., an image of mapping the unit circle into $\mathbb{R}^3$, "represented properly" would mean that it never touches simplices of dimension ≤ 1 and the intersection of the file with a 3-simplex should never consist of more than one connected component. One may hope that this enables us to compute. A knot will then be just a sequence of 3-simplices with common adversely oriented boundary. The sequence will indicate over which bounding triangle/sub-simplex the wire or file of the knot runs.

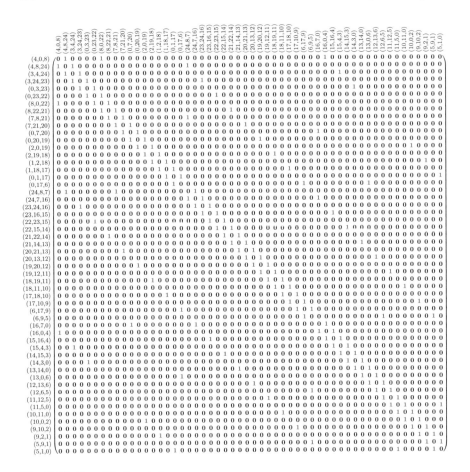

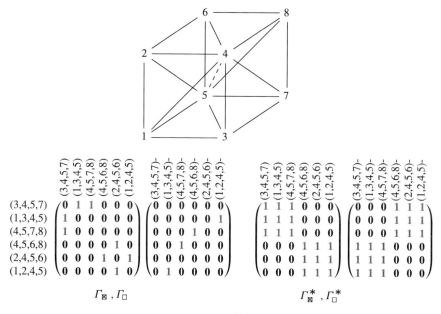

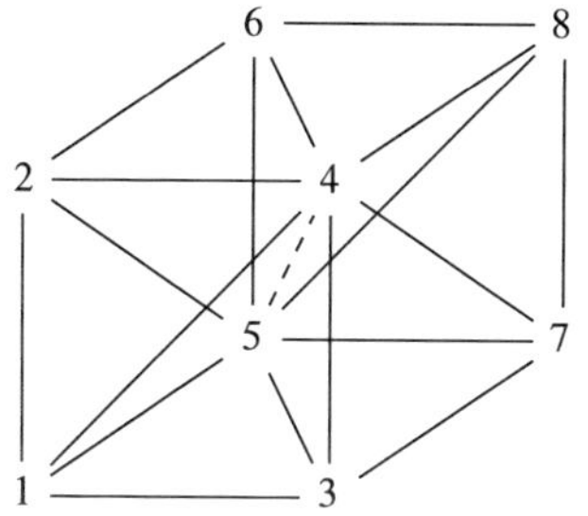

	(3,4,5,7)	(1,3,4,5)	(4,5,7,8)	(4,5,6,8)	(2,4,5,6)	(1,2,4,5)	(3,4,5,7)-	(1,3,4,5)-	(4,5,7,8)-	(4,5,6,8)-	(2,4,5,6)-	(1,2,4,5)-
(3,4,5,7)	0	1	1	0	0	0	0	0	0	0	0	0
(1,3,4,5)	1	0	0	0	0	0	0	0	0	0	0	1
(4,5,7,8)	1	0	0	0	0	0	0	0	0	1	0	0
(4,5,6,8)	0	0	0	0	1	0	0	0	1	0	0	0
(2,4,5,6)	0	0	0	1	0	1	0	0	0	0	0	0
(1,2,4,5)	0	0	0	0	1	0	0	1	0	0	0	0

$\Gamma_{\boxtimes}$, $\Gamma_{\Box}$

(3,4,5,7)	(1,3,4,5)	(4,5,7,8)	(4,5,6,8)	(2,4,5,6)	(1,2,4,5)	(3,4,5,7)-	(1,3,4,5)-	(4,5,7,8)-	(4,5,6,8)-	(2,4,5,6)-	(1,2,4,5)-
1	1	1	0	0	0	0	0	0	1	1	1
1	1	1	0	0	0	0	0	0	1	1	1
1	1	1	0	0	0	0	0	0	1	1	1
0	0	0	1	1	1	1	1	1	0	0	0
0	0	0	1	1	1	1	1	1	0	0	0
0	0	0	1	1	1	1	1	1	0	0	0

$\Gamma_{\boxtimes}^{*}$, $\Gamma_{\Box}^{*}$

Fig. 10.24 3-dimensional cube triangulated; $\Gamma_{\Box}^{*} \subseteq \overline{\mathrm{II}}$ indicates orientability

The most trivial "unknotted" knot consisting of just a circle around the dashed space diagonal would then be represented as the cyclic sequence

$$(1,3,4,5), (3,4,5,7), (4,5,7,8), (4,5,6,8)\text{-}, (2,4,5,6)\text{-}, (1,2,4,5)\text{-}$$

of simplexes.

Császár Polyhedron The Császár polyhedron is highly remarkable and possibly uniquely determined by its properties. Too much could be said about that polyhedron to be presented here, so that we better refer to

https://en.wikipedia.org/wiki/Csaszar_polyhedron.

There an animated visualization can be found. In brief, the Császár polyhedron resembles the complete graph on 7 vertices and all the 21 lines are edges of the polyhedron. It can be cut down along triangles/sub-simplices to 7 tetrahedra. We show here that the Császár polyhedron is topologically equivalent with the torus by presenting Fig. 10.25.

There are two ways to look at Fig. 10.25: One may consider it being a 2-dimensional simplicial complex or a 3-dimensional one. The orientation for the 2-dimensional case is easily derived counter-clockwise from the upper left picture.

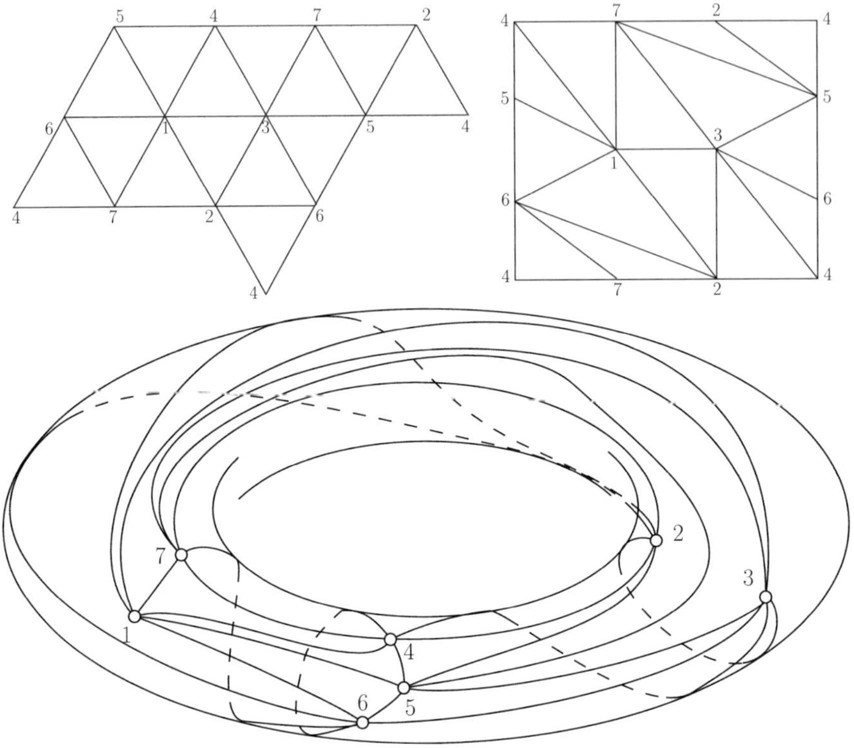

Fig. 10.25 Different presentations of the Császár polyhedron

Far more involved is it to find orientations for the 3-dimensional complex. Some help is provided by the realization on a torus.

The Császár polyhedron is according to $\Gamma_{\square}^{*} \subseteq \overline{\mathbb{I}}$ of Fig. 10.26 orientable.

The Császár polyhedron has been included in order to present something that may not be broadly known, but is nevertheless interesting. It gave the opportunity to show a non-trivial 3-dimensional simplicial complex.

$B^{\Uparrow}{}_{\square} =$

	(2,3,6)	(3,4,6)	(1,2,5)	(2,3,5)	(1,2,4)	(2,3,4)	(1,3,4)	(1,2,3)	(2,4,7)	(2,4,6)	(2,6,7)	(4,5,6)	(4,6,7)	(4,5,7)	(1,5,6)	(1,6,7)	(5,6,7)	(1,3,5)	(1,3,7)	(1,5,7)	(3,5,7)
(2,6,3,4)	0	0	0	0	0	1	0	0	0	1	0	0	0	0	0	0	0	0	0	0	0
(5,1,2,3)	0	0	1	1	0	0	0	0	0	0	0	0	0	0	0	0	0	0	0	0	0
(4,2,1,3)	0	0	0	0	0	0	1	1	0	0	0	0	0	0	0	0	0	0	0	0	0
(4,7,2,6)	0	0	0	0	0	0	0	0	1	0	0	0	1	0	0	0	0	0	0	0	0
(4,6,5,7)	0	0	0	0	0	0	0	0	0	0	0	0	0	1	0	0	1	0	0	0	0
(1,5,6,7)	0	0	0	0	0	0	0	0	0	0	0	0	0	0	1	1	0	0	0	0	0
(1,3,5,7)	0	0	0	0	0	0	0	0	0	0	0	0	0	0	0	0	0	1	0	1	0

$B^{\Downarrow\Uparrow}{}_{\square} =$

	(2,3,6)-	(3,4,6)-	(1,2,5)-	(2,3,5)-	(1,2,4)-	(2,3,4)-	(1,3,4)-	(1,2,3)-	(2,4,7)-	(2,4,6)-	(2,6,7)-	(4,5,6)-	(4,6,7)-	(4,5,7)-	(1,5,6)-	(1,6,7)-	(5,6,7)-	(1,3,5)-	(1,3,7)-	(1,5,7)-	(3,5,7)-
(2,6,3,4)	1	1	0	0	0	0	0	0	0	0	0	0	0	0	0	0	0	0	0	0	0
(5,1,2,3)	0	0	0	0	0	0	0	1	0	0	0	0	0	0	0	0	0	1	0	0	0
(4,2,1,3)	0	0	0	0	1	1	0	0	0	0	0	0	0	0	0	0	0	0	0	0	0
(4,7,2,6)	0	0	0	0	0	0	0	0	0	1	1	0	0	0	0	0	0	0	0	0	0
(4,6,5,7)	0	0	0	0	0	0	0	0	0	0	0	1	1	0	0	0	0	0	0	0	0
(1,5,6,7)	0	0	0	0	0	0	0	0	0	0	0	0	0	0	0	0	1	0	0	1	0
(1,3,5,7)	0	0	0	0	0	0	0	0	0	0	0	0	0	0	0	0	0	0	1	0	1

$\Gamma_{\boxtimes}$

	(2,6,3,4)	(5,1,2,3)	(4,2,1,3)	(4,7,2,6)	(4,6,5,7)	(1,5,6,7)	(1,3,5,7)
(2,6,3,4)	0	0	1	1	0	0	0
(5,1,2,3)	0	0	1	0	0	0	1
(4,2,1,3)	1	1	0	0	0	0	0
(4,7,2,6)	1	0	0	0	1	0	0
(4,6,5,7)	0	0	0	1	0	1	0
(1,5,6,7)	0	0	0	0	1	0	1
(1,3,5,7)	0	1	0	0	0	1	0

$\Gamma_{\square} = \Gamma^*_{\square}$

	(2,6,3,4)-	(5,1,2,3)-	(4,2,1,3)-	(4,7,2,6)-	(4,6,5,7)-	(1,5,6,7)-	(1,3,5,7)-
(2,6,3,4)	0	0	0	0	0	0	0
(5,1,2,3)	0	0	0	0	0	0	0
(4,2,1,3)	0	0	0	0	0	0	0
(4,7,2,6)	0	0	0	0	0	0	0
(4,6,5,7)	0	0	0	0	0	0	0
(1,5,6,7)	0	0	0	0	0	0	0
(1,3,5,7)	0	0	0	0	0	0	0

$\Gamma^*_{\boxtimes}$

	(2,6,3,4)	(5,1,2,3)	(4,2,1,3)	(4,7,2,6)	(4,6,5,7)	(1,5,6,7)	(1,3,5,7)
(2,6,3,4)	1	1	1	1	1	1	1
(5,1,2,3)	1	1	1	1	1	1	1
(4,2,1,3)	1	1	1	1	1	1	1
(4,7,2,6)	1	1	1	1	1	1	1
(4,6,5,7)	1	1	1	1	1	1	1
(1,5,6,7)	1	1	1	1	1	1	1
(1,3,5,7)	1	1	1	1	1	1	1

Fig. 10.26 Orientability of the Császár polyhedron

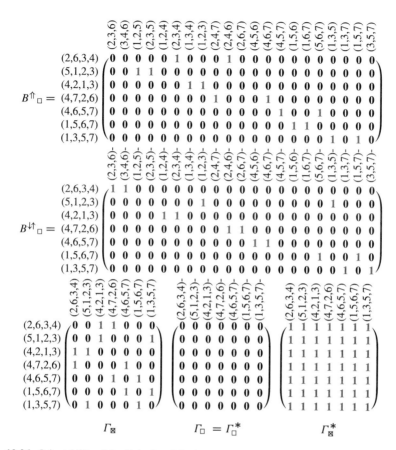

Concluding Remarks

In the present work, we have for the first time presented a thorough relational and algebraic treatment covering the broad range of such concepts as topology, proximity, nearness, apartness, contact, closure, frames, and finally orientability of simplicial complexes. Much of the impetus to perform all these computations came from the intention to sharpen the relational tools. In the mean time, we have reached a status from which it seems possible to classify what can be achieved relationally and what not.

Another stimulus for this research was the idea to solve practical problems computationally. For several of the topics mentioned, it seems that this is possible. The implementation of relational methods as with RELVIEW http://www.informatik.uni-kiel.de/~progsys/relview/ has gained substantial power. The merely term rewriting TITUREL system http://www.titurel.org/TituRel/indexTituRel.html proved versatile enough to underpin all the formulae with the examples presented. Its powerful unification mechanism did also work safely with products, quotients, membership relations, residuals, Kronecker products, etc.

TITUREL as well as RELVIEW brought, however, two further and not unimportant contributions. Quite often the matrices produced when computing examples gave intuition how a solution might look like. But there was also another frequently occurring effect: This text contains a multitude of sophisticated formulae which are uncommon to the average reader. More than once the one initially assumed to hold was erroneous. Any attempt to prove such a formula was thus bound to fail; but too detailed work on it could usually be avoided by looking at the relations produced when investigating intermediate steps.

A third stimulus for the present authors was the intention to unify concepts. There exist many internally coherent groups of researchers which do not cooperate with each other. Young scientists in particular contribute their excellence when working on a specific topic. But when still young their knowledge is sometimes restricted to the fields they have seen so far. Contrasting this observation, we have here been

G. Schmidt, M. Winter, *Relational Topology*, Lecture Notes in Mathematics 2208, https://doi.org/10.1007/978-3-319-74451-3

looking at ever new and at first sight different fields that later turned out to be cryptomorphically the same.

Another topic that more or less obviously lends itself to being treated relationally are matroids and their exchange property. It would be highly desirable to find a point-free as well as quantifier-free relational form of the respective axioms. In the literature they are mostly given with counting arguments.

It has been a particular concern to identify those topics where one inevitably has to use *points*, in the relation-algebraic sense, and where one may get along without. The evasion to pointwise reasoning, much in the same way as in the sharpness problem in the early 1980s, could widely be avoided.

Having mentioned relational work with computer help, we should add yet another remark. Typing often gives a good guess how a suspected formula should look like. However, typing in existing proof assistants seems not to be sufficiently developed so far. Dependent types in particular, such as the quotient according to an equivalence, e.g., require to prove the property thereby requested. This is usually not considered being part of the typing system. So these types may not be handled in a way to fit into every unification procedure. In this regard, further work needs to be done.

References

[AN98] Samson Adepoju Adeleke and Peter M. Neumann. *Relations Related to Betweenness: Their Structure and Automorphisms*. Number 623 in Memoirs of the American Mathematical Society. American Mathematical Society, 1998.

[Aum69] Georg Aumann. *Reelle Funktionen*, volume 68 of *Grundlehren der mathematischen Wissenschaften*. Springer-Verlag, 1969. 2nd Edition.

[Aum70] Georg Aumann. *Kontakt-Relationen*. Sitzungsberichte der Bayer. Akademie der Wissenschaften, Math.-Nat. Klasse, 1970.

[Aum74] Georg Aumann. *AD ARTEM ULTIMAM — Eine Einführung in die Gedankenwelt der Mathematik*. R. Oldenbourg München Wien, 1974. ISBN 3-486-34481-1.

[BD07] Brandon Bennett and Ivo Düntsch. Axioms, Algebras and Topology. In Marco Aiello, Ian E. Pratt-Hartmann, and Johan F.A.K. van Bentham, editors, *Handbook of Spatial Logics*, pages 99–159. Springer-Verlag, 2007.

[BHSV94] Rudolf Berghammer, Armando Martín Haeberer, Gunther Schmidt, and Paulo A. S. Veloso. Comparing two different approaches to products in abstract relation algebra. In Maurice Nivat, Charles Rattray, Teodore Rus, and Giuseppe Scollo, editors, *Algebraic Methodology and Software Technology*, Workshops in Computing, pages 167–176. Springer-Verlag, 1994.

[BStV01] Douglas Bridges, Peter Schuster, and Luminiţa Vîţă. Apartness, topology, and uniformity: a constructive view, 2001. Submission found in the internet.

[Des99] Jules Desharnais. Monomorphic characterization of n-ary direct products. *Information Sciences*, 119:275–288, 1999.

[Die74] Jean Alexandre Dieudonné. Sollen wir "Moderne Mathematik" lehren? In Michael Otte, editor, *Mathematiker über die Mathematik*, pages 402–416. Springer-Verlag, 1974.

[DL12] Ivo Düntsch and Sanjiang Li. Extension Properties of Boolean Contact Algebras. In Timothy G. Griffin and Wolfram Kahl, editors, *RAMICS 2012*, number 7560 in Lect. Notes in Comput. Sci., pages 342–356. Springer-Verlag, 2012.

[Dob15] Ernst-Erich Doberkat. *Special Topics in Mathematics for Computer Scientists — Sets, Categories, Topologies and Measures*. Springer-Verlag, 2015.

[dRE98] Willem-Paul de Roever and Kai Engelhardt. *Data Refinement: Model-Oriented Proof Methods and their Comparison*. Number 47 in Cambridge Tracts in Theoretical Computer Science. Cambridge University Press, 1998.

[DV06] Georgi Dimov and Dimitar Vakarelov. Contact algebras and region-based theory of space: A proximity approach — I, II. *Fundamenta Informaticae*, 74:209–282, 2006.

[Eng78] Ryszard Engelking. *Dimension Theory*, volume 19 of *North-Holland Mathematical Library*. North-Holland/Polish Scientific Publishers, 1978. ISBN 0-444-85176-3.

G. Schmidt, M. Winter, *Relational Topology*, Lecture Notes in Mathematics 2208, https://doi.org/10.1007/978-3-319-74451-3

[Fab59] Georg Faber. *Mathematik*. C. H. Beck'sche Verlagsbuchhandlung München, 1959. Sonderdruck aus Geist und Gestalt, Biographische Beiträge zur Geschichte der Bayer. Akademie der Wissenschaften; Sonderdruck aus dem zweiten Band Naturwissenschaften.

[Fra60] Wolfgang Franz. *Topologie I*. Number 1181 in Sammlung Göschen. Walter de Gruyter, 1960.

[GW14] Manas Ghosh and Michael Winter. Refinements of the RCC25 Composition Table. In Peter Höfner, Peter Jipsen, Wolfram Kahl, and Martin E. Müller, editors, *RAMICS*, number 8428 in Lect. Notes in Comput. Sci., pages 379–394. Springer-Verlag, 2014.

[Hus77] Taqdir Husain. *Topology and Maps*. Mathematical Concepts and Methods in Science and Engineering. Plenum Press, 1977.

[HW41] Witold Hurewicz and Henry Wallman. *Dimension Theory*. Princeton University Press, 1941. Nineth Printing 1974.

[KS00] Wolfram Kahl and Gunther Schmidt. Exploring (Finite) Relation Algebras With Tools Written in Haskell. Technical Report 2000/02, Fakultät für Informatik, Universität der Bundeswehr München, October 2000. http://titurel.org/Papers/RelAlgTools.pdf, 158 pages.

[Men28] Karl Menger. *Dimensionstheorie*. B. G. Teubner, 1928.

[NW70] Somashekhar Amrith Naimpally and Brian D. Warrack. *Proximity Spaces*. Cambridge University Press, 1970.

[Per40a] Oskar Perron. Modulartige lückenlose Ausfüllung des R_n mit kongruenten Würfeln I. *Mathematische Annalen*, 117:415–447, 1940.

[Per40b] Oskar Perron. Über lückenlose Ausfüllung des n-dimensionalen Raumes durch kongruente Würfel. *Mathematische Zeitschrift*, 46:1–26, 1940.

[Per40c] Oskar Perron. Über lückenlose Ausfüllung des n-dimensionalen Raumes durch kongruente Würfel II. *Mathematische Zeitschrift*, 46:161–180, 1940.

[Per41] Oskar Perron. Modulartige lückenlose Ausfüllung des R_n mit kongruenten Würfeln II. *Mathematische Annalen*, 117:609–658, 1941.

[RK07] Karin Reich and Alexander Kreuzer, editors. *Emil Artin (1898–1962 — Beiträge zu Leben, Werk und Persönlichkeit*, number 61 in Algorismus. Dr. Erwin Rauner Verlag, Augsburg, 2007.

[Sch11] Gunther Schmidt. *Relational Mathematics*, volume 132 of *Encyclopedia of Mathematics and its Applications*. Cambridge University Press, 2011. ISBN 978-0-521-76268-7, 584 pages.

[Sch11a] Gunther Schmidt. Partiality I: Embedding Relation Algebras. *Journal of Logic and Algebraic Programming*, 66(2):212–238, 2006. Special issue edited by Bernhard Möller; https://doi.org/10.1016/j.jlap.2005.04.002.

[Sch12] Gunther Schmidt. Partiality II: Constructed Relation Algebras. *Journal of Logic and Algebraic Programming*, 81(6):660–679, 2012. Special Issue edited by Harrie de Swart, http://dx.doi.org/10.1016/j.jlap.2012.05.005.

[Sie79] Carl Ludwig Siegel. On the History of the Frankfurt Mathematics Seminar — Address Given on June 13, 1964, in the Mathematics Seminar of the University of Frankfurt on the Occasion of the 50th Anniversary of the Johann-Wolfgang-Goethe-University Frankfurt. *The Mathematical Intelligencer*, 1(4):223–230, 1979.

[SS89] Gunther Schmidt and Thomas Ströhlein. *Relationen und Graphen*. Mathematik für Informatiker. Springer-Verlag, 1989. ISBN 3-540-50304-8, ISBN 0-387-50304-8.

[SS93] Gunther Schmidt and Thomas Ströhlein. *Relations and Graphs — Discrete Mathematics for Computer Scientists*. EATCS Monographs on Theoretical Computer Science. Springer-Verlag, 1993. ISBN 3-540-56254-0, ISBN 0-387-56254-0.

[SW14] Gunther Schmidt and Michael Winter. Relational Mathematics Continued. Technical Report 2014-01, Fakultät für Informatik, Universität der Bundeswehr München, April 2014. http://arxiv.org/abs/1403.6957.

[VDDB02] Dimitar Vakarelov, Georgi Dimov, Ivo Düntsch, and Brandon Bennett. A proximity approach to some Region-based theories of space. *Journal of Applied Non-Classical Logics*, 12:527–559, 2002.

[vdV93] M. L. J. van de Vel. *Theory of Convex Structures*, volume 50 of *North-Holland Mathematical Library*. North-Holland, 1993.

[Vic89] Steven Vickers. *Topology via Logic*, volume 5 of *Cambridge Tracts in Theoretical Computer Science*. Cambridge University Press, 1989.

[vQ79] Boto von Querenburg. *Mengentheoretische Topologie*. Hochschultext. Springer-Verlag, 1979. Zweite, neubearbeitete und erweiterte Auflage.

[Win83] Glynn Winskel. A note on powerdomains and modality. Technical report, Carnegie Mellon University, 1983. http://repository.cmu.edu/cgi/viewcontent.cgi?article=2481&context=compsci.

[Zie88] Hans Zierer. *Programmierung mit Funktionsobjekten: Konstruktive Erzeugung semantischer Bereiche und Anwendung auf die partielle Auswertung*. PhD thesis, Fakultät für Informatik, Technische Universität München, 1988.

[Zie91] Hans Zierer. Relation algebraic domain constructions. *Theoret. Comput. Sci.*, 87:163–188, 1991.

Index

G. Schmidt, M. Winter, *Relational Topology*, Lecture Notes
in Mathematics 2208, https://doi.org/10.1007/978-3-319-74451-3

LECTURE NOTES IN MATHEMATICS Springer

Editors in Chief: J.-M. Morel, B. Teissier;

Editorial Policy

1. Lecture Notes aim to report new developments in all areas of mathematics and their applications – quickly, informally and at a high level. Mathematical texts analysing new developments in modelling and numerical simulation are welcome.

   Manuscripts should be reasonably self-contained and rounded off. Thus they may, and often will, present not only results of the author but also related work by other people. They may be based on specialised lecture courses. Furthermore, the manuscripts should provide sufficient motivation, examples and applications. This clearly distinguishes Lecture Notes from journal articles or technical reports which normally are very concise. Articles intended for a journal but too long to be accepted by most journals, usually do not have this "lecture notes" character. For similar reasons it is unusual for doctoral theses to be accepted for the Lecture Notes series, though habilitation theses may be appropriate.

2. Besides monographs, multi-author manuscripts resulting from SUMMER SCHOOLS or similar INTENSIVE COURSES are welcome, provided their objective was held to present an active mathematical topic to an audience at the beginning or intermediate graduate level (a list of participants should be provided).

   The resulting manuscript should not be just a collection of course notes, but should require advance planning and coordination among the main lecturers. The subject matter should dictate the structure of the book. This structure should be motivated and explained in a scientific introduction, and the notation, references, index and formulation of results should be, if possible, unified by the editors. Each contribution should have an abstract and an introduction referring to the other contributions. In other words, more preparatory work must go into a multi-authored volume than simply assembling a disparate collection of papers, communicated at the event.

3. Manuscripts should be submitted either online at www.editorialmanager.com/lnm to Springer's mathematics editorial in Heidelberg, or electronically to one of the series editors. Authors should be aware that incomplete or insufficiently close-to-final manuscripts almost always result in longer refereeing times and nevertheless unclear referees' recommendations, making further refereeing of a final draft necessary. The strict minimum amount of material that will be considered should include a detailed outline describing the planned contents of each chapter, a bibliography and several sample chapters. Parallel submission of a manuscript to another publisher while under consideration for LNM is not acceptable and can lead to rejection.

4. In general, **monographs** will be sent out to at least 2 external referees for evaluation.

   A final decision to publish can be made only on the basis of the complete manuscript, however a refereeing process leading to a preliminary decision can be based on a pre-final or incomplete manuscript.

   Volume Editors of **multi-author works** are expected to arrange for the refereeing, to the usual scientific standards, of the individual contributions. If the resulting reports can be forwarded to the LNM Editorial Board, this is very helpful. If no reports are forwarded or if other questions remain unclear in respect of homogeneity etc, the series editors may wish to consult external referees for an overall evaluation of the volume.

5. Manuscripts should in general be submitted in English. Final manuscripts should contain at least 100 pages of mathematical text and should always include

   – a table of contents;
   – an informative introduction, with adequate motivation and perhaps some historical remarks: it should be accessible to a reader not intimately familiar with the topic treated;
   – a subject index: as a rule this is genuinely helpful for the reader.
   – For evaluation purposes, manuscripts should be submitted as pdf files.

6. Careful preparation of the manuscripts will help keep production time short besides ensuring satisfactory appearance of the finished book in print and online. After acceptance of the manuscript authors will be asked to prepare the final LaTeX source files (see LaTeX templates online: https://www.springer.com/gb/authors-editors/book-authors-editors/manuscriptpreparation/5636) plus the corresponding pdf- or zipped ps-file. The LaTeX source files are essential for producing the full-text online version of the book, see http://link.springer.com/bookseries/304 for the existing online volumes of LNM). The technical production of a Lecture Notes volume takes approximately 12 weeks. Additional instructions, if necessary, are available on request from lnm@springer.com.

7. Authors receive a total of 30 free copies of their volume and free access to their book on SpringerLink, but no royalties. They are entitled to a discount of 33.3 % on the price of Springer books purchased for their personal use, if ordering directly from Springer.

8. Commitment to publish is made by a *Publishing Agreement*; contributing authors of multiauthor books are requested to sign a *Consent to Publish form*. Springer-Verlag registers the copyright for each volume. Authors are free to reuse material contained in their LNM volumes in later publications: a brief written (or e-mail) request for formal permission is sufficient.

Addresses:
Professor Jean-Michel Morel, CMLA, École Normale Supérieure de Cachan, France
E-mail: moreljeanmichel@gmail.com

Professor Bernard Teissier, Equipe Géométrie et Dynamique,
Institut de Mathématiques de Jussieu – Paris Rive Gauche, Paris, France
E-mail: bernard.teissier@imj-prg.fr

Springer: Ute McCrory, Mathematics, Heidelberg, Germany,
E-mail: lnm@springer.com

Printed in the United States
By Bookmasters